THE POLITICS OF DIALOGUE

BORDER REGIONS SERIES

Series Editor: Doris Wastl-Walter

In recent years, borders have taken on an immense significance. Throughout the world they have shifted, been constructed and dismantled, and become physical barriers between socio-political ideologies. They may separate societies with very different cultures, histories, national identities or economic power, or divide people of the same ethnic or cultural identity.

As manifestations of some of the world's key political, economic, societal and cultural issues, borders and border regions have received much academic attention over the past decade. This valuable series publishes high quality research monographs and edited comparative volumes that deal with all aspects of border regions, both empirically and theoretically. It will appeal to scholars interested in border regions and geopolitical issues across the whole range of social sciences.

The Politics of Dialogue

Living Under the Geopolitical Histories of War and Peace

RANABIR SAMADDAR
Director, Calcutta Research Group
Kolkata

ASHGATE

Published by
Ashgate Publishing Limited
Gower House
Croft Road
Aldershot
Hants GU11 3HR
England

Ashgate Publishing Company
Suite 420
101 Cherry Street
Burlington, VT 05401-4405
USA

Ashgate website: http://www.ashgate.com

British Library Cataloguing in Publication Data
Samaddar, Ranabir
The politics of dialogue : living under the geopolitical histories of war and peace. (Border regions series)
1. Dialogue - Political aspects 2. Boundaries - Political aspects 3. Peace-building 4. Partition, Territorial
5. Boundary disputes
I. Title
303.6'9

Library of Congress Cataloging-in-Publication Data
A catalog record for this book is available from the Library of Congress

ISBN 0 7546 3607 0

Printed in Great Britain by
Antony Rowe Ltd, Chippenham, Wiltshire

Contents

It is understandable that some people should weep over the present void and hanker instead, in the world of ideas, after a little monarchy. But those who, for once in their lives, have a found a new tone, a new way of looking, a new way of doing, those people, I believe, will never feel the need to lament that the world is error, that history is filled with people of no consequence, and that it is time for others to keep quiet so that at last the sound of their disapproval may be heard...

'Masked Philosopher' (interview with Michel Foucault by Christian Delacampagne (translated by Alan Sheridan) in Lawrence D. Kritzman (ed.), *Michel Foucault - Politics Philosophy and Culture: Interviews and Other Essays* (New York: Routledge, 1988) p. 330.

Acknowledgements

This is a book on the political question of dialogue. But I write here less on the formulations that this question may have in terms of political theory, and more on its practices as we find them in our current histories of war, conflicts, and peace. In commenting on current events I knew I was burning my hands, particularly as the political situation in Sri Lanka and Nepal started changing rapidly. But I tried as much as possible in taking current history as trends only, and refrained from making judgement. Thus while in the three years in which the book has been in the making, some of the events narrated in these pages have changed, the main arguments worked out here remain the same. These arguments are basically products of my readings of a very contentious history, but reinforced by comments, views, and advice of a number of colleagues, friends, and scholars in the field. This is a long list, but that is how I made this work.

I am grateful to Benedict Anderson of the Cornell University (retired), Rajeev Bhargava of the Jawaharlal Nehru University, Stefano Bianchini of the University of Bologna, Sanjay Chaturvedi of the Panjab University, Klaus Dodds of the University of London, Lev Grinburg, David Newman, and Oren Yiftachel of the Ben Gurion University of Negev, Tamar Hermann of the Tami Steinmetz Center for Peace Research at the Tel Aviv University, Ivan Ivekovic of the American University of Cairo, Rada Ivekovic of Paris University 8, Paul Joseph of Tufts University, Jyrki Kakonen of the University of Tampere, Ram Narayan Kumar, the indefatigable human rights investigator on Punjab and biographer of the Punjab insurgency, Alexander Motyl of Rutgers University, Gurharpal Singh of the University of Hull, and Daniel Warner of the Graduate School of International Studies at the University of Geneva.

Etienne Balibar and Charles Tilly gave me some of their current writings, which would have remained otherwise inaccessible to me. Readers will find my debt to them in many places in this volume. Similarly inspiring for me has been my annual visits to the book-laden small room of Anaour Abdel Malek in Paris. He will know on reading this book, how much I have drawn upon my discussions with him.

Friends and well-wishers provided equal support, some in fact in abundant measure throughout these three years. Kanti Bajpai, Paula Banerjee, Sabyasachi Basu Ray Chaudhury, Francois Bernard, Pradip Kumar Bose, Indrani Chatterjee, Roberta Cohen, Samir Kumar Das, Ghislaine Glasson Deschaumes, Meenakshi Gopinath, Frederic Grare, Goran Fejic, Meghna Guhathakurta, Danielle Hass-Dubosc, Pieter van Hensbroek, Christophe Jaffrelot, Amitabh Mattoo, Gautam Navlakha, Francisco Privitera, I.A. Rehman, Helmut Reifeld, Cathy Schneider, Anita Sengupta, Hari Sharma, Jeevan Thiagarajah, Marie-Claire Caloz Tschoppe, Hubertus von Welck, and Stephen Wright, encouraged me with comments and suggestions, sometimes with good-humoured tolerance when I bombarded them with rough drafts of essays. Itty Abraham, Maurice Aymard, David Newman, Jean

Luc Racine, Mike Shapiro, and Doris Wastl-Walter in particular, took interest in my work, especially in this manuscript. I am still overwhelmed by their kindness.

Most of the chapters in their original essay form were written in the last three years during my tenure as director of the peace studies programme at the South Asia Forum for Human Rights (SAFHR) in Kathmandu. I am grateful to colleagues there, particularly its Secretary General, Tapan K. Bose, because they bore with the odd man out. My insistence on reflecting critically on the issues that a human rights forum has to engage with in its daily life and on the way it engages was tolerated with courtesy, if not always encouraged. My gratitude is also to the South Asia Regional Office of the Friedrich-Naumann Stiftung in New Delhi, the sponsor of my position there. Visiting professorships to the Maison Des Sciences De L'homme in Paris, particularly my attachment to its International Programme for Advanced Studies gave me time to write some of the chapters. Similarly, the fellowship in the advanced studies programme of the Columbia University Institute for Scholars at Reid Hall gave me the time and space to revise the work. Irene Khan, now the Secretary General of the Amnesty International, earlier working in the UNHCR, enabled me to visit Geneva and consult the documentation centre of the UNHCR and the documents available at the UN library at the Palais de Nations.

Some of these essays were prepared as papers and lectures in seminars and conferences, also used occasionally as teaching material, in Jawaharlal Nehru University, Benaras Hindu University, Calcutta University, University of Zurich, University of Groningen, Jadavpur University, Rabindra Bharati University, International Peace Research Association Conference at Tampere, The European Masters Programme on Human Rights, Constitution, and Peace at the University of Sarajevo, Masters Programme on Humanitarian Action in the University of Geneva, International Border Studies Conference and the Dialogue on Geopolitics and Peace at the Panjab University, Paris University 8, Sciences Politique in Paris, Guahati University, Ben Gurion University, University of Hyderabad, Women in Security, Cooperation and Management of Peace in New Delhi better known as WISCOMP, the Strasbourg Conferences on Regional Systems for the Protection of Human Rights hosted by the Council of Europe and organized by the Brussels Office of the Friedrich Naumann-Stiftung, the Afro-Asian Peoples' Solidarity Organization conference on Dialogue of Cultures in Cairo, the UN Working Group on Minority Rights in Geneva, and the South Asia Peace Studies Orientation Courses and regional dialogues on human rights organized by SAFHR. My gratitude is to all the participants in these meets who commented on and at times forced me to reconsider my arguments.

The chapter on the line of control in Kashmir has a witty career. Part of the chapter is shamelessly drawn from Paula Banerjee's controversial essay, "Tenuous Control over the Line of Control" published in *Peace as Process – Reconciliation and Conflict Resolution in South Asia*, co-edited by Helmut Reifeld and myself, and published by the Manohar Publications (2001). Some of the ideas therein were integrated later in an essay written by me jointly with Paula Banerjee, and finally these ideas were reworked by me in form of a note for a round table contribution in *Security Dialogue* (Vol. 32, No. 1, March 2001), and now they find their place in the present form of a chapter. The essay on peace accords was first brought out as a SAFHR paper, and was then substantially revised and enlarged for publication in *Hagar – International Social Science Review* (Vol. 1, No. 2, 2000). The chapter on power, fear, and ethics began as a short essay on method in refugee studies titled "Inside/Outside" in a volume on *Refugees in West Bengal – Institutional Practices and Contested Identities*, edited by Pradip Kumar Bose and published by Calcutta Research Group (Calcutta, 2000), then was revised substantively as a contribution to the Eminent Persons' Group (EPG) meeting on forced migration hosted by the UNHCR in Colombo, 2001, and was then again revised for a collective work of the European Centre for Conflict Prevention (ECCP). The first draft of the essay on partition under the title, "The Last Hurrah that Refuses to Go Away", was written for *Transeuropeannes* (Nos. 19/20, 2001, the French version translated by Richard Crevier and Ghislaine Glasson Deschaumes), and was then used as the raw material for *Partitions – Reshaping States and Minds*. Part of the material on the politics of ceasefire was published in *La Mazarine*'s special issue on post-coloniality (*Le post-colonialisme, d'une revolution a une autre,* 2001), and part of the material on the politics of globalisation was published in *Himal*, the South Asian magazine as an essay titled, "Crime and Punishment in Genoa". The Network Institute for Global Democratisation, jointly run from the Nottingham Trent University and the Finnish Institute of International Affairs, published part of the material on autonomy in their collection of papers on *Politics of Civil Society – A Global Perspective on Democratization*, edited by Heikki Patomaki (Volume 2, 2000). Jeevan Thiagarajah and Fiona Blyth-Kubota were responsible for bringing my arguments on autonomy to the notice of the seventh session in Geneva of the UN Working Group on Minority Rights, which gracefully offered to put my arguments on minimal justice in form of a UN paper (E/CN.4/Sub.2/AC.5/2001/CRP.2) notwithstanding its "philosophical" intonations on one of the most contentious issues of international politics. The essay on peace as a political value was published in the volume *Peace as Process – Reconciliation and Conflict Resolution in South Asia*, co-edited by Helmut Reifeld and me, and published by the Manohar Publications. Sanjay Chaturvedi, Jyarki Kakonen, Bishnu Mohapatra and Neerja Gopal Joyal inspired my arguments on friendship and dialogues, which later on appeared as a special essay in the *Economic and Political Weekly* (Vol. 36, No. 10, March 2001). The chapter on received histories of war and peace has a similar animated career. It began with a short essay by the name "Long and Short of Peace" as an attempt to forge a new critique of John Lewis Gaddis' "Long Peace" and was written at the behest of the Department of Political Science of the University of Calcutta for the special number of the *Calcutta Journal of Political* Studies (1999), and then was revised as a lecture in the context of the theme of globalisation and history, for the Afro-Asian Solidarity Conference in Cairo. The lecture has now found a place in *Space, Territory and the State* (Orient Longman, 2002). Francois Bernard published parts of my arguments on globalisation in an online French dictionary on globalisation (translated by Francesca Gee and François de Bernard) run by the French newspaper, *Liberation*, and GERM, and these found place in the *Le Dictionaire Critique de la Mondialisation* (Paris, 2001). Now of course the essay has undergone again substantial reworking and enlargement in the context of the theme of this book. To translators, editors and publishers of these essays I acknowledge my debt. Similarly my editorship of *Refugee Watch*, published by the South Asia Forum for Human Rights and the Calcutta Research Group, forced me to reckon with many of the issues discussed in these pages. My debt remains to them also.

Finally, my gratitude to Krishna can only increase. At home my work was done at her expense. And, when I had to leave everything and stay away, I could only count on my increasing debts while writing this book there.

Ranabir Samaddar
Kathmandu

Chapter 1

Introduction

Dialogue and the Continuum of War and Peace

This is a book that describes the politics of dialogue, congealed in its practices amidst conflicts, war, and peace. In the chapters that follow these practices are brought out in various forms – failures and relative successes, forms of accommodation and negotiation – and in their bearings on democracy. These dialogic practices, the book finally argues, are conditioned by the geopolitical history of our time. Geopolitics influences in a large way the dynamics of the dialogic self, because democracy by itself is parched on our geopolitical destiny. Democracy and geopolitics are, thus, the two themes that continuously make their appearance throughout this volume in discussions on the politics of dialogue, though I do not discuss these two themes independently. Their presence remains like the ever-present shadow. In this introductory chapter therefore I intend to spend some lines on the three-way relation between dialogue, democracy, and geopolitics.

Similarly, though the sub-title of the book refers to war and peace of our time, it does not describe any war; it does not wish to prescribe also how to bring in peace. In stead, the book inquires into conditions of our existence under these situations, and our dialogic practices in the milieu of wars, fragile peace, endemic and structural violence, and a pervasive condition of no war-no peace. Rulers, members of the political class, strategists, and even otherwise perfectly decent individuals are feeling claustrophobic – either wars should be allowed to produce clean results, or the peace should allow the rulers to carry on with their rules to pursue their policies for governance, development, and glory. The book is a comment on the tension, the dynamics of no war-no peace situation, about low-level fratricide, about the act of conducting war while conducting peace – about some of the less noticed aspects of statecraft in South Asia.

Yet there is a defence about this sub-title. After all modernity along with its many gifts has given us the complex of no war-no peace situation which from the time of Machiavelli and more particularly Kant has been marked as one of strategic ambiguity, where and when the politics of carrying out the policies of one situation (war/peace) through the politics of its reverse situation (peace/war) is mapped out. If our claim of belonging to an age of criticism is true, we must turn our attention to this situation of hybridity, claustrophobia, and this all-pervasive desire to 'come out or break clean' of a situation that one of the modern gurus of imperial power, Zbigniew Brzezinski describes in a related context as 'a black hole situation'.[1] The language of strategic studies fails in describing and analysing this. The language of international politics or regional politics similarly fails, for their nose extends very little beyond States and official multilateral institutions, and therefore fails to discern the mode of power being exercised in post-colonial regions like South Asia – the political technology of rule in this region. That is where criticism must begin its task, critiquing existing knowledge and languages, and pointing out the fatality of the philosophy that guides our understanding of the post-colonial politics of war and peace.

Dialogues are the marks of this hybrid condition. Plural dialogues carry the signs of a capillary situation of power, resistance, and contested conversations in politics. In itself no virtue, dialogue in all its complexities waits to be analysed as a pointer to how war and power grow out of dialogue in as much as plural dialogues may lead to accommodation, reconciliation, and peace. In making the politics of dialogue the subject of inquiry, the book in fact asks, what questions do the practices of dialogue present before the politics of our time, and what are the issues that it asks of politics today? In my study of issues and events such as partitions, lines of control, negotiations over ceasefires and peace accords, dynamics of separation and friendship, interventions by citizens' groups in the form of dialogues with the governments and militants, new forms of autonomy that can accommodate conflicting claims in a dialogic existence, and the need for conversation on issues of care and justice hitherto dealt in a unilateral manner in the field of humanitarian law and institutions, and the limits of a dominantly constitutional mode of accommodation, I have focused therefore on modes of dialogue. The aim is not to set up a grand theory of dialogue, but to find out how the human practice of dialogue is forcing the present time to face certain issues of contentious politics and demanding answers.

But why is it that studies of the politics of post-colonial regions never took up the theme of dialogue in a serious way? Why is such a major strategy for war or peace left unconsidered? We can think of several reasons. First, with the assumption of state power the political class that had been dialoguing within itself, among its own members, and was conversing with the leaders of communities and parties on broad anti-colonial issues, stopped conversing with groups, communities, and the nation. On the assumption of power after de-colonization, the political society intended to be an extremely reserved club; one could gatecrash, but never got entry purely by appeal or by sheer indulgence and goodwill of the members inside. Second, the reasons of state dictated that the citizens were to be governed well so that they became loyal to the state, and the administration could improve; also with each area of life open to rule, governance, and mediation, citizen and the state would enter in a transactional relation. Conversation would not be essential, because conversation would not be necessary except on specific occasions. The agenda of dialogue would be limited. Third, with the post-colonial state embodying the highest will of the citizens, and thus the political truth of the time, there was no particular imperative to dialogue. Finally, the transition from colonialism to independence gradually signalled that the nation no longer required to feel itself bound with the memory of anti-colonial politics (in case of countries like Nepal with the age of pre-parliamentary democracy), so much so that dialogue, that traditional artefact of conversing with memory, became of very little consequence. All in all, silence and a kind of unilateral self replaced dialogic relation. Therefore, politics went back on its earlier modes of self-knowledge. 'Dialogue as a method of discovering truth' almost disappeared.[2] The imperial tradition of command, listening to the petitioner in silence, and the pervasive use of a juridical language that examines issues from the vantage point of an onlooker sums up the reality today of the post-colonial politics within and among nations. The pedagogy of the judge and the master has replaced that of the conversation.

Yet, as I try to show in these essays, the other side of silence is also there waiting to spread out its wings. One of the purposes of this exercise is to throw light on the practices of accommodation, reflection, and reconciliation, shaped as they are in the midst of conditions of war and peace. My argument in this case is motivated by a simple desire. Can we escape the binary of war and peace, and thus escape the fate of being a victim of the exclusivist discourse of either militarism or banal pacifism, and bring to light the conditions of war and peace, their alterability and mutuality, so that we can think of the conditions of justice, dialogue, and therefore sustainable peace? In this enterprise I have heavily drawn from existing narratives of several contentious situations of dialogue, for these situations as events produce a kind of abundance from within that they as events cannot exhaust. In retrieving that unaccounted space, we can only follow the technique of opening up a series of alternative possibilities from within the existing situations, possibilities that seemed to be closed. Thus we do here no comparison, make no analogy, no functional classification – in other words, we do not erect another system counter to the present states system producing violence and the no war-no peace condition in this region. Our task is to subject the received arguments and accounts to endless interrogations to show the unlimited range of possibilities other than the outcomes that these accounts record, contain and give seal of approval, so that we can see the dialogic possibilities within several historically contentious situations, appearing before us in forms of both war and peace. The binary of war and peace has been a curse for peace studies, precisely because the latter accepted the binary beyond criticism. Peace studies never agreed to become a study of flux. It never wanted to become political.

My argument is however not only methodological. Its existential concern is also clear. The discourse of civil society with various cultural attendants was the major site of dialogue in colonial times. In that enterprise the motto was to acquire identity and sovereign-hood. With the birth of the post-colonial state this was abruptly terminated.[3] State discourse of constitution, parliament, election, party, law, and above all national security, replaced the earlier association-oriented conversation in politics. As against this, those who wanted to keep the earlier site of freedom intact and those who against the post-colonial state (for instance India) wanted to have their own State took recourse to a discourse of rights. Now it is clear that this post-colonial state like the earlier colonial state is determined to be the most vigilant guardian of privacy of the political society that thinks of the state as its private wealth. Various classes, communities, sections and individuals want an entry. Rights still remain the language of claim.

But in this world of entrance, exit, and counter-voice, more and more the terrain is shifting from claims and rights to justice. To claim is not enough. To show that the claim is just is equally important. The challenge to State from within and without is now on this new ground, which has the advantage of forcing the most reluctant power to converse and to communicate, also the advantage of sidestepping the awesome task of mounting challenge to the state with an army that can be equipped with only 3.3 rifles, at times with AK 47 and AK 56 rifles, hand grenades, some rocket launchers, in some cases with high frequency communication sets, and nothing more. As if the message from those who yearn for justice is, *we are not out to destroy the world; all we want is justice, speak to us, give us political recognition.* Such a notion of dialogic justice has the added advantage of crossing the limits of retributive justice by aiming to be restorative. In terms of political technique, it inverts the moral fabric of the reasons of the state. And in terms of a political ethic, this inversion is explosive. It replaces the state agenda with the agenda of expanding political community, violence with accommodation, revenge with reconciliation, and rights with justice. In this double inversion, nothing new is invented. The replacements appear possible with the re-emergence of other prospects in the wake of the near exhaustion of state discourse. The possibilities of these replacements show that the architecture of politics of a post-colonial region such as South Asia has consumed whatever there was as sustenance in form of an anti-colonial legacy in the 'new' post-colonial politics consequent to de-colonization. With the exposure of the full form of its finitude, democracy must have other forms beyond constitutionalism, ethnic accords, militarism, security pacts and alliances, and an increasingly tattered electoral system. The significance of partitions, wars, borders, revolts, constitutionalism, accords and the elaborate system of law, order and security in post-colonial democratic politics is not only in the fact that these can co-exist with democracy as a complex reality in this region (to more or less degrees), but that in this expanding milieu of mass political culture, these have to prove themselves 'democratic'. The official discourse on these has to explain the legitimacy of State policy of renouncing various forms of dialogue, trans-border experiences, and the ethic of reconciliation. The structure of democracy in South Asia is now open before our eyes with its deficits. This book intends to argue that the issue is not that we are not democratic, or that our politics has certain anti-democratic features, but that our democracy is producing its deficit. Inquiry of the practices of dialogue therefore leads us on several occasions in this book to a discussion on democratic theory, specifically on how constitutional democracy is producing the deficit in democracy, a phenomenon known as the 'democratic deficit'.

Dialogue, Power, and the Reworking of the Arguments of Liberalism

How can a condition that is supposed to be marked by return of democracy (after 1989) be characterized by absence of dialogues? The paradox is intriguing, and we should take a look into this phenomenon.

We are living in a strange time when the original challengers to state theories of dominance have laid their weapons down, and those who have become the victors are trying to absorb the message of the challenge. Therefore liberals who won the great battle of ideas of the last century are, so to say, withdrawing from their positions inch by inch to reformulate their positions to go ahead. The lessons are to say the least instructive. In almost each of the positions worked out in this book therefore, I try to begin with an examination of the liberal position, which is being re-worked for reconciling the irreconcilable such as state/community, universalism/particularism, globality/territoriality, nationalism/global morality, capital/labour, citizenship/alien-hood, and destruction/peace. I am not suggesting that there is nothing insightful in these reworking of positions. On the contrary I try to show, if one were to be critical enough, one would have to see if these reworked positions are adequate to meet the fallacies and inadequacies of the realist assumptions in international politics, and whether in the process of taking into account the Marxist formulations in international politics and the Left-wing challenges to realist-liberal paradigms, the liberals are not reconciling themselves to realist and neo-realist positions. Of course, we have to remember that there is no solid liberal camp engaged in the reworking of the problematic. There are all kinds of social theorists and social scientists who are not satisfied with the ruling neo-liberal strategies that rule international politics today. But the fact this endeavour is coming from various quarters as dispersed say from a figure like John Rawls to a more critical thinker like Jurgen Habermas shows the power of the revisionist current. Its fallacies, fatalities and misplaced hopes are therefore more liable to producing grave errors.

Rawls speaks of 'Laws of Peoples'[4] where his enterprise is to get over the distinction between what he thinks to be authoritarian polities and liberal polities in the interest of peace. Communitarians are pointing out that the abstract and apparently rootless individuals who figure in much liberal theory have to be reconciled with community identity, and hence group emotions, sentiments, and loyalties. If nationalism and ethnic allegiances are producing appalling bloodshed and devastations, then in the interest of multicultural harmony based on group rights an alternative form of loyalty and harmony has to be theorized which will be compatible with universal norms and yet capable of attracting people of a particular territory.[5] Jurgen Habermas speaks of constitutional patriotism to 'fight nationalism on its own ground'.[6] Michael Walzer revived few decades ago the Kantian position of just wars.[7] In all these revisionist exercises, there is an attempt to reconcile morality and the facts of life through dialogic tropes – dialogues between the contradictory features in place of their antinomies mentioned above. And, in all these attempts there is a remarkable silence about the question of power.

Another variant of the liberal revival is the attempt to grasp current international politics through institutional approach, which is supposed to epitomize the post-1989 dialogic politics. Under the influence of great institution setting in Europe through European Union, Council of Europe, Organization of Cooperation and Security in Europe, the European Court, various treaties, conventions, and not the least NATO, the institutional argument has gained strength, and seems to suggest almost universal efficacy of the formula that institutional processes despite lack of consensus on critical issues of political life can lead to sustained cooperation and dialogue among nation-states under conditions that would be fairly defined. Indeed, to carry the institutional argument little more, institutional processes can define the conditions where they will work.

Thus there was a near unanimous opinion after the Balkan tragedy, that new institutions should be set up, institutional processes and supervision would have to be strengthened, and erring nations would be gradually trained and disciplined in institutional behaviour, and cooperation and dialogue. Of course, the 'curious history of Europe', to borrow these recent words of the historian Hobsbawm, has been the particular backdrop against which hope on institutional efficacy has soared, also the anxiety of Europe about wars and disorders has led to an institutional emphasis and reinforcement. But the resonance has been widespread. With World Bank-IMF (here the Asian Development Bank also) emphasis on reconstruction, reforms, institution setting, 'putting house in order', and linking regional cooperation with reconstruction, we have now a great revival of institutional wisdom. Of course one aim is to defend against what the economist Paul Krugman calls, 'the return of depression economics',[8] but the connections between backing away from the economy of chaos and savagery and return to the economy of discipline and prosperity, between the institutional prescriptions and neo-liberal theology, indeed between neo-liberal politics and neo-liberal economics are too pronounced to be ignored. The difference between the current phase of institutionalism and the earlier ones is that, the reformers today prefer to infuse notions of dialogue and peace in their prescriptions. Since war has come home, it is not something that is taking place some thousands of miles away, but here – at the backyard, almost at the doorstep, known as the Balkans, the institutional pundits are no longer Reagan, Thatcher, or Friedman, but more trained in social realities, more literate. Reconstruction, the concern that led to the setting up of premier international financial institutions after the Second World War, is again the watchword with few more words adopted as companion terms such as poverty, race, social safety net, social capital, new democracy, dialogue, and good governance. The institutional insight and wisdom is at its worst an ancillary of the Fund-bank regime, at its best a reflection of new workings in international relations theory.[9] If we have to see dialogue in the context of the incipient forms of justice and democracy, we must also see it in the context of power, whose one form is the institutional management of conflicts.

Each of the re-workings mentioned above has insights, withdrawals from earlier positions in political philosophy in order to suggest new positions from which current reality can be negotiated. And each then supplements realities of inequality, asymmetrical power positions, and new wars of conquest with ideal positions. But these revisionist attempts also indicate that we are on the verge of a conceptual break in the realm of political strategies in the quest of equality and justice. If the new ideas and practices are not be lost in the quagmire of a neo-liberal theory, and this possibility is irrespective of the wishes of the theorists (for example, Rawls is in clear pursuit of an enriched liberal position, while Habermas is concerned with the central issue of negotiating the spectre of fascism with the treasure of enlightenment thought, and his critical insights can be scarcely designated as a re-worked liberal position), then a critical argument has to go deeper and show how the ubiquitous problematic of power still eludes the revisionist arguments, and predicates the relation between democracy, dialogue, and geopolitics.

This is what I indicate by invoking the notion of democratic deficit, for the deficit is not produced by a void existing outside, in other words from inadequate democratization and inadequate institutionalization of democracy. The *lack* is constitutive of democracy. It is produced by the very democratic logic where fundamentally liberal arguments meet with 'non-democratic' people. Thus democratic theory is eternally concerned with 'transition', getting over 'inadequate democratization'. That the demos is concerned with power, that this demos may be an organic community intolerant of others, that classes, masses and ideas constitute a field that may be beyond the negotiating ability of a democracy, and that successful democracies being inherently 'organic/organicist' produce conflicts where the poor and the weak suffer and justice is given a go bye – is not something that the democratic theory can admit. Its business is too serious, stakes too high to admit its own constitutive deficit or lack. One of the intentions of this book in interrogating the practices and possibilities of dialogue is to point out, how democracy remains bound with the geopolitical realities of our time, how in order to make sense of the three way game between dialogue, geopolitics, and democracy, our political theory has no choice but to be intensely critical.

The practices of democracy, in as much as the practices of dialogue, are therefore my concern in these chapters. The particular method of writing and the themes that occupy my attention here – both are dictated by these concerns.

The Dangerous Liaison of Globalization and Dialogue and the Shrinking Space of Dialogue

The need to disentangle the story of alternative possibilities of dialogue and reconciliation from ruling accounts of contentious events is now more in the context of what has been described as the 'dangerous liaison' between globalization and the current discourse of democracy and development.[10] The political technology of modern rule, which has refused for the most part to give any seat to the principle of reconciliation, accommodation, and understanding towards achieving what I term as 'minimal justice', and discuss later in details, has its own history. This history has shown how war and development have been always connected as phases in this history. Michel Foucault in one of his last writings had remarked,

> ...The French Revolution gives the signal for the great national wars of our days, involving national armies and meeting their conclusion or their climax in huge mass slaughters. I think that you can see a similar phenomenon during the Second World War. In all history it would be hard to find such butchery as in World War II, and it is precisely this period, this moment, when the great welfare, public health, and medical assistance programs were instigated. The Beveridge program has been, if not conceived, at least published at this very moment. One could symbolize such coincidence by a slogan: Go get slaughtered and we promise you a long and pleasant life. Life insurance is connected with a death command.[11]

The dramatic improvements in the technology of rule have come in the wake of global developments in technologies of social administration, which base themselves on the increasing capacity to govern and have nothing to do with dialogic politics. Ivan Ivokevic, the Balkan political analyst, in a recent essay on different stages of social structures of accumulation has shown how development has induced underdevelopment, the inadequate resolution of the peasant question has provoked neo-patriarchal clan wars, and the institutional dynamics of accumulation today faces three traps – the Malthusian trap, the technology trap, and the ecological trap. The severity of these three is compounded by the illusion that while keeping huge sections of population at subsistence level (the level itself is revised from time to time), accumulation can continue unhindered on the basis of institutional management of these three issues of population, technology and environment.[12] Ivekovic also shows how patriarchy, neo-patriarchal convulsions and political violence are linked in this process. However, as we shall see, it will not be enough to say, the 'patriarchal/neo-patriarchal convulsions of decaying peasant communities are the main cause of today's 'ethnic' conflicts'.[13] To understand the pattern of today's violent non-dialogic politics, we shall have to inquire more.

Besides the well-known factors such as inequitable access to land reforms, patriarchal residues after 'modernization' with its attending features of urbanization and migration, and the production of déclassé elements in such modern society,[14] Ivekovic draws our attention to the crucial development of neo-patriarchy, that is 'patriarchy distorted by modernization'. Therefore we have one on hand what Valentine Moghadam calls the classic patriarchy. Classic patriarchy in her words is where 'the senior man has authority over everyone else in the family, including younger men and women subject to distinct forms of control', and the 'key to the reproduction of classic patriarchy lies in operations of the patriarchic extended household, which is also commonly associated with the reproduction of the peasantry in agrarian societies'.[15] On the other hand today we have the situation, where with new technology and new process of accumulation, household mode of production faces grave threat. Poorly paid family labour, and the eternally marginal existence of household unit in the world of advanced capitalist production produce neo-patriarchy, where women are compelled to be the lowest-paid labour in the household to maintain the house as an economic unit. The family organization of survival and work and the patri-linear mode of appropriation are now the direct result of globalization where guest-workers, family workers, marginalized peasant groups, small producing units, operatives in service sector, and various categories of 'frontier workers' have to survive by constantly negotiating globalization. Women form the core of the 'surplus-army' now employable, now unemployable, but always the underclass of the present age. States are being re-formed, re-organized, and regrouped and re-arranged on the basis of neo-patriarchy not only in an economic sense, but directly politically. Women symbolize most the phenomenon of citizen-alien, the insider-outsider, and the worker-family keeper. Eternally moving from one 'own home' to another, across families, villages, towns, and then international boundaries, they are not only the 'un-homed', but crucial in homing the modern state in an age of globalization. Neither our realists and neo-realists nor our moralists have looked into the one of the fundamental phenomena in international politics – the role of women in the reorganization of the world power structure today. It should not astonish us therefore, that the demands of justice and the desire to transcend the language of rights with a language of justice, propelled by the women's desire everywhere for dignity and recognition (for example in South Africa, Bosnia, Palestine, Guatemala, Chile, and in our very own South Asia), has not been theorized. Almost everywhere massive numbers of women are confronting dominant power structures with demands for dialogue on all contested issues. One of the aims of critical theory is to acknowledge the massive existence of women as subjects of world politics today. It will be one more step away from the futile language of conflict moderation towards one of democracy, dialogue, and peace.

In short, the reflections in this book on practices of dialogue amidst war and peace, and the conditions of democracy and territoriality, are an attempt to achieve two things at the same time. The first is to secure for critical political studies an analytical strategy that brings out alternative possibilities existing within these histories of conflicts and wars. The second is to throw the light on the State again, much in the same way political scientists had called some years ago for 'bringing the state back in' political discourse.[16] For precisely by throwing light on the State we can study the dynamics of the self in post-colonial politics caught in the three way relation between democracy, territoriality, and dialogue – the self that demands the right to determine its destiny, the self that demands autonomy, the conditioned nature of this autonomy, and the self that quarrels, dialogues, accommodates, and repeatedly dies to live anew. Practices of dialogue, as readers will find this book arguing repeatedly, are contained both in the narrative of alternatives (hence always half-formed, existing only in the realm of possibilities) and in the themes of contests, dialogues, and accommodation in which the post-colonial political self engages.

Peace Question and the State

The return to the theme of the State, specifically to the way in which the State governs by conducting war and making peace, provides us the clue to the significance of no war-no peace situation, mentioned earlier, in a study of dialogue. As wars form post-colonial states-system of a region such as South Asia, peace politics also enables such a system to stabilize. Therefore, in a chapter on peace accords in this volume, I show the other side of Charles Tilly's powerful hypothesis (war and state-formation), namely, the relation of peace and dialogue to the formation of state structure. Wars and war-peace continuum have played a key role in the intermeshing of polities to form a complex whole. Contrary to Rousseau, unthreatened little societies never merge of free will to promote the happiness of their citizens. This is too happy a scenario to believe in. Confederacy as was of the Zulu state in South-Western Africa, German politics of the nineteenth century (Frankfurt parliament of 1848, restored confederation of 1850, North German confederation of 1866, and then the surrender of confederate power to imperial form in 1871), or in our own region of South Asia – it is the war-peace continuum that carries within its dark ambiguity the truth of state-formation in modern age.[17] As the constitutions of Sri Lanka and India show, it was a 'fordism' of a third world variety, namely amidst overall backwardness there were an expansion of intensive mass production techniques, wage increase in organized sector, legitimacy of large organized unions, state/corporate subsidy to wage earners, spread of education and the reproduction of a skilled working force, finally continuous 'deferred wages' in terms of fringe benefits, retirement benefits, and increasing emphasis on literacy and skill – in brief, a milieu of peace and development in which the post-colonial State formed itself. The political consequences of the growth of a paternalistic (in Hindi it is called the 'mai-bap sarkar') and developmental state on subsequent patterns of conflict and dialogic politics have been to say the least enormous,

'Peace and development' resulted in both India and Sri Lanka in the most unlikely republican resistance to the oligarchy presiding over the post-colonial development. The phenomenon that was termed in the West as 'flower children', took shape here with students and youth revolutionaries flanking up with organized unions; and the militants thinking that their counter-power was so strong that the revolt could proceed uninterrupted, and the economies were so powerful that the workers could demand full reparations from galloping inflations.[18] But what was most remarkable of this period of state-stabilization in this region, was that corporate South Asia, at least corporate India, could pretend and make others believe that social spending in the post-colonial developing epoch had been imposed on the nation by 'irresponsible leftists'. What came to be known as the 'fiscal crisis of the state'[19] had to lead sooner or later to two phenomena – the emergence of the much spoken of 'casino society' and 'crony capitalism', and second, increasing political conservatism, authoritarianism, and militarism developing into 'permanent war-alertness', in war and no-war condition alike. Political accounts of post-coloniality did not pay attention to conditions of war and peace because, they were so much concerned with the notion of 'development' in both appreciation or criticism that they forgot that as in other regions, war had brought development in this region also – development in structures of rule, of taxation, of other monetary and fiscal measures, and in due course welfare measures. A new deal to people has always been actually a 'war deal'[20] – deal, for example in India, through/after the first Kashmir war, the border wars of 1962, 1965, more particularly 1971. It is a war deal that brought peace dividends after war and catastrophe developed administration, made the citizenry quiet, improved social security measures, and made the agenda of development possible as a legitimating factor in political life. It needs an extra-ordinary insensitivity to the politics of our time to ignore the connection between development and militarism, to praise the spread of education in Sri Lanka and forget in the same breath the spread of civil war that this very developmental state in Sri Lanka imposed on society, the debilitating impact that the combined phenomena of war and development has had on the dialogic culture of a society.

War and development signalled in a democratic framework an all enveloping and all consuming notion of 'gain', and consequently a majoritarian polity. International relations theory failed to understand the impact of war and development on patterns of dialogue, because it too immersed itself in the neo-classical theory of utility and gain. Thus, it had nothing to offer as insight on the dynamics of dialogue, an enduring feature of world relations in the second half of the last century. It had no answer to the following posers: What does the mediating agency of dialogue speak of the war-peace continuum that it straddles? More important, is dialogue reducible to a status of being the language of negotiation? All that the international relations theory could do was to explicate various scenarios of bargaining in terms of rational choices. Thus, its own explanation of the dynamics of dialogue in terms of comparison of utilities, the 'Pareto optimality', distributive bargaining, and integrative bargaining was of no relevance to the historical situation developing, particularly in the post-colonial world, amidst cold war, neo-colonialism, proxy wars, irreducible poverty, great power domination, and an increasingly global/fragmented scenario. The theory of rational bargaining tells us that each party has stable preferences, which are arranged along a single line of homogenous dimension utility. It of course acknowledges that though, 'the bargainers (may) differ in their stockpiles of utility so that one is far wealthier than the other, this will have no impact on their evaluation of further units of utility, (and though) an additional unit will not be more valuable to the poor bargainer than to the wealthy one.'[21] Thus notions of justice, independence, equality, reparations, and human rights were not considered as calculable units of utility; yet, as we know, these have persistently influenced the fates of attrition, co-existence, and peace, and have provoked major demands for dialogue. We can follow the question further. Thus, for example, what again if parties do not have stable preferences either due to information deficiencies or changing perceptions because these parties as new born states and nations are in that seemingly endless twilight zone of 'transition'?[22] While conflict theorists will put these to 'limited perceptual capabilities of bargainers in actual conflicts' and 'goals comprised of independent, and even conflicting constraints as opposed to homogenous utilities (that) cause some of the difficulties',[23] the post-colonial experiences lead us to factors such as ideologies, values, group ethos, which remain as powerful as the principles of calculation laid out by the theory of bargaining – an woefully inadequate supplement for a theory of dialogue.

This is one of the reasons why readers will find that in several chapters I examine the dynamics of dialogue in events such as ceasefire, peace accords, conversations for popular coalitions, inquiries into massive human rights abuses, failed relations between partitioned states and peoples, and try to show that beyond the rational account of gains, there remains the history of the contending notions of justice and rights in which dialogic practices have shaped up. And again to me it seems that if we have to step beyond the theory of bargaining in our effort at understanding the politics of dialogue, to the key such understanding is a rigorous probe into the war-peace and conflict-peace continuum.[24]

To be sure, one point of beginning such an investigation would be to inquire into the conditions and consequences of constitutionalism on patterns of conflict and accommodation in this region. After all constitutionalism is the brightest flag of peace, and its shine hides the reality of the continuum I speak of.

In trying to understand the extremely volatile field of dialogic acts, for instance in the context of minority rights, existent forms of autonomy, or the pattern of peace accords, I have therefore in course of my arguments gone back time and again to the issue of constitutionalism – the single most important site governing the relation between the state and the self in this region. Needless to say, international relations theory never understood its significance. And only when reconstruction of the state through the mechanism of restorative constitutions became an international agenda in the last ten years, that international politics started reflecting on law, legality and constitutionalism. It is in order therefore to make few remarks on the significance of a critique of constitutionalism in a new reading of post-colonial politics. And though in this case my arguments are based primarily on the experiences of India and Sri Lanka, we shall need a more rigorous theoretical explanation of the link between constitutionalism and conflicts, of the phenomenon of constitutions producing conflicts.[25]

Constitutionalism and Dialogue

In South Asia constitution has long ceased to be a matter just for constitutionalists. Legal philosophy, legal theory and legal deliberations are now continuously subjected to contending selves who are putting the bicameral character of constitutional law, namely its factual nature and ethical underpinnings, to severe test. As conflicts conflagrate into civil wars, constitutions have to explain if they are freedom guaranteeing or coercive. In other words, constitution, hitherto enjoying a validity that stems from its origins in a colonial power, and therefore substantively free from popular deliberations, now needs to self-explain – is it a collection of norms backed by threat of state sanction or norms whose validity does not primarily stem from the state, but from the fact that these norms guarantee the autonomy of all legal persons equally? And either way, the answer faces the next question, what grounds the legitimacy of rules that can be changed at any time by the political lawgiver? What Habermas terms as 'strategic approach', that is considering law as commands in the sense of factual constraints on scope for action, and 'performative attitude', that is viewing law as norms to be respected for being valid precepts – these two increasingly refuse to be reconciled, as modern citizens in their role of subjects of law increasingly cannot see themselves as authors of law as well. In terms of legal theory, modern legal order therefore faces increasing challenges, based, as this order has to be on the legitimacy of self-determination. Oscillating between *facts and norms*, the legal notions of rights and duties, entitlements, autonomy, equivalence, and subject-hood, belie the demands of both facticity and validity. People start speaking of democratic self-legislation, which comes to mean increasing deliberative procedures for producing law by the legal subjects. This presents for the constitutional state a greater problem. Constitution speaks of rights. But rights are against the state, and as norms against a state, arise 'co-originally' with constitutional principles of legality. Thus rights supposed to deepen constitutionalism weaken the regime by constant recourse to legitimacy from some higher-ranking moral law. Wars, particularly the declared and undeclared civil wars in Sri Lanka, India, Nepal, often become in such situation the expression of the demand to frame other procedures for will-formation.[26] Constitutionalism bereft of a dialogic context faces the limits to juridical truth.

In the context of different readings of the legacy of the French revolution, Jurgen Habermas speaks of the abiding legacy of the Revolution in presenting popular sovereignty as procedure.[27] Yet, in the long explanation that he offers, he does not ask, why could only a *form of violence* offer popular sovereignty as procedure, a new mode of practice and self-determination, a new mode of communication, legitimation, and thus a new mode of 'communicatively generated power'?[28] In other words, the question that Habermas does not reflect on relates to the ability of war to present itself as a procedure of deliberations. What happens when and where war forces constitution to accept deliberations, or presents a new set of political lawgivers as deliberators of sovereignty? Or, what happens when a constitution cannot produce deliberative procedures, which originate from another set of political practices outside its precincts, namely accord, compromise and reconciliation? Clearly we are in a situation today when constitutional legitimacy is giving way to other legitimating modes of democracy and rule.

The question to be asked will be, will this entrenched constitutional culture allow new practices a broader site? And will this entrenched culture of constitutionalism facilitate the emergence of a political society that by broadening its sphere will help the new set of practices to gain legitimacy? Going by the indications available now, the answer unfortunately seems to be negative. Not only constitutions everywhere are failing to be effective conflict-resolution mechanisms, they are on the contrary exacerbating these; for constitution as legal subject in post-colonial politics loses no time in demanding that the society has to acknowledge constitution as an autonomous political actor, and often even before receiving an answer sits on the top of political society. The political society does not form the constitution, but the constitution increasingly becomes the power to form the political society. The deliberations in the Indian Constituent Assembly carry the traces of various alternative possibilities and alternative communicative modes between different political actors, which the constitution itself was to give a deep burial. Similarly, the long history of accords and amendments in Sri Lanka was ultimately only a sub-player in a constitutional game that could survive only by turning its face against other possible rules of conversation with the dissenters. In any case as the saying goes, *Inter Arma Silent Leges* – in times of war, the laws are silent. And not only the present juridical situation is shaken in the wake of national, social and political wars, even the formation of future laws for the purpose of codified rules of negotiation of differences are also suspended. In such a context, to put emphasis on the public discourse of constitutional democracy is to interrogate in the first place, how this 'public is constituted', and how this 'discourse of democracy' is a constitutional essentialism that needs to be critiqued by a philosophy of alternative conversations.[29] Habermas has admitted that,

> After a century that, more than any other, has taught us the horror of existing unreason, the last remains of an essentialist trust in reason have been destroyed. Yet modernity, now aware of its contingencies, depends all the more on a

> procedural reason that is on a reason that puts itself on a trial. The critique of reason is its own work: this double meaning, first deployed by Immanuel Kant, is due to the radically anti-Platonic insight that there is neither a higher nor a deeper reality to which we could appeal – we who find ourselves already situated in our linguistically structured forms of life.[30]

The trouble with such a faith in procedure is that, reason places itself as *the* procedure and does not allow other conversations to take place. Law always places itself as *the* mode of order. Constitutional communication always places itself as *the* only way to discourse. In such a situation, the critique of reason is not produced by reason but by romance, the yearning to converse anew, atypically. The romance of the democratic voyage begins first by stepping beyond what reason has laid down as the procedure of democracy – namely law, constitution, and state. If the political experiences of the post-colonial times mean anything in term of adding to the corpus of democratic learning, it is a chronicle of how reason is pushed aside often by romance, reason (of the state) by dialogue, rights by justice, and attrition by accommodation. No reason suggested this replacement. Indeed its limits produced romance, and therein, lay hopes for the next turn in the democratic journey. It is essential then, that in order to understand other ways of conversation, that the romantic tradition in communication be re-affirmed to counter corporate and state control of communicational rationality and to animate the renewal of public voice in society. Also it is important to remember that the alternative in the procedural world emerges not because of the foundations and not because of advent of new technologies. Nor does it arise out of consumer demand. Rather at distinct points of historical rupture, radical activists seize the moment to create a new medium and a new form of conversation. Alternatives arise out of dramatic struggles of the visionaries in communication.

Communicating as a Dialogic Act

In any case communication of war and peace today in this region as elsewhere is the result of a packaging of the public space as a homogenous space to be served by a standardized *medium* called the *media* – a distinct result of the eighties of the last century when the quasi-autonomous public space was taken over worldwide by the Reagan-Thatcher revolution whereby powerful corporate groups won control of the state and mainstream media. If in the USA that led the monetarist revolution, all three television networks were taken over by major corporate conglomerates (ABC was taken over by Capital cities, NBC by GE, and the CBS was taken over by the Tisch financial group. ABC and NBC sought corporate merger, a trend that took ten years to complete with the proposed merger of AOL and Time-Warner), in the last five years the same trend is visible here. Whoever witnessed the media coverage of the Kargil border war between India and Pakistan[31] will understand how with this has passed into myth old innocent accounts of 'national democratic urges' on which the constitutional-republican (in another version civic-republicanism) spirit would thrive, and has come into existence the age of support from the world of communications to conservative policies, military adventures, more important simulated elections and democracies.

Since communication specialists often cite the Gulf War as the inauguration of the reception of the war spectacle by the public in the drawing room, there will be a natural tendency to begin an analysis of the public arena with focusing on the spectacle of this communication. All the more therefore, we shall have to focus on the *economy* of communication and the *economy* of this public. These two tell us of the process of production of the public – the public in war and the public in peace – for there is nothing called communication if it is not produced with *news* and flowing through cultural artefacts. That awareness will show, that the integrated narrative about the world of communication has now become a myth, probably it was always so. With the public being actually many publics making their varying senses of war and peace, the study of the degree or extent of parallel communication becomes significant. For, the singularity of dissent tears up the theme of one public, constitutes several publics, not one public space but many public spaces; and in this plural universe the romantic communicator makes his/her presence felt. The critical exercise of engaging in an interrogation of everyday practices of politics is a romantic task, for it has to work against hegemony, find out the fault-lines, and help spaces of dissent and autonomy make themselves present through the fault-lines in this supposed integrated structure of communication. There is no doubt that the politics of moral dissent leads the romantic communicator to opt out and creating the dissenting niche in the social structure.

Technologies of communication come and go; romance remains. A CD-ROM can transport an encyclopaedia, but the technology of the self, nurtured on interaction is more than that. It is the care of the self that mounts the utopia to become the opposition to a world in which communication as circulation of news/views and as a specific political technology has been never dialogic. Creative way of communicating is romantic and based on dialogic response, because it aims to transgress the given structures of communication with a critical re-arrangement of signs – satirical, funny, grim, dry, authorial, objective, pictorial, verbal, relative and absolute. More than a political declaration, the re-affirmation of the romantic tradition in communication is the first step in going beyond the given world of communication with the help of critical tools. In the claustrophobic time of no war-no peace condition (and sometime war), the attempt to understand how we live amidst this claustrophobia, and as political subjects how we converse, and cope with our geopolitical fetters, is like an act of little distancing, and asking ourselves in an almost imaginary fashion, what it was like to have glimpsed, by accident and through privilege of living through dissenting time, the aesthetics of passionate but critical communication – a strange feeling of witnessing the romantic communicators passing through the screen. There should be therefore no hesitation in admitting that reasons of State, war, conflict, and domination, can be at times countered by only dialogic imagination – an imagination that is at once critical and romantic. In order to unearth alternative rationalities, alternative reasons, alternative conversations, in short to find out other possibilities besides those that have been historically realized in form of current wars, conflict and domination, we must begin by accepting the 'language to infinity'. It also means that our aim in capturing the dialogic truth in a contentious situation, we cannot be just some modern day 'shaman' whose vocation is to expose the pathologies of what goes on by the name of international politics. The need to engage calls for the combination of the critical and the romantic, the dual awareness of the need to critique the current politics of the international and recognize at the same time that the struggle for justice and dignity is forming within this terrain that is known by the name as international, yet opting out of it in terms of meaning and rights.

In this project, dialogue occupies a place of critical importance as a strategy. There are several reasons for that, of which let me tell the most important one.

Collective memory of the victims of all the injustices of the past century seems to have become a 'problem' in a situation termed as 'post-conflict', where it is assumed that in this situation of post-conflict trauma, the problem becomes more pressing and needs healing through effecting instant judgment with the help of dialogue and reconciliation. The past, it is held, must be made transparent, everything must be told out, and we cannot allow forgetting and cannot wait or strive for a more just process of healing. The pressure for acceleration of the speed of reconciliation is the new device. The urgency to put the ethic of dialogue on the holy alter increases as the divides mentioned above increase, and the different speeds of different groups of people, communities, and states become a problem for a strategy to rule. The theory of democratic restitution glosses over the realities of divides. Central to it becomes 'some form of dealing with the past' – a form in which truth is necessary and beneficial for it heals wounds, thereby allowing reconciliation and preventing conflicts from recurring. But what when this truth and truth-telling are a problem, when many prefer silence to telling and accepting the divides, and realities of power overwhelm the realities of the discourses of dialogue? Also central to such supposition is the faith in democratic peace theory, which holds that constitutional orders by themselves have the capacity to heal.[32] But what happens when the requirements of justice and reconciliation call for new modes of dialogue beyond constitutional prescriptions for mediation, compromise and restraint? And the important question associated with this, what happens when constitutionalism as the dialogic site of transition fails?

The Multiple Sites of Dialogue

In order to conduct a concrete analysis of that great failure, I had to abandon here the centrality of sovereignty in the study of war and peace. The theory of sovereignty lends a centralized form to power ignoring the myriad ways power forms to dominate and suppress people, and turn subjects into victims. To go back briefly to the issue of constitution, the Indian Constituent Assembly debates were clear as to how power was being formed. Groups, classes, estates, rebellious nationalities were not fighting each other to death, 'War of all against all' was in fact suspended in favour of 'measuring of all against all' to found the commonwealth. It is a record of a will to compose various forms of power, of the postponement of that final fight, because society had to be run, citizens had to be ruled, and therefore the act of suspending bellicosity was scripted in the political document (the classic no war-no peace document). All classic components of sovereignty while being included were tacitly questioned here – territory, citizenship, agency to command life and death, origin of sovereign power, nature of the commonwealth, and the supreme capacity to rouse fear and care. In this diffused scenario where power has been most effective by spreading out to all layers of society, we must ask few questions, namely – To what extent are wars and conflicts moderated by constitution? Or, to look at the question from the other angle, how much has constitutional essentialism to do with the process of warfare? How much do the contentions and struggles flow from the constitution? Finally, to what extent are political institutions guided by constitutional notion of sovereignty in their involvement in war and peace?

It seems to me that in view of the answers that my inquiry provides in the following pages, we had been barking up the wrong tree. Constitution helps, but is never the principal dialogic site, simply because it does not encode what will be called 'sovereign power', and dialogue undermines that fiction of a monolithic institutionalization of power. Beneath the compact text of a constitution, one does not have to stretch ears far to hear the drum of wars of all kinds (caste, ethnic, religious, class), the violence, passions, enmities, revenges, also strains of illusive peace, desperate truce, reconciling acts, precarious friendships promising both war and peace, and finally the tensions that cut social body, the decisive battles for which all are preparing. Sovereignty is not the explanatory principle. Its role is to provide an image in which processes of war appear to be under control, processes of peace appear as harmless enough so as not to consume every organ with the fire of justice. Constitution and the fiction that it has given birth to and goes by the name of sovereignty aspires for the juridical-philosophical universality that lawmakers and philosophers have always dreamt and aspired for – the position between adversaries, the position of the centre and yet above the adversaries, the agency that imposes the armistice, the order that brings reconciliation. Dialogue by bringing forth its huge treasure of practices (including constitutional dialogues) poses before such a politics that had anointed constitution to that impossible seat questions about ways of power, the visible brutality of law, and about possibilities of imagining new ways of conversation. In this way it sheds light on forms of power, processes of war and peace, and processes by which peace and reconciliation become issues to be governed, its political-historical account is therefore darkly critical and intensely anti-mythical.

This book therefore will not discuss the virtues of dialogue, but dialogue and its place in politics. This as we know is one of oldest themes in politics, and now after many, many centuries it has become the fundamental theme in international studies today – the politics of dialogue as its practices bring out. Therefore I try to recount how sometimes themes, countries, and groups are excommunicated from the dialogic universe, because they have become so infamous – they are impugned, shunted out of the dialogic act; how nations of the region talk with each other to talk more, to renew fight, or to make peace; how state talks with the rebels; how rebels talk with the state; how groups talk with each other; how a single theme usurps places of all other themes of conversation; and how an ethnic time and space is created conditioning friendship, dialogue, and forms of accommodation, and conversely how its other is located in which friendship renews. Dialogue becomes a field of power, government, and challenge, in which even humanitarian politics plays itself out; whence themes of love, hospitality, sympathy, friendship, and trust, which make the anthology of human existences and of noble souls, cry out in anguish in face of the impersonal power that is shaping up in the world today in form of humanitarian politics, globalization, and the new world order.

I try to show repeatedly in my accounts of the political practices of dialogue, if at all communicative rationality has to be understood, we are not to do it in terms of its approximation to an 'ideal speech type situation' where better argument wins, but in terms of finding out the way, that great mystery, by which the practices build their own rationality – the rationality of a war-peace situation. There is no rational choice, but 'truth games' of bewildering variety, where parties are defining their mutual position only to discover that they have reached a situation, which they must circumvent, and that while they thought that they were relying on the power of arguments, they suddenly find out that they actually were arguing about power. Therefore to the question, 'does arguing really matter?',[33] we must answer not by saying yes or no to the query, but by showing how factors such as existence of common life-world, high degree of institutionalization, a fantasy of narration and dialogues, uncertainty of knowledge, and awareness of hierarchic international institutions, make such question possible – a question that only indicates, beneath the ordered, calm, and the stylized discourse of diplomacy, traces of outrageous, enigmatic, incisive, angry, reconciliatory, appeasing, and fraternal gestures, as if all along these were in 'a free state' deciding from time to time to mark their presence in the narrative of dialogue. On one hand the 'international public sphere' of UN, demarches, conferences and commissions, visits, on the other hand the practices of dialogue by those who are not subject of international law, refusing to be civilized, living an existence of fables and legends that by their very marginal existence to the international are continuously affecting it, unsettling it, and scaring it with their anonymity and modest identity like the lightning's occasional bolts. Whence the agenda to civilize these existences, teach them to talk, be civil, sit around tables and politely converse, and come out of scandalous existences – the agenda to make them truthful, contrite, and reconciliatory.

Obviously I am not suggesting that these are two absolutely marked out different worlds; these are intermeshed worlds precisely because these are different and the international regime of power rules over both universes, at times one to be found therefore within the other. In the quotidian life of protest, murder, anger, solicitousness, gesture of amity, shout of justice, grief, remonstrance, revolt, coalition, altercation, consideration, empathy, hatred, friendship, trust, hope, and hospitality, we have endless traces of the dialogic act –consequently an immense possibility of a discourse of dialogue. Some knowledge of the quotidian life has as much a part in its origin as special knowledge of communication, international relations, and above all anthropology.[34] One day possibly the need for all this special knowledge in order to promote the culture of dialogue would be swept away. Acts performed at the level of every day life will no longer wear the robe of a monarch distant, special, omnipotent, and capricious – sort of a magical authority that the oracles of truth telling have bestowed upon it. The economy of the world of dialogue will shed at least much of its mystery, in other words its discovered and (thus) prescribed rules and orders will be wrestling in open with promise of transparency, that is to say, banality.

The theme of dialogue is, thus, is set here amidst a thick description of post-colonial conditions, and this has broader significance in the world once we look at it in tandem with the twin themes of democracy and geopolitical histories of war and peace. In this context, perhaps the reader should be cautioned at the outset that in suggesting a broader significance of our geopolitical histories in this global age, this book does not attempt a theory of dialogue to be fit for the current age of globalization. In my discussion on the politics of dialogue, not only I have tried to resist the temptation of resorting to any overarching argument resting on a global explanation,[35] as I try to summarize in the penultimate chapter, the examinations of the dynamics of dialogue are intended to present dialogic acts as specific political interventions, which bring out some of the most definite limits of current international politics, also limits of what goes on by the name of conflict and conflict resolution studies. The aim is thus partly to examine the received notion of the international itself by a scrutiny of the boundary of what is considered as international in post-colonial regions through the looking glass of the quotidian – hence the repeated return to the theme of the relation between geopolitics and democracy.

In this relational context new ideas of rights and justice are emerging; they form the terrain where new dialogic acts are shaping up, old divides are being interrogated, and quotidian conversation wrestles with the exceptional. The question therefore is: how can we investigate the truth of the daily practices of dialogue in our historical context? The only available help here, it seems to me, is an awareness of the wide gulf that separates the pasts and futures of friendships and accommodations – pasts of friendships that have left no testaments for us, and futures of friendships that bear no immediate record. Therefore, the importance is of studying dialogue in the sense of current history. Hannah Arendt begins her exercises on political thought of contemporary times by quoting the French resistance poet, Rene Char, 'Our inheritance was left to us by no testament'.[36] It is the same difficulty that marks our discussion on the politics of friendship and enmity. The discussion that follows in these pages therefore has several overlapping aspects – some deliberately brought out to show the complexities of dialogue as current acts; some are intended to show the role of dialogue as resistance acts steeped in ethical connotations, some show dialogue as strategy of rule, and some come out un-intended, they are the result of the unpeeling of layers in the relation between the three tenses, and the three notions of dialogue – namely, geopolitics, conversations on justice and rights, and democracy.

Hence the book is plagued by an awareness of the need for a new language to catch this overlapping world. Such a language is meant to serve narratives of alternative happenings, that is to say other possibilities, always suspended during the enactment of current history. Foucault in one of his intriguing essays, 'Language to Infinity' (1963) refers to the story, 'The Secret Miracle' of Jorge Borges who tells (the story) of a condemned writer to whom God grants, at the precise moment of execution, another year of life to complete the work he had begun.[37] The suspended but impending death is the life of the alternative. Readers will find that what is placed in the following chapters is actually an endless series of possibilities, construction of an impossible history of the politics of South Asia, a site that is nowhere – a series whose each part can only survive by murmuring to death to postpone its arrival. And, since possibilities appear only around happenings or events, such a language to infinity can build only on those events. In other words, the alternative history of friendship and accommodation in this region resides mostly in events of collision and conflict. I have therefore written these chapters around happenings such as partition, borders and border wars, evocation of self-determination, ceasefires, fruitful and fruitless dialogues, and great population flows of massive and mixed nature. Instead of attempting the great mythical book of South Asian relations, it is time we start writing a chronicle of all that could have been said in these fifty odd years, initiating ourselves into that impossible story of what we could have been through knowing what we are today.

Dialogue always brings to our mind the suggestion of a path to peace. Here I show not only that to be not true, my account of contentious dialogue and contested forms of accommodation begins with an inquiry into the supposedly unproblematic nature of the idea of peace based on harmonious conversation. From there I proceed straight to narratives of contentious dialogues and contested forms of accommodation – narratives in which themes of difference and understanding constantly make their appearance, and clamour for attention.

In that landscape, laid bare in the following pages, we shall discover our common sense of justice and dialogue, but in such an unrecognisable form, that we shall be able to imagine the possibility of making that the basis of an alternative argument of democracy.

Notes

1 Zbigniew Brzezinski, *Out of Control - Global Turmoil on the Eve of Twentieth Century* (New York: Collier Books, 1993), p. 163.
2 Michel Foucault, 'Technologies of the Self' in Luther H. Martin, Huck Gutman and Patrick H. Hutton (eds.), *Technologies of the Self - A Seminar with Michel Foucault* (Amherst: University of Massachusetts Press, 1988), p. 31.
3 Based on my experiences of Bangladesh, I write on this in detail in 'Leaders and Publics - Stories in the Time of Transition', *Indian Economic and Social History Review*, 37 (4), October-December 2000.
4 John Rawls, 'Laws of Peoples' in Samuel Freeman (ed.), *Collected Papers of John Rawls* (Cambridge, MA: Harvard University Press, 1999).
5 Charles Taylor, 'Why Democracy Needs Patriotism' in Martha C. Nussbaum *et al*, *For Love of Country – Debating the Limits of Patriotism* (Boston: Beacon Press, 1996); Taylor, 'Cross-purposes – The Liberal Communitarian Debate' in Nancy L. Rosenblum (ed.), *Liberalism and the Moral Life* (Cambridge, MA: Harvard University Press, 1989); see also in this context, Alasdair MacIntyre, 'Is Patriotism a Virtue?', E.H. Lindley Memorial Lecture (Lawrence: University of Kansas, 1984).; for a critical summary of the views, Margaret Canovan, 'Patriotism is not Enough', *British Journal of Political Science,* 30, 2000, pp. 413-432.
6 Jurgen Habermas, 'Citizenship and National Identity' in Jurgen Habermas, *Between Facts and Norms – Contributions to a Discourse Theory of Law and Democracy* (Cambridge, MA: The MIT Press, 1996), appendix 2.
7 Michael Walzer, *Just and Unjust Wars – A Moral Argument with Historical Illustrations* (New York: Basic Books, 1977). For a critique of Walzer's exoneration of Israel's aggressive wars, Richard Falk, 'The Moral Argument as Apologia', *The Nation*, 25 March 1978, pp. 341-343; Noam Chomsky, 'An Exception to the Rules', *Inquiry*, 17 April 1978, 23-27; and Edward W. Said, *The Question of Palestine*, new edition (New York: Vintage Books, 1992), pp. 205-207.
8 Paul Krugman, *The Return of Depression Economics* (London: Allen Lane The Penguin Press, 1999).
9 The most well-known representative work in the institutional approach, R. Keohane and S. Hoffman (eds.), *The New European Community – Decision-Making and Institutional Change* (Boulder: Westview Press, 1991), then, R. Keohane, J. Nye, and S. Hoffman (eds.), *After the Cold War – International Institutions and State Strategies in Europe, 1989-1991* (Cambridge, MA: Harvard University Press, 1993); also Stefano Bianchini and Robert Craig Nation (eds.), *The Yugoslav Conflict and its Implications for International Relations* (Ravenna: A. Longo Editore, 1998). For a critique of the new institutional approaches that base themselves on an argument of 'decline of state capacity' to eliminate poverty and moderate conflicts, and the consequent need for new institutions to tackle poverty and racism, see Louis Kushnck and James Jennings (eds.), *An Introduction to Poverty – The Role of Race, Power, and Politics* (New York: New York University Press, 1999) – particularly two essays by David C. Ramney, 'Class, Race, Gender, and Poverty' (Ch. 2, pp. 39-56), and by Michael B. Katz, 'Reframing the Underclass Debate' (Ch. 3, pp. 59-74).
10 The democratic theorist Claude Ake calls the current interface of globalization and democracy as one of 'dangerous liaisons' in 'Dangerous Liaisons: The Interface of Globalisation and Democracy' in Axel Hadenius (ed.), *Democracy's Victory and Crisis* (Cambridge: Cambridge University Press, 1997), pp. 282-296.
11 Michel Foucault, 'The Political Technology of Individuals' (1981), in *Technologies of the Self – A Seminar with Michel Foucault*, p. 160.

12 Ivan Ivekovic, 'Contemporary Ethnic Conflicts in a Comparative Perspective' in Ursula Oswald-Spring, *Peace Studies from a Global Perspective* (Delhi: Madhyam Books, 1999), pp. 186-251.

13 Ibid. p. 224. See also David C. Ramney, 'Class, Race, Gender, and Poverty' in *A New Introduction to Poverty*.

14 On the relation between new ethnicity and masculinity, brought out so vividly by among others the Balkan wars, see Cynthia Enloe, 'All the Men are Militias, All the Women are Victims – The Politics of Masculinity and Feminity in Nationalist Wars' in Lois Ann Lorentzen and Jennifer Turpin, *The Women and War Reader* (New York: New York University Press, 1998), pp. 50-62.

15 Valentine Moghadam, *Modernizing Women – Gender and Social Change in the Middle East* (Cairo: The American University in Cairo Press, 1993), p. 104.

16 Theda Skocpol, 'Bringing the State Back In: Strategies of Analysis in Current Research' in Peter Evens, Dietrich Rueschmeyer and Theda Skocpol (eds.), *Bringing the State Back In* (Cambridge: Cambridge University Press, 1985).

17 See in this context an unusual narrative of state-formation, Jared Diamond, *Guns, Germs, and Steel* (New York, W.W. Norton & Company, 1997), pp. 265-292.

18 I discuss some of these aspects in my previous work, *A Biography of the Indian Nation 1947-97* (Delhi: Sage Publications, 2001), Chapters 2-3.

19 We are indebted for the insights on this to James O'Connor, *Accumulation Crisis* (New York: Blackwell, 1984).

20 Michael Harrington in *The Next Left – The History of A Future* (London: I.B. Tauris, 1987) has written of Fordism, 'And finally it was of course the 'War Deal', not the New Deal, that put an end to depression'. p. 29.

21 Charles Lockhart, *Bargaining in International Conflicts* (New York: Columbia University Press, 1979), p. 5.

22 In fact Lockhart cites the work of Prichard Walton and Robert Mckersie, *A Behavioral Theory of Labor Negotiations – An Analysis of Social Interaction System* (New York: McGraw Hill, 1965), to show how rational bargaining leading to expectations of 'distributive bargaining' (division of spoils) or 'integrative bargaining' (upgrading of common interest) are prevented by intra-party disagreements that make 'commitments fuzzy and hinder integrative bargaining', thus making 'attitudinal structuring' (influencing perceptions of the opponent) – one of the main goals of bargaining – almost impossible. See, *Bargaining in International Conflicts*, pp. 13-15.

23 *Bargaining in International Conflicts*, p. 43.

24 In my *A Biography* (Chapter 8) I discuss two instances of how bargaining becomes the form of contradictory and yet complimentary processes of relations between post-colonial nations marked by the war-peace continuum. Irving Louis Horowitz had shown many years earlier (1957) how in philosophical and social thought the continuum had been an important object of inquiry. See his *War and Peace in Contemporary Social and Philosophical Theory* (London: Souvenir Press, 1973).

25 On some of the issues concerned with constitutionalism and the debate on the relative importance of the procedural and substantive aspects of deliberation, see Jack Knight, 'Constitutionalism and Deliberative Democracy' in Stephen Macedo (ed.), *Deliberative Politics – Essays on Democracy and Disagreement* (New York: Oxford University Press, 1999), pp. 159-169. Knight writes commenting on Amy Gutmann and Dennis Thomson's joint work, *Democracy and Disagreement* (Cambridge: Belknap Press of Harvard University Press, 1996), 'The pervasive role of democratic deliberation in the interpretive process challenges some of our fundamental ideas about constitutionalism and the rule of law.' (p. 168).

26 Jurgen Habermas in *Between Facts and Norms – Contributions to a Discourse Theory of Law and Democracy*, pp. 447-462. I am pushing here Habermas' argument to the extreme.

27 'Popular Sovereignty as Procedure', Lecture in 1988 on The Ideas of 1789, published as Appendix 1 in *Between Facts and Norms*, pp. 463-490.

28 Ibid. p. 483.

29 The African political scientist Julius Ihonvbere notes the rise of new constitutionalism in several countries of Africa. He describes how earlier even if states had constitutions, the leaders whom he calls 'the big man' did not care for these constitutions, or were simply ignorant of the existence of such basic laws. There was no popular deliberation, and constitution was irrelevant to the social and political conflicts and turmoil in the country. He then describes how in many of these countries new rules are being forged through popular deliberations or at least popular coalitions are trying to formulate new basic laws, moved by experiences of conflicts and wars. Julius O. Ihonvhere, *Towards the New Constitutionalism in Africa* (London: Centre for Democracy and Development, 2000); also his lecture, 'Engaging the Leviathan – Constitutionalism and the New Politics in Africa', Annual Minority Rights Address, organized by the International Centre for Ethnic Studies (Colombo), Geneva, 18 May 2001.

30 *Between Facts and Norms*, preface, p. xli.

31 For such documentation, interested readers may see Sabyasachi Basu Ray Chaudhuri and Shahid Fiaz (eds.), *The Kargil War – From the News Files*, SAFHR Paper 6 (Kathamndu: South Asia Forum for Human Rights, 1999).

32 Paul D. Sense in his essay, 'Democracy and Maturity – Deciphering Conditional Effects on Levels of Dispute Intensity', *International Studies Quarterly*, 43 (3), September 1999, pp. 483-502, writes, 'At the core of democratic peace theory is the notion that democracies behave in accordance with norms of mediation, compromise, and restraint in their interaction with one another' (p. 483). He notes that while there is no agreement on this and no clear empirical evidence, on the other hand with 'relative frequency democratic disputants select hostile strategies sometimes stopping just short of war'. (p. 500). See also Ralf Dahrendorf, *The Modern Social Conflict – An Essay on the Politics of Liberty* (New York: Weidenfeld and Nicholson, 1988), where he speaks of the democratic agenda in form of new social contract and speaks of the democratic capacity and necessity to depart from policies of unilateralism. (pp. 166-194)

33 Thomas Risse, ''Let's Argue!' – Communicative Action in World Politics', *International Organization*, 54 (1), p. 20.

34 A glimpse of such a twilight world, Pradip K. Bose, 'Anthropology of Reconciliation' in Ranabir Samaddar and Helmut Reifeld (eds.), *Peace as Process – Reconciliation and Conflict Resolution in South Asia* (Delhi: Manohar, 2001), pp. 45-60.

35 On this see, Justin Rosenberg, *The Follies of Globalisation Theory* (London: Verso, 2000); Rosenberg however does not specifically address the question of global reach of power, His concern is to show the weak foundations of the theory of globalization, particularly its overemphasis of time-distance axis.

36 Hannah Arendt, *Between Past and Future – Six Exercises in Political Thought* (London: Faber and Faber, 1954), p. 3.

37 'Language to Infinity' in *Michel Foucault – The Essential Works*, Vol. 2, *Aesthetics, Method, and Epistemology*, ed. James D. Faubion, trans. Robert Hurley and others (London: Allen Lane, The Penguin Press, 1998), p. 91.

Chapter 2

Peace as the Political Question of Our Time

As the introductory chapter tried to explain, in the post-colonial conditions of war and peace dialogue is not an uncontested act, nor is it inherently pacific in the ethical sense of the term. Deployed by rulers as a strategy of rule, mounted by the ruled as the ethical plank of justice and rights, used by different actors as trope, dialogue carries within it marks of contest and contention. Yet, as we know, dialogue in history has been traditionally associated with peace, for while war is seen is conflict-centric, peace is seen as born from different degrees of harmony. Dialogue is seen as mark of an attempt to resolve differences and found relations of harmony. Dialogue, therefore, in the contemporary world of contention is viewed as the way to peace. But does this view throw light on the contentious nature of peace itself, which can throw light on the contentious nature of the dialogic act? What shall we say, when we see that peace as much as war – in essence the war-peace continuum – is bound by geopolitical reasoning and our geopolitical history? We must, therefore, begin our account of the politics of dialogue by interrogating the supposedly unproblematic nature of peace. And though we know that philosophy has tried to come to terms with the problematic of peace, there cannot be a better way than taking a leaf out of our bloody past of partition to see the contested nature of peace on which dialogue is perched.

Contesting Visions of Peace

In 1960, thirteen years after the partition of the Indian sub-continent, India's first prime minister Jawaharlal Nehru reflecting on the consent of the Indian National Congress in 1947 to the decision of the colonial rulers to partition of the country is said to have told Leonard Mosley, a British author, 'The truth is that we were tired men, and we were getting on in years too. Few of us could stand the prospect of going to prison again – and if we had stood out for a united India as we wished it, prison obviously awaited us. We saw the fires burning in the Punjab and heard everyday of the killings. The plan for partition offered a way out and we took it'. When the countdown to partition had begun that year, Gandhi was in the riot torn area of Noakhali in the eastern part of Bengal, and he wrote in his diary on 2 January of the eventful year, 'Have been awake since 2 A.M. God's grace alone is sustaining me. I can see there is some grave defect in me somewhere which is the cause of all this. All around me is utter darkness. When will God take me out of this darkness into His light?' And then few months later declining an invitation to join Nehru and Patel in Mussoorie he wrote on 25 May, 'I am quite well though in boiling heat. I must not think of Mussoorie or any other similar climate', and then, 'I see no place for myself in what is happening around us today...I have no wish to live if India is to be submerged in a deluge of violence, as is now threatened', and still again, 'I am in the midst of a raging fire. Is it God's mercy or irony of fate that the flames do not consume me?' Sardar Baldev Singh, the defence minister in the Interim Government who was reasonably confident that large scale troop deployment would ensure peace during division of the country, had endorsed the plan of partition amidst massive violence particularly in Punjab, with these words, 'It does not please everybody, not the Sikh community anyway. But it is certainly worthwhile. Let us take it'. The Commander-in-Chief of the armed forces Field Marshal Claude Auchinleck concerned himself so much with the safety of the British in the sub-continent after the British would withdraw on 14-15 August rather than concerning himself with saving human lives amidst that unprecedented destruction and carnage, that the General Officer Commanding of the Eastern Command could only forecast in disgust 'a bloody massacre in Punjab'.[1]

Fatigue, fear, escape, illusion, despair and the desire to commit suicide, choice of a lesser evil, self-preservation – all kinds of feelings gave form to the desire for peace and led to adoption of policies that to the actors of that time appeared as capable restoring sanity. It is a different matter that peace came much later after many, many lives were gone, even many more were rendered homeless and displaced, while some were not touched at all in this all-consuming fury. Peace has thus many visions, probably as many as life offers it in form of way out of contentious politics. Peace is no less a bright mirror also, in which even day light murders play their selves out.[2]

The question therefore is: how does a particular politics of peace achieve universality, in the process predicating or almost determining the dynamics of dialogue? And is this not what happened when discussions on Cabinet Mission proposals were held and failed in the Indian sub-continent – a particular way of holding dialogue and achieving peace became the universal consensus on partitioning the country? Or, how does peace, which is seen only in the light of war, become so general and elemental that it loses its opposition to anything (including war) and thus loses its speciality, even hiding at times the reality of war? In other words, the inquiry is not about a generalized politics of peace, but the particular issues that the notion of peace poses before the politics of our time, especially for a dialogic politics. Let us for once then begin with tales of those questions around peace, and not war. Needless to say, as happened during the months of our partition, since the early time the tale of peace has been interwoven in that of war, and arguments for peace have been derived from arguments of strife, thus making accounts of peace always conditional and unclear. It is necessary, therefore, to read these tales of peace as readings of the various ways of enmity and accommodation, accounts of failed and un-failing politics of friendship – of arguments embedded in histories of varying kinds, including histories of power, war and geopolitics. Let us start that enterprise by taking a brief look into different readings on peace.

Peace in Different Mirrors

Quarrel and Fear

The causes of war Thomas Hobbes had reminded us was 'quarrel', and he had three origins of quarrel in mind. These quarrels were *first*, a result of competition or pursuit of gain, *second*, diffidence or a desire for safety, *third*, glory or a desire for recognition. Quarrel thus signified a broader thing, *war of every man against every man*. Self-instinct for survival leads to the quest for power. Mankind is therefore perpetually compelled to seek power because each has to go to war for self-preservation. Honour, dignity, science, technology of control, peace – are all to be understood as expressions of power. The phenomenon of power is thus basic to humankind's experience of war and peace. The 'fear of death' leads man to strife, war, survival and compacts for peace.[3]

But if the instinct of self-preservation leads to war, alternatively to peace, how can mankind be sure that peace may be durable, for man by *nature* has to seek power, and therefore may relapse into war? We may have three solutions to this conundrum. First, the very aspects of human nature that cause war can be pressed into the service of peace, and this is what Hobbes seemed to have in mind. Second, human nature can be changed, at least partly. We are reminded of Plato's *Dialogues* in this context. One can also remember in this context Marx, though in reference to any such possibility, he never spoke of any abstract human nature. Human nature depended on the relations that man entered into, and the change of human nature depended greatly on the changes the structure of relations underwent. Third, as the *Federalist Papers* conveyed the idea, a strong sense of rights and constitution can put the concern of society above individual concern, and thus can stay war. War and peace are thus interchangeable. They defy solutions. They can be stayed, but as products of assumptions about power, they remain. Only 'fear of death' keeps the war at bay. Madison who had said, 'ambition must be made to counteract ambition', went further. There were two types of war, flowing from the will of the government, and that of the society. By changing the structure of the government, the former could be prevented. And, while the second could not be prevented, those who declare war would have to submit to law that those who warred must pay for it.[4] Thus, fear, desire, reform, and cost and compensation principles – are the elements that build peace that may any time slide into war. In this unavoidable quandary, we are told by political philosophy that, war and peace reside. We have no moral propositions here to discuss and decide. This is a world of the rational abound with assumptions of power, fear and desire. The aim of dialogue, we may deduce, is to make the world rational, or to confirm the rational and make it explicit.

Unfortunately, there hangs a tale about this rationality, rather about its imperfection. The point is, if war and peace are interchangeable, then why is it that only peace condition can be visualized as *absence of war*, and not war as absence of peace? If peace is only in the mirror of war, shall we say then that war is the natural condition, and peace is what authority brings about? From Hobbes to the arguments of *Pax Americana*, this seems the suggestion, though there is no answer to the inquiry as to why conflict, that product of power, fear and desire, should *naturally* lead to war and not adjustment, negotiation and understanding. The imperfection of inter-changeability thus raises the basic question: must peace be seen in the light of war?[5] Are these two concepts at all connected? Can there be a positive definition of peace that does not link to war for its own conceptualization? Revolt and rebellion have legitimacy in society because they override the binary, even when ushering war they become widely acknowledged as a pathway to reconciliation and peace. The rational line of argument will however say that this should be seen as the 'vigorous way to peace', in much the same way Hegel had spoken of the 'ethical health of the people' corrupted by 'prolonged calm', 'corruption in nations...the product of prolonged peace'. (Section 324, *Philosophy of Right*). This was like Immanuel Kant saying that long peace brought about a predominant commercial spirit and along with that 'low selfishness, cowardice, and effeminacy'.

Desire and Good Life

Since the rational argument makes peace an ancillary of war, it is understandable why the thinking on peace whenever it appeared in its 'original form' had taken the line of beauty, happiness, and goodness. Particularly religious philosophy in its many forms has never held *reason* as a positive attribute. It is neutral faculty that may serve wicked wills. Therefore the theme of peace is lodged in a 'longing', 'desire' for the good, the beautiful. Saint Augustine had told, 'There will be a peace so good that no peace could be better, a peace so great that a greater would be impossible.' This phenomenology of the desire for peace subsumes within it the practice of war also. 'Notice that there can be life without pain, but no pain without some kind of life. In the same way, there can be peace without any kind of war, but no war that does not suppose some kind of peace. This does not mean that war as war involves peace. But war, insofar as those who wage it or have it waged upon them are beings with organic natures, involves peace – for the simple reason that to be organic means to be ordered, and therefore, to be in some sense, at peace.' But appeal to nature does not end with order. The immanent goodness is also crucial. We are told that 'anyone who grieves over the loss of peace to his nature loves itself, (because)...there can be no nature completely devoid of good'. 'If there were no good left, there would be no one to lament the good that has been lost'.[6]

So far, so good, but how does man feel happy? What is there in happiness that creates this longing for peace, a pacifist disposition, an inclination not to hurt but to understand, to appreciate, to approach quarrel with mediation? If appeals to immanent goodness have to be supplemented with logic that we have left behind, we cannot be sure that peace will be inherent to society. This is an admittedly a closed circle, we have to try to break out. The consequence of this closed circle for a theory of dialogue is obvious. Dialogue cannot be considered as an act in poetics, its aesthetic dimension is subject to its rational dimension.

Conflict

The closed circle is obvious when we see that commentators like Kenneth Boulding though not always sympathetic with the rationalist-realist argument, are forced to concentrate on the dynamic nature of peace, and therefore emphasize that peace is ultimately a state of condition in which war is absent.[7] In his words, 'All non-conflict is peace, but conflict can be divided into war and peace'.[8] And what then is conflict? Boulding defines conflict as, 'a situation of competition in which the parties are *aware* of the incompatibility of potential future positions and in which each party *wishes* to occupy a position that is incompatible with the wishes of the other' (emphasis author's).[9] In this given asymmetry of consciousness and desire ('awareness' and 'wish'), rational arguments about peace have to depend on, what Boulding calls, an 'ordered set of positions', which are fragile, for they depend on the 'boundaries of possibilities'.[10] In other words, this is not a stable peace, but limited peace, because of 'rapid succession of situations', only 'partial equilibrium, existence of 'hostility and friendliness as fundamental variables' defining 'system boundaries'.[11] Therefore, in Boulding's terse words,

> This is a very important phenomenon in the conflict processes of the real world, where arms race frequently leads to war, industrial disputes to strikes, marital discord to divorce, verbal quarrels to fisticuffs, theological disagreements to separations, and party strife to dictatorship...

Sometimes the system never recovers after a breakdown; one country is destroyed by the other in war; the firm is bankrupted or the union is destroyed in a strike, one party is killed in a fight or duel, or a divorced couple never see each other again. On the other hand, it not infrequently happens that, after a breakdown and an interlude of open conflict, the old system re-forms, perhaps in a slightly different pattern. Nations renew relationships after a war, a firm and a union continue after a strike, friendship survives a quarrel, and divorced people marry each other. This alteration of systems is one way in which systems that have no true point of equilibrium, or an equilibrium beyond the limits of tolerability can persist.[12]

Thus we have here the admission that conflict theory cannot provide permanent guides to rational solutions to conflicts spread large on a canvas that ranges from quarrels between nations to those between a couple. Dialogue is then perennially contingent on the nature of the flux in the midst of which it is being conducted. The outcome of the 'game' (all conflict theorists urge us to look conflict-outcomes as games) is uncertain, the payoffs limited, the payoff matrix defying predictability, and the bargaining power curves inviting countless modifications.[13] Boulding admits at the end of his massive work on conflict, that the 'critical boundary' of conflict poses the greatest problem, because it may be consisting of several discriminate elements such as geographical boundaries or legal boundaries, and comments, 'Reconciliation will be presumably easier if reconciliation itself is highly valued as a process by the contending parties'.[14] In other words, peace and reconciliation is not reducible from conflict; there is something more fundamental at stake.[15]

But if reconciliation and dialogue are values by themselves and not deducible from conflict, then why do we find a marked lack of dialogic and reconciliatory spirit in our politics and daily life, stretching, as Boulding reminded us, from marital discord to national wars? Obviously, what Kenneth Boulding did not engage in was an analysis of the element of power, more necessary than an analysis of the phenomenon of conflict. In fact, in 1962, the same year when Boulding had published *Conflict and Defense – A General Theory*, the noted behavioural scientist Amitai Etzioni had written, *The Hard Way to Peace*, where he had noted an escalation-effect in the deterrence strategy and had argued that 'multi-deterrence is not only utterly incapable of averting these dangers, it actually serves to increase them'.[16] He termed it as 'mad rationality'.[17] Also in arguing for a new political strategy for peace, he defended the idea of expanding the moral community,[18] because he said that mere disarmament would not do, as 'history teaches us that after disarmament frequently comes, not peace, but rearmament',[19] and 'the confusion of arms-control with disarmament is not completely unintentional',[20] because the purpose may be to prepare public opinion for armament again. Boulding himself had to admit later on that, 'correlates on the incidence of war and peace are elusive', that cause and effect angle was useless, and that there were too many dynamic and structural variables in strain and strength functions of the war-peace systems.[21] He said that justice and notions of justice formed an important segment of a war-peace system, and a peace system perceived by increasing number of participants to have elements of injustice will be subject to increasing strain'.[22] He commented,

> I have often compared the peace movement to the labors of Sisyphus – we push the stone uphill and continually it breaks away from us and rolls down again and we have to start all over again. But the hill is not infinite and it has watershed, and one day the stone will roll over the watershed and we will be chasing it instead of pushing it up.[23]

Not that conflict-centred peace studies (conflict-resolution) did not analyse power, but because it accorded centrality to conflict, power remained in the background, foreshadowing many things, but as if it was a monster, better left untouched, un-addressed. Thus, with all the writings on games, the guru of games cautioned his readers, 'The only guarantees of peace are compatible self-images. These have been produced largely, one suspects, by accident, though occasionally by an almost sub-conscious series of policies'. He further noted the 'increasing spiral of illusion and folly' and the 'essential fragility of taboos in the war situation'.[24] Thus, there was inadequate study of mode of power at global level, the nature of informational, referent, legitimate, expert, strategic, rewarding, and coercive influences that went into making the form of global power – in short quantitative and qualitative aspects of power. Anatol Rapaport, one of the well-known observers of the conflict scene in the 1960s and 1970s wrote, 'Search for solutions of specific problems led to so-called unsolved problems', and 'an enlargement of the repertoire of concepts and the formulation of entirely new classes of problems inconceivable in the old framework of thought.'[25] Conflict specialists had to justify their trade by tying all analysis of peace and dialogue to the immediate issue of conflict – a bizarre return to Say's law – the merchandize beginning with the Cuban missile crisis. The Indian and Pakistani civilian militarists displayed the same pattern of response after the nuclear tests by the two countries in 1998. What Paul Swingle wrote in 'Dangerous Games' is even more relevant in the context of our discussion and is worth quoting at some length,

> Beyond the clear set of objections...I have pointed out over the years, there is one that ahs seemingly escaped everyone's attention in the past (including my own) – namely, the role of war game theory as a legitimate device for whatever crude military strategy has been decided upon.

A tautological aspect thus emerges: If the decision to blockade Cuba is taken, war game theory is appealed to as the ultimate arbiter; if the decision to lift the blockade is taken, the same appeal to war gaming is made; and since any complete holocaust would 'terminate the game' and 'eliminate the players' there is no real possibility of disconfirming the 'theory' on which the decision is ostensibly reached. Under such a wonderful protective covering of *post hoc* (italics author's) legitimation, and with every strategic decision confirming anew the worth of war game theory, it is extremely difficult to reach any final estimate of the theory as such; for this reason, the examination of real events – particularly military fiascos such as the Bay of Pigs landing or the Gulf of Tonkin retaliation – may be the clearest way open to analysts for evaluating the potency, or as is more usually the case, the paucity of war game strategies.

In part, what has been described in this paper belongs to history rather than to politics. And this is as it should be, since when a particular strategy becomes elevated to the level of military theology, the clear and present danger to human survival becomes apparent. And in the shock surrounding the Cuban missile crisis – the delayed awareness that the world stood still for a week while games of strategy were permitted to run their course – war game theory had its proudest moment and yet its last moment.

It was not long after the Great Missile Crisis that the 'game of chicken' was abandoned in favour of conventional forms of political accommodation. This came about through the mutual realization of the Soviet Union and the United States (especially the later) that Cuba was not a pawn or an ace-in-the-hole, but a sovereign power in its own right. The Castro Revolution was both national and hemispheric; it evolved its own brand of socialism to meet the challenges of a single crop island economy. Thus the Cuban regime was a system that had to be dealt with in traditional political term of how sovereign states with differing political structures relate to each other. When this dawning took place the Cuban 'crisis' was really solved, precisely by surrendering the notion that this was a behavioural situation reducible to the moves and countermoves of the world's big two military powers. Yet, as long as such repudiation of strategic thinking remains informal, the dangers in a repetition of such forms of crisis management through games of chance remain ever present. And what first appeared as a tragedy may return not so much as comedy, but rather as absurdity – in this instance, the absurdity of total mutual annihilation.[26]

Moving away from conflict theory (and not contest) in analysis of peace politics is thus not a tactical move. It implies fundamental reconsideration of how we are to think of peace – again not as an abstract moral, but one based in time, in the continuum that we have spoken of above. Given the fact that contradiction is not an optional mode of social existence, the activity of peace invariably goes back to the issue of choice also, which includes preference for a life that denotes satisfaction about values, in other words, a choice relating to *living how*, the happiness of having a vision, and therefore a capacity to exchange, interact, align, ally, and befriend, in short to understand *difference*. It includes issues of ethics, and as I show in the following pages, these issues are always historically predicated. In an age of nuclear menace, fierce civil wars, massive migration, destitution, and a resurgence of statism, *living how* is a crucial lead to the inquiry of dialogic existence, and peace. If indeed, we are to go beyond conflict-centric analysis, and take a look into the ethical-psychological aspects of the question, we have to pass not only the modern exponents of the theory of security, but also move beyond Saint Augustine.

Security and the Mirror of Behaviour

Yet it will be worth our while to see how the peace argument fares when it is perched on the security tree. It begins with realist assumptions and then goes into a fierce psychological tailspin. To say the least, the result is intriguing. The realist argument will predictably say that to have peace is to actually have workable peace; to have workable peace, peace has to be sited in security considerations. Security implies security of one's body, body of the nation/state, and in a milieu of mutually accomplished total destruction security of the body of the international. I leave aside for the time the possible objection to defining the *international* – what this includes and excludes. But we can at least agree that security implies threats to security, consequently perceptions of threats to security. Perception is a matter of psychology, so the security argument pitches the issue of peace on the ground of mutual invasion of *peace of mind*. After this twist, then, we find the peace argument taking a bizarre form. We are told that peace is achievable only when we do not have peace of mind. By the same token, we can dialogue, only there is a complete end to dialogue.

The argument of *peace as security* is premised on conflicts. It is true that conflict studies in the form of conflict-management and conflict-resolution studies have contributed to security studies that are assumed to be part of peace studies as well. One of the notable phenomena in peace studies has been the rapid development of conflict resolution studies.[27] Studies of dialogue, bargaining, and diplomatic negotiations and treaties are on one hand parts of what is known as diplomatic history, on the other hand they are considered falling in the genre of conflict and security studies. The theory, practice and skills of conflict resolution have taken off from behavioural and anthropological studies, and system-building programmes in the wake of managerial development in the field of arbitration, mediation, adjudication, and the practice of evaluation research. Starting as a managerial tool in the field of social conflicts, it became a matter of academic pursuit. Conflict-resolution programmes based in social and behavioural sciences and sometimes in liberal arts, spread in form of seminars, workshops, collaborative research, and foundation activities. Teaching conflict resolution turned an amorphous practice into a discipline. Curriculum, simulation, instrumental dissemination, and research combined into the making of a community of practitioners who became a part of the community of strategic analysts, security specialists, and conflict moderators. Conflict resolution could not keep itself distinct from conflict management and conflict moderation. The liberal arts and sociological dimensions of the theory and skills of conflict resolution and dialogue were subsumed in the behavioural studies of diplomacy and statecraft. This is of course not to deny that conflict resolution approach in many cases has led to the growth of peace studies. Experiences in community leaving, the academic anchorage to themes of social justice, and the emphasis on the training of mediation techniques have developed a field where peace has become important element. But, at best this has given academic recognition to peace studies in the times of cold war and the subsequent years of the Washington consensus, and at worst it has become an appendage of conflict studies that in its turn had grown as an ancillary of cold war studies. In the development of disciplines, power, after all, has exercised a tremendous role.

The result is that even when peace studies have attended to security concerns through the route of conflict resolution and conflict moderation studies, studies of bargaining, interacted with strategic studies, and have attempted to engage the security lobbies in a dialogue – dialogue between peace and security – the result has not been encouraging. The security studies lobby looks to peace studies academics and practitioners as upstarts or gatecrashers, men (these days some women) with soft belly, too utopian, too amorphous, too impractical. The security specialists think in this way, more because, they now occupy important positions in the portal of power, and they have developed an entrenched interested in prioritizing security among all concerns. Studies of security are less a matter of study as an academic vocation, more a matter of getting job. But this is a special job, one of being hired as consultant, writing reports and joining disarmament and security conferences, acting as negotiator in arms control dialogue, and finally becoming the truly multinational, because the more one is specialist in security matters, the more one becomes global. In this hierarchy of disciplines and fields where the leitmotif is power/knowledge, peace studies have been the victim both as knowledge and as a depository of power. It is important therefore to know at least briefly the nature of security criticisms against peace studies. This will show whether the security route to peace studies is viable or not.

George Quester in an essay 'International-Security Criticisms of Peace Research' neatly outlines these criticisms. Peace researchers are too optimistic and unrealistic. They tend to assume that good things must go together. They think that domestic and international improvements are always linked, and that disarmament serves all the purposes of humankind. They redefine peace to include justice, and social improvement, and whatever other causes they favour. The result is that they remain confused about which particular problem to be solved at a particular time. And all these they do or think because they hold that others, especially the security-specialists do not understand the desirability of peace.[28] Quester has written,

> The assumption that domestic reform will lead to international peace or that one kind of international improvement will always support another kind of international improvement has emphatically not been endorsed by traditional European analysts of power politics...When asked how they would choose between preventing nuclear war and preventing conventional war, many people who style themselves peace researchers will thus simply refuse to comment, regarding it as a moral betrayal even to acknowledge that such a choice might ever have to be made, arguing that there must somehow be a solution that would never giving up on either gain. Similarly, if asked to choose between eliminating poverty and eliminating war, the same researchers will again run away from the question, as if it were some kind of trap.[29]

Thus peace is security and must not be related to end of poverty, or any 'domestic' reform. By the same token, nuclear security is peace; conventional disarmament is a false pursuit. Yet this is not all.

The peace studies approach seems to be on the verge of blaming our problems on an insufficient aversion to war among government officials and the establishment, and it extends this accusation to the people who work on national security, international security, or arms control...The international-security community would normally counter that the disagreement is not about values or awareness or paradigms but about analysis...For example, would the total nuclear disarmament of the Soviet Union and the United States make it less likely that a Qaddafi would seek nuclear weapons of his own, or would make it more likely? And what are we then to do about vertical and horizontal nuclear proliferation if there is no preordained linkage that all of these threats to peace recede in the same move?[30]

And then the final complaint that peace studies does not given enough importance to self-interest. Self-interest makes the 'fashionable' distinction between 'negative' peace and 'positive' peace meaningless. Attempt to make such distinction,

> ...is to use these words of our English language in a manner that substantially underrates the human priority of eliminating warfare, whatever its causes and whatever the remedy...The crucial accomplishment in the advancement of arms-control reasoning comes with the open acknowledgement that now we must always care about the self-interest and motivations of our adversary. [31]

The comments are familiar. The underlying assumptions are worth noting. In order to have peace we need to have a demon from whom we need (security and thus) peace. We must have therefore nuclear weapons, because Qaddafis may have them. But making a demon is not enough as a peace strategy. It is necessary to believe in the pacific nature of state, officialdom, military officials, arms-control experts, and security specialists. Finally, peace is a phenomenon that is not rooted in any process, 'causes', and 'remedy'. In the security argument we have thus a return to psychology – we need justification for war in order to justify peace. In this play of psychology, to which all behaviouralism relapses, the realists are the most surreal. They inhabit the dual world of reality and fantasy. In fact, the real can appear as real only by being unreal.[32] In this surreal mirror, peace is transformed as security, conversation is reduced to bargaining, relation is transformed into behaviour, and existence collapses into self-preservation. The implications of the multiple series of metamorphosis for a culture of dialogue are to see the least enormous.

Peace as a Discipline

The disciplinarian argument is thus about disciplining peace with the argument of security. The security *mantra* is however is only one way of disciplining peace. What is significant is that peace, as an educational pursuit, has to be disciplined. Yet a mere acknowledgement of this is not enough. It is important to understand why there is the need to discipline peace.

The uniqueness of peace studies at the present moment is that the pursuit has originated from a conjuncture of circumstances. Disciplining means cutting out traces, links, and silencing the multiple, sometimes discordant and cacophonous, voices that arise from the circumstance of conjuncture in which peace studies as a vocation is situated. That is how disciplines originate; the effervescent field of differential traces is cemented into a solid structure of knowledge and an intellectual tool of power. Peace studies imply peace research, peace education, and peace action together. At a minimum level it means simply stopping violence, and particularly war, and at a positive level overcoming social injustice. Its locomotive had been earlier the traditional knowledge of society and increasingly now the newer ideas of, and movements for justice. Finally, studies of peace depend on the knowledge produced from the local (sometimes called the grassroots) movements, and as a concept it relies on global frameworks. Peace vocation is thus at, what Chadwick Alger calls, the 'crossroads'.[33] Can one merely study peace? Can one be content with prevention of war? Or, perhaps, can one be satisfied with local knowledge while working for peace? These questions are invitations to take the political character of peace studies with more seriousness.

If technologies of rule and violence are getting more and more complex, peace actions need a more secure foundation in studies, and peace studies needs more researches into these technologies. To Johan Galtung, peace research is crucial, because,

> The danger is not that researchers also are interested in education and action. They should be, and thereby get some important feedback from the real world, from people who challenge the 'findings' and from a complicated reality that refuses to respond to actions in the way suggested by the research. Much more dangerous than this would be those who engage in education and action without any research basis and for that reason have a tendency to repeat their cherished ideas, whether those of the establishment or the anti-establishment, trying to shape the world according to their dogmas.[34]

The problematic is then clear: can peace programmes be deepened with research, but without being turned into another sterile pursuit or the professional pursuit of an exclusive community? The skills of community organizers, social scientists, mediators, lawyers, pedagogues, teachers, and elders of society, go into the enrichment of peace education. But peace produces its own practitioners; they pose a problem for peace, for peace is essentially a reflection of the conjuncture of circumstances and roads. It is therefore essentially an experiential act – an act of challenging values, a moral act. Dialogue sits therefore at the heart of peace. The contested nature of the peace question reflects in the contested nature of the dialogic act. On the other hand, the experiential nature of the dialogic act predicates the various ways in which social constituencies appreciate the question of peace.

The political trend of turning peace into a fund-driven, curricula-based, university-incorporated vocation is against such moral-experiential act, though ironically the moral act sets off the political agenda. In such institutionalization of peace, we have the seeds of a merger of peace studies and conflict-based programmes. In the conflict scenario, as we have noted, endemic features are not issues of concern, only the cataclysmic form known as conflict is the concern. Peace, therefore, in such disciplinarian form, defines itself as an end of a conflict, and not an engagement with endemic matters. Also we have noted, conflict studies is based on behaviour-centric analysis. Thus violence is seen mostly a matter of human behaviour, of pathology, in other words endemic. It has to be cured. The stress is on non-violent behaviour, and a non-violent resolution or moderation of conflict. In such lore of non-violence, interrogation of issues of justice becomes secondary. In well-intentioned academic programmes on peace, themes of non-violence, human nature, feminist education, pathological cure, and community restructuring, form the rainbow of peace. Birgit Brock-Utne in *Educating for Peace: A Feminist Perspective* shows how in educational practices, behavioural aggression, competition, and oppression of women are fostered in boys, and female-male differences make this world violence-prone.[35] Thus, even though in this case, peace studies speak of *peace with justice*, peace is taken to be fundamentally *pacific behaviour*. The themes of *security as the core of peace* and of *right behaviour as the core of peace* are thus two sides of the same problematic of defining peace in an institutional format. The complex and the contested nature of peace, and its character as a political question thus go out our sight. Consequently in the disciplinarian mirror of peace studies, dialogic act becomes essentially a matter of correct and appropriate behaviour.

The problem is made acute when the behavioural is supplemented with the sociological. Thus, it is said, we need peace because there is too much 'gruesome interracial and interethnic bloodshed'. As one sociologist says, 'Something has gone awry in centre-periphery relations', there is 'collective violence among people who are not aliens but enemies intimately known'. 'Ethnic conflict is a major reality of our time'...thus upheavals...have become climactic in recent years, among them (are) the Sinhala-Tamil war in Sri Lanka, the Sikh-Hindu and Muslim-Hindu confrontations in Kashmir...the Chakma-Muslim turmoil in Bangladesh....Most of these conflicts have involved violence, homicide, arson, and destruction of property...The classic definition of the state as the authority invested with monopoly of force has become a sick joke'. It is further said that easy access to technology of warfare and international fraternization among resistance groups, who have little in common save their resistance to the status quo in their own countries and who exchange knowledge of warfare, has increased violence.[36] In this scenario the pathological problem is shown with the addition of sociological explanations. Thus in this landscape violence is civil. State does not feature, and even if it does, it does in a minor way. In the rose-tinted word processor every conflict is ethnic. Thus there is a 'Chakma-Muslim' problem in the Chittagong Hill Tracts in Bangladesh, similarly a 'Hindu-Muslim' problem in Kashmir. The nation-state or statism is not the prime root of violence, but ethnicity is. The road to peace is primarily through a negation of particularities; the state may rest in comfort while peacemakers go about extinguishing fires of ethnicity.

In the context of such banality, we find increasing spread of the idea of 'powerful peacemaking'[37] that comes from a concept of experiential education. Experiences of community living, democratic decision-making, handling of ethical issues and organized protest against injustices are reconciled with social analysis, technical reasoning, and historical research. The experiential nature of peace makes dialogue a living education. Experiential education is thus not about the theme only, it is also about the way. 'There is no way to peace, peace is the way.' It is thus a matter of moral imagination – working against the themes of discipline and power with the imaginative tools of morality. Non-violence can be thus only one aspect of this experience; other important aspects are engagement, dialogue, and non-compliance with the rules of domination.

History, Contests, Geopolitics and Dialogue

Peace question, therefore, finds itself in the complex and contested arena of power, injustice, structural violence and war – all that make our history. If this history has in it the threat of an impending nuclear catastrophe provoking the desire for peace, it also includes a growing understanding of human rights as a critical factor in making peace a political issue. Awareness of a whole range of civil, political, and economic rights, rights pertaining to minorities, language, education, and work, has either forced the States to open up the political sphere at least to some extent), or have broken the social peace imposed on societies by the States. With the growing awareness of rights, the social peace imposed by regimes in countries such as Turkey, South Korea, Singapore, Brazil, or Nigeria – and there are many more – has increasingly faced protest. Today's peace question is thus unique in the history of peace. It builds not on social peace, but on contesting that peace.

The historicity of peace is thus to be noted. Our discussion on the politics of dialogue in subsequent chapters will be in this historical context.

First, in post war era international institutions providing some legal clarity to the notion of peace have grown.

Second, the older definitions are becoming antiquated in the context of the discourse of new generations of rights. Redefining peace is thus becoming an imperative.

Third, the relation of peace to democracy is becoming complex as the conduct of liberal democracy in matters of war and peace becomes more and more non-ideological, and purely a matter of convenience.

Fourth, causes and phenomena hitherto considered 'external' and internal', distinct from each other, are collapsing into one zone of unrest whose singular feature is the breakdown of states.

Fifth, restoration of state and restoration of peace have become a single enterprise wherein peace building and state building have become the two inter-linked aspects of peacekeeping.

Sixth, like the conduct of war, the conduct of peace politics is in a distinct geopolitical setting marked by globalization, collapse of several States in wide areas, emergence of zones of unrest, the phenomenon of new contests around issues of environment, resources, trade, and information, and the seemingly unrivalled military power of the United States.

And, finally, all these take place in a historical situation that begins with the close of Second World War; and then passing through the cold war reaches the time of *Washington consensus,* when the world is supposed to have reached the end of a conflict-ridden history and arrived at a consensus on the desirability and universal validity of liberal democracy. The political nature of today's peace question thus rests in its historically embedded relation with democracy, particularly liberal democracy.[38]

We shall return to some of these themes in chapter 11 by way of concluding our discussion on the complex politics of dialogue, in order to show how the received histories of war and peace influence the rise of new issues in dialogue, and shape our consciousness as actors labouring within and outside the given world history. But here, without anticipating our later remarks, we must pose at least in some details the critical aspects of current historical scenario in order to realize how the question of peace and dialogue has evolved in the context of certain developments in international democracy, such as evolutions in institutional definition of peace and human rights. This will help us to realize why the issue of dialogue has been placed in the dual context of geopolitics and democracy.

The preamble of the Universal Declaration of Human Rights (1948) in the first paragraph says, 'Recognition of the inherent dignity and of the inalienable rights of all members of the human family is the foundation of freedom, justice and peace in the world'. And peace (and also freedom and justice) can be ensured when, the Declaration says in the third paragraph, 'human rights' are 'protected by the rule of law', and in the fourth paragraph, there is development of 'friendly relations between nations'. Article 3 says, 'Everyone has the right to life, liberty and security of person'. Art. 28 says, 'Everyone is entitled to a social and international order in which the rights and freedoms set forth in this Declaration can be fully realized'. And, Article 30 proclaims, 'Nothing in this Declaration may be interpreted as implying for any State, group or person any right to engage in any activity or to perform any act aimed at the destruction of any of the rights and freedoms set forth herein'. Peace as the space to enjoy the rights is thus a universal claim. The General Assembly in a resolution on 'Right of Peoples to Peace' adopted on 12 November 1984, declared, 'The peoples of our planet have a sacred right to peace'. Further, 'the preservation of the right of peoples to peace and the promotion of its implementation constitute a fundamental obligation of each State'. If this resolution is read with the injunction in the same resolution (A/Res/39/11) on the states to behave peacefully in resolving conflicts, the right to peace becomes clearer in import.

Yet in terms of the historical development of the notion of rights, peace defined as a right (to a space) to enjoy human development poses few problems. Proceeding from Locke's notion of natural rights, rights have come to occupy primarily a juridical-political meaning, especially in the last two centuries. Yet we retain the notion of inherent rights and inherent justice. Even customary rights retain meaning though in a very reduced way. In such situation the notion of peace as right is perched precariously on the branch of justice and custom. If peace as right is justifiable legally, then the instruments of compulsion are not yet developed. The legal obligation of states is yet confined to rhetoric. The Helsinki Accord of 1975 is not a treaty law; respective countries have not ratified the specific obligations of states under it or under the Assembly resolution. The international law of security speaks of *jus ad bellum* that implies anticipatory self-defence, collective self-defence, reprisal as defence, and intervention. We have thus a quaint situation here. The first generation of rights was badly executed; the expectations that might have been fulfilled by a rigorous application of these, led to the development of the second generation of rights. Now this too remaining half-executed, the expectations have generated the need for another generation of rights. The category of third generation of rights to which the right to peace belongs thus hangs suspended between the traditional desire for end to *hostilities* that destroyed towns and villages, and affected human life, the new trend towards valorizing rights with a comprehensive notion of good life, and an accompanying low level of applicability of rights. Conceptualized in a non-juridical way, right indicates a democratic desire for a higher quality of life, and it also shows how much the later generations of rights depend on the applicability of the preceding ones. Thus peace as a comprehensive concept of disarmament, social justice, and democracy spans all the generations, and it seems that like all rights this one too depends on how much the history of popular struggles would make of it.[39] Also it will depend on the way in which the tensions in the current history of globalization resolve.

The current history of global politics has encouraged some to speak of 'a revolution in international affairs' whereby the principle of sovereignty, the principle of international order and the right to self-defence have to come to occupy the heart of the political storm worldwide.[40] Revolution or not, there is no doubt that globalization has made the issue of peace an irreducible historical-political question. Moralists or realists, all would have to decide on the entire discursive tradition of *just war* the ghost of which has resurfaced in the context of the current debate over international military intervention on 'humanitarian ground' and has engulfed the peace question. We are now in a situation where sovereignty has been internationalized, and 'order' is no longer taken as the natural adjunct of peace unlike in the Westphalian scenario where the capacity of a state to maintain order at home and remain disciplined abroad was recognized acknowledged as being peaceful. The attitude of the United Nations has changed notwithstanding Article 51 of the Charter that upheld the just war tradition. The trouble with the current scenario is that it does not allow us a clear picture of this supposed revolution. The UN rarely intervenes in difficult situations. The USA does, or the NATO may do. The legitimacy of the state in claiming to be peaceful by maintaining order at home and observing discipline abroad is not grated anymore, for such conduct is not taken to be necessarily pacific. On the other hand, the interventions by private clubs of world powers on 'humanitarian grounds' can also be treated (indeed they are by many) as equally illegitimate. This is a situation where the just war principle is invoked on both sides – the state defending self-interest and its 'right of self-preservation' and the club defending 'humanity'. This has far-reaching implications in terms of the evolving scenario of peace politics. According to Article 51 of the UN Charter, self-defence is the basic right of any state. Indeed, this can be said to be a *jus cogens* principle of international law, one that cannot be superseded by treaties or written legal provisions. But Article 51 does not anticipate the lack of 'an internationally agreed normative template'[41] necessary to acknowledge and define the new concept of threats to international peace and order. Indeed, if we posit *jus ad bellum* (rightness of going to war) with *jus ad bellow* (rightness of actions in war) when thousands have died in humanitarian war, dispensed as 'collateral damage', economic wars whereby generations of children have become permanently weak and impaired, and depleted uranium used in tactical war weapons, both *jus ad bellow* and *jus ad bellum* become big question marks for peace politics. Indeed as the debates on NATO's war over Kosovo show, the issues of 'legitimate authority', 'just cause' and 'right intention', the three elements of the just war theory, become new points of contest for the establishment of an internationally acknowledged arena where the fulfilment of these criteria can be discussed, debated, dialogued, and where sanctions can be decided for beaches of the norms.[42]

Democratic Inadequacy, the Inadequacy of Defined Positions, and Dialogue

The history of the institutional definition of peace shows in this way the perennial inadequacy of defined positions in the contentious politics of dialogue. Indeed, dialogue begins with the implicit acceptance of the uncertainty of defined positions. The implication for geopolitics is that, dialogic politics not only challenges positions embedded and fixed in territoriality of one kind or another, it raises the possibility of further exploring the geopolitics of flux. The reason why peace appears as belonging to a new generation of rights in stead of belonging to considerations of security or balance of state interests or order maintained by an imperial power, is the inadequacy of some the earlier notions related to the concept of peace, that a dynamic vision of rights and justice (which are trans-national and trans-territorial) brings out. As the political consciousness about democracy deepens, the supposed universality and particularity of the instruments of human rights and peace seem inadequate. We can cite two cases of inadequacy of the available instruments of human rights – rights of the refugees and minorities.

The Convention of 1951 relating to the status of refugees recognized the obligations of the contracting states to grant protection and set standards for treatment of refugees in the asylum states. It also established the principle of *non-refoulement* according to which persons may not be returned against their will to a territory where they may be persecuted. The institutional governance of the convention was assigned to the UNHCR that was to encourage governments to subscribe to the convention, and make regional conventions also. It was to promote the grant of asylum to refugees. It was to secure due process of law in examining refugee applications and enforce international standards in the enforcement of legal status. And the UNHCR was to secure durable solutions also through voluntary repatriation or eventual grant of nationality of the country of residence. The protocol of 1967 enlarged the scope of the convention. The functions of the UNHCR increased. A separate category of internally displaced persons was recognized. The UN human right machinery seemed to be paying greater attention to the condition of the refugees. The convention also led to regional initiatives. The Council of Europe recognized in 1994 that refugee question called for a regional approach. Methods such as information-sharing, parliamentary attention, defining acceptable humanitarian standards for acceptance of refugees emerged as parts of a regional framework of refugee protection in Europe. The Organization of African Unity convention of 1969 broadened the scope of definition of a refugee. The 1984 Cartagena Declaration of the Organization of American States took note of 'generalized violence' which may produce refugees. It seemed that the traditional arrangement was working well.

Yet, these developments indicate on the contrary, that the system has virtually collapsed in face of the unprecedented population flow in the nineties. The massive refugee flow, forced repatriation, the decline of liberal asylum practices, the creation of *non-entrée regime* indicating the existence of a 'fortress Europe', and the security mentality overwhelming the refugee care regime have made the vision of 1951 inadequate for negotiating the population flow in the world of today. In any case the convention of 1951 was intended for Europe, and did not consider the millions that were fleeing from homes in the wake of de-colonization, partition, and the emergence of new states. It is still unable to administer quick and proper care of the millions fleeing from ethnic violence and state terror. It is meaningless in face of the structural adjustment policies that produce what the historian of indentured labour Hugh Tinker calls 'new slavery'. The phenomenon of forced migrants, unwanted migrants and migrants of the underworld has made the demand of a more humane protection regime insistent.

Similar is the situation with the international law on the minorities. Though the protection of ethnic, religious, and linguistic groups has been one of the oldest concerns of international law on human rights, the task of ensuring minority rights through arrangements of autonomy has become only more and more complex. Without going into the details of the issue that we discuss in chapter five, suffice it to note in the context of old definitions and new realities, that for long there was no specific provision even in international human rights law as to what constituted a minority and the protection of minority rights. At least three definitions are before us, each reflecting a different facet of the reality. Art.

27 of the UN covenant on Civil and Political Rights mentions minorities in terms of rights. The Capotorti definition emphasizes the feature of 'non-dominance'. The Deschemes definition refers to a 'collective will to survive and whose aim is to achieve equality with the majority in fact and in law'.[43] So who is a minority? Is it a group, which is simply non-dominant? Or, is it a group that which not only suffers from domination, but also has a will, or, that enjoys certain rights? Does it need protection? Or, does it self-determine? But self-determination is a right of *peoples* and not *minorities* (Again this is a distinction that has implications for a dialogic politics, and will be discussed in chapter 5). Several international treaties erected the League system, and the UN started building on that. Treaties on minorities in Poland (Versailles, 1919), Romania (Paris, 1919), Greece (Sevres, 1920), Upper Silesia (Geneva, 1922), Turkey (Lausanne, 1923), and many others wrestled with ideas and clauses of cooperation, separation, self-determination, peaceful solution and guarantee. Yet events overtook these efforts in less than twenty years. And though the UN learnt from the preceding failures and made the obligation to protection wider through provisions of non-discrimination, the institution of a sub-commission on prevention of discrimination and protection of minorities, prohibition of genocide, the covenant on civil and political rights, particularly Article 27 of the covenant, and the human rights committee (Articles 28-45 of the ICCPR), yet minority rights seem to be as unsafe as they were, and the arrangements still inadequate. Indeed, the Universal Declaration of Human Rights, the most significant document in the field of human rights, when passed in 1948 was conspicuous by its omission of a 'minorities article'.[44] Obviously the idea was that with universality of rights established, special care was secondary. That was non-discrimination. The crisis of the liberal state, wars, extermination of minorities, nationalist majoritarianism, ethnic cleansing, proxy insurgency, and great power backing of ethnic groups as means of waging the cold war made the UN system of universality fragile, if not a mockery. Group rights, called the second-generation rights, have created a paradoxical situation. On one hand the discourse of group rights has enriched the notion of human rights, on the other hand this discourse reflects and produces the ethnic futures which have disfigured the notion of toleration and justice by turning themselves over to the global power grid. Thus, as the 'cunning of history' would have it, group rights become in many places the new front-paw of great power politics whether in erstwhile Yugoslavia or in Afghanistan. Clearly, conflicts overwhelm the structure of rights and peace. New discourses emerge, new generation of rights surfaces, only to become the terrain of new contests between democracy and politics of domination. The new discourse of minority rights therefore does not signify the arrival of the *final mechanism*. Given the various definitions and provisions available to us reflecting different facets of a complex reality, namely weak capacity and resource of certain groups for political, cultural, and economic survival, domination and powerlessness, hierarchical representation, and discrimination, the situation only indicates that ensuring fairness and justice, which lies at the core of the issue of protection of minorities, is a process linked irrevocably with the process of power, and conversely with that of dialogue and democracy.

The minority question like the refugee question presents before us the inadequacies of existing protection mechanisms of human rights in enhancing peace as a political value, or in addressing the issue of peace as the principal political question of our time. Events such as the bombing of Afghanistan and the consequent massive refugee exodus, or the genocides and ethnic cleansing in various parts of the world making minority populations the permanent victims of States and majoritarian democracies, testify to the inadequacy of the current institutions of human rights. They also explain, why dialogic politics in challenging the State and other national and international authorities, carries the promise of challenging the current institutional limits of rights and justice. Contest here too remains the inevitable historical agenda.

It is important to take pause here and think: why does dialogue tend to defy the institutional limits or forms of democracy, and raise new issues beyond their pale? One reason mentioned in the preceding chapter is that though rights and justice are closely inter-related concepts, rights belong more to the domain of law, while justice is more closely associated with dialogue. What begins as a demand for justice can end in the codified form of a right. The democratic discourse builds on the arguments of rights, and not on justice. Justice in fact expands their frontier. Therefore, though democratic theory may anticipate dialogue, and should anticipate it, the politics and practices of dialogue contest the given institutional limits of democracy; to be truthful, it challenges the democratic form, because dialogic imagination is fired less by the clamour for rights, more by the passion for justice. From this we can advance the second reason. The demand for justice leads dialogic politics to inquire into democratic inadequacies, summed up in the phrase 'democratic deficit', to the extent that dialogic politics in the form of peace politics begins questioning the hitherto accepted congruence of democracy and peace. Indeed the issue of democratic inadequacy assumes critical importance in view of the increasingly raised question, are democracies pacific? Are they inherently dialogic, or are they war-like as any other political ideology or system?

Democrats from Woodrow Wilson to William Jefferson Clinton (and with a change in party nomenclature, George W. Bush, Jr.) have reminded us that democrats are peace loving, and therefore a world of democracy is inherently peaceful and safe for the people on earth. Immanuel Kant found a link between rights and peace – a view that had distinct traces of the Lockean attitude. The UN charter also declares in Art. 55, that human rights are essential for friendly relation among nations. But what happens when democracies commit aggression, or invade other countries, or conduct wars of subversion for 'humanitarian reasons' (Kosovo), to overthrow autocracies (Grenada and Panama), to roll back communism (Nicaragua), or continuously arm themselves with nuclear weapons? To keep peace, democracies wage war. If democracies do not fight wars among themselves, they combine to fight war against others. In fact, this is almost a law of democracy that an ideological Calvinism goads democracies to launch wars against others. These wars may be overt, but more often covert. These covert wars aim at overthrowing democratically elected governments (Mossadegh in Iran, Lumumba in Congo, Arbenz in Guatemala, Allende in Chile), or governments with very strong legitimacy (Castro in Cuba or Ho Chi Minh in North Vietnam). Thus existence of civil and political rights to a considerable extent may not be a barrier to violence. On the contrary, the (alleged or real) absence or relative absence of civil and political rights in a country may become an occasion for other countries to wage war (the five-year war waged in Russia by the allied interventionist army after the Bolshevik revolution). But again, ironically when a democratic government is being overthrown by a fascist coup, and the grossest abuse of human rights is taking place, democracies may choose to remain neutral (Anglo-French neutrality in Spanish civil war). The club of democracies is indeed an elite club. Therefore, ideology, forms of power, forms of weaponry, techniques of war (like covert operations, development of small arms, intelligence gathering), exclusivity/hierarchy of certain rights, economic system, and location, are as important as the formal presence of democracy. Democracy as a form of dialogue and peace is therefore not without problem; on the contrary dialogue as a form of democratic practice brings out the problematic inherent in democracy, namely democracy as a system championing rights, and democracy as politics negating many social values, namely, justice, tolerance, and reconciliation. Indeed, as we can see, beneath the apparent smoothness of the theory of 'democratic peace' is a consensus on a neo-classical model of asocial, utility maximizing, rational actors interacting in a milieu of shared values and consensus about certain priorities.[45]

In the following pages I try to deliberate on this paradox, and the related questions in the context of geopolitical histories of war, conflict, and peace in South Asia. In the process of thinking beyond the received wisdom on the relation between democracy and peace,[46] the following pages try to argue for the need for a critical theory of dialogue. Because, such a critical theory will not be satisfied with some supposed mystical power of dialogue in stopping violence and healing wounds; it will investigate into the geopolitical contours of democracy, democracy's compulsive relation with wars in the last half century, and dialogue's own vulnerability in front of governing and warring techniques whereby dialogues may become part of preparing for war or domination.

We may now re-present the paradox in slightly different terms. Democracy begins with a politically-territorially bounded population who then decides how to get on with life. But what happens if that basic is contested, if, as some claim, the boundaries are never founded with democratic procedures, and therefore, democracy has to begin by re-defining, who 'we the people' are, and not be satisfied with the given bounds of the 'people'?[47] This inquiry was so long theoretically considered as impossible in terms of the democratic discourse; but now with this inquiry becoming historically possible with the emergence of new demands for justice and rights, which question the given bounds, both democracy and peace are being re-imagined in the mirror of a dialogic politics.

People as subjects constitutive of democracy are an imagined configuration, as the classic struggles of democratization by women and lower classes have shown.[48] Democracy in the form of a national community can also be imagined, much in the way as Benedict Anderson suggests; but as I try to show in *A Biography of the Indian Nation, 1947-97*, its passage to statehood must be concrete. Democracy facilitates the passage from the imagined features of a collective community-hood to statehood by a double method. It resorts to concretization of the features of a political community by laws and institutions; it also makes this political community of democracy a matter of imagination, almost beyond contest. An imagined democracy then becomes intolerant of others, aggressively defines and protects its boundaries, and most significant, includes and excludes others emphatically from the democratic nation, just it had at one time excluded and later included women and lower classes. As I show repeatedly in the subsequent chapters here, democracy in such incarnation does not help in conflict resolution; being a matter of the imaginary, it aggressively involves itself in conflicts. In such context, democratic resolution of conflicts may not mean fundamentally a broadening of dialogue, but a revision of the rules of the game whereby moderate elites of both sides try to marginalize old rules and old actors, and resultant process of negotiation is arrived with hardly any public political debate. If old agenda of contest was inspired by a mythological discourse of a democratic community, the new methodological politics is similarly a product of imagined consensus, which shuts out issues of justice, which indeed is built on excluding these contentious issues. The Israeli peace researcher Lev Grinberg has remarked of the conundrum, 'imagined democracy imagined peace.'[49] While the political strategy becomes increasingly one of trying to democratize a polity through peace, and using democracy to reinstate peace, the fundamentals of inclusion/exclusion remain – the contradictory goals of including the excluded but keeping them at the margin, transforming them into legitimate political actors, yet ensuring their externalization. To quote Grinberg, 'Imagination is often an efficient way to mobilize the weak in their struggle, but usually it is also an effective means for legitimizing the rule of the powerful'.[50]

The demands of peace and dialogue based on new demands of justice thus lay bare the imaginary of democracy. Imagined democracy makes us aware of the need for a critical theory of dialogue, that does not take democracy as a formal set of institutions only, but as a regime, a form of power relations, a system capable of waging war, and making limited peace. The process of dialogue is thus of contest, contesting the politics of exclusivity, raising the standard of democracy against democracy. But is that not what the history of popular politics is all about, *democracy as the politics of the ruled?*[51]

Embedded Histories of Contests and Peace

The task of peace and dialogic studies thus lands itself in the deep, in the inner region of the problematic of formation state and particularly the formation of the modern polities. Peace as I have tried to suggest in this chapter is a matter of history of a process linked to the process of power, also to the politics for justice, equity and plurality. This implies that peace remains inseparable from the problematic of power and authority that has always provoked the struggles for justice and a participatory politics. And here lies the irony of modernity. For, present in the architecture of that desire, known by the word *peace*, and provoked today by the most ruthless demonstration of power and authority ever in the history of mankind, is an important binary element called anarchy/order. Order produces authority, authority produces violence, and violence provokes the desire for peace. But violence breaks down authority also, and the breakdown of the historically admissible form of authority leads to a collapse of the state. This is recognized socially as anarchy, and the desire for peace gets the phantasmagoria of a desire for lost authority, order, and thus the state. Thus in Afghanistan, Rwanda, Cambodia, Somalia, Lebanon, erstwhile Yugoslavia – in all these countries peace has been sought to be restored by restoring the state. The usual process has been promulgation of a 'peace constitution' (much like reconstructing post-war Japan), talks, interim power-sharing arrangement, a multilateral peacekeeping army, elections under international supervision, and the *restoration of the state*. The breakdown of the state is a universal phenomenon, so is its restoration. I have termed this elsewhere as 'conditions of globalisation and conditions of peace',[52] and I return to this theme at the end of this book. Restoration of peace and the restoration of state have therefore become the twin tasks of liberal democracy today.[53] Together they present the strange spectacle of the politics of the modern. But probably the more significant conclusion that we can draw from this is that, since peace is restored through restoration of the state, and not primarily through plural dialogues, what we have is the return of the state on a weak and fragile basis, backed by a narrowly based consensus, and certainly not a return of the dialogic politics/polity. In other words, modernity is restored in politics only by severely curtailing its dialogic base.

The fact that the process of peace is now governed by today's conditions of globalization has one more implication. If technologies of war have become global, so have become technologies of peace. Modes of intervention by 'civil society' (for want of better word or phrase), the democratic agenda now more and more seen as linked with issues of human rights and peace, new identities, such as the regional identity, as a path to conflict-resolution, normative issues articulated as issues of peace – these are *political* techniques of reinforcing the dialogic practices. The question of peace, in this way, has become intensely political, indeed to a degree unseen before, just as war. If war is a matter of state, the working of the modern state has ensured that peace too becomes a matter of politics, in fact its *raison d'être*. Constituencies of peace emerge, as new tools of peace make such constituencies viable. Peace in short becomes a durable programme, not because men (and women) are peace loving, but relations and technologies make peace constituencies durable features of society. That is why peace studies is perched on a critique of the condition of the state, of patterns of domination, of global hegemony, and of technologies that make dialogue a sustained demand in today's global world. Strategies of peace are therefore based on the 'capillary' condition of dialogue today.[54]

Therefore, we shall have to pay attention to the contests amidst which peace as a specific political value and dialogue as a specific political practice are shaping up.

As we have already noted, the implications of globalization on the politics of peace have become significant more than ever. A range of new issues has emerged from the interaction of economic globalization, new technologies, and rapid political change, and has affected the political and normative conceptions of the state system with which the question of peace had been hitherto tied. Even we keep aside the debate on whether the orthodoxy on globalization exaggerates the demise of domestic power of the state and the disintegration of national economies,[55] it will be reasonable to argue that state authority has undergone radical changes in the face of trans-national markets and footloose companies, and the politics of dialogue and peace has to now not only engage with the violence of the state, but the realities emerging out of the rearrangements in geopolitics (control and distribution of territory)[56] and what the French social theorist Paul Virilo has called 'chrono-politics' (the control and distribution of time).[57] With the redistribution of both these resources, we are witnessing trans-national (that is, escalating across nation-state boundaries) conflicts, enhanced vulnerability of states and regions such as sub-Saharan Africa perishing from unequal flow of trade, and structural adjustment programmes and re-location of investment in the world-economy. The United Nations Human Development Report of 1996 confirmed that aid to sub-Saharan Africa had fallen as fell the support of the World Bank in terms of loans and credits. The 'new geographies of money lending'[58] have increased tensions in South Asia as well as East Europe. The immediate result for peace politics has been that it will have to work out its way within a more complex global setting, marked by re-constitution and redefinition of state power and state-society relations.

This reconfiguration of the peace question in the form of emerging structures of international justice in the new global and regional setting has two meanings. On one hand there is a globalization of protests against massive human rights violations, and demands being placed on global power structures to agree to dialogue with constituencies of protesters on new issues of justice; on the other hand, the human rights question is framed in the discourse of 'humanitarian intervention' by great powers, though such interventions lack consensus.[59] Old dictators are being let off, because they had fought communism; new dictators are hauled, because they represent rogue states. Carlos Nino considers these 'dirty wars' that the dictators fought against the trade unions, leftist political parties, and church as not simple evil, but 'radical evil'.[60] Throughout Southeast Asia and Latin America in the decades of the seventies and eighties of the last century thousands and thousands of men and women were subjected to torture, kidnapping, extortion, forced disappearance, and on many occasions, execution. Leaders of severely oppressive regimes used the legal systems to justify and legitimize the policies of violence. At the same time, women remained vulnerable to a range of dangers that are still deemed to be outside the scope of most national legal systems. While the language of human rights has been used by the great powers to re-establish control over vast territories, in particular the post-colonial areas, issues of rights are being deployed in another contest also of different nature – this is the contest against global power structures, in which the demand for dialogue occupies a place of critical importance.

Yet, we can see, how the situation is further complicated by the use of dialogic strategy by states to quarantine the dirty wars of the past through peace process and peace accords in order to stabilize the polities. Though I shall come back to the issue of accords in a later chapter, we can note here at least another face of the quarantining strategy. This strategy not only quarantines past dirty wars, it also prioritizes issues of dialogue in such hierarchic order that, virtually some of the aspects of modern warfare will not be discussed, while some others will be brought under collective scrutiny, and quarantined. Here, dialogue is a means of disciplining some, also a means of ensuring liberty and private power for others. Since, massive expenditure in the arms race of the sixties and seventies of the last century raised worldwide protests in form of peace movements, the calculus of arms reduction is now stalemated through development of new arms technology that has brought about the so-called 'revolution in military affairs' (RMA). This has meant politically that while weaker nations cannot have landmines or other weapons of defence, thus all kinds of conferences and arms control discussions go on to quarantine the military capacity of the weaker nations, superior nations can have missile shields and means of long distance electronic warfare. Thus while the RMA can say that weaker nations need not spend on the military, the relations between development and defence have now taken an even more unsettled form. Thus many of the old arguments on the cost-benefit balance sheet of production of civilian and military goods, stock of labour and capital, rate of depreciation, budget deficit and defence welfare trade-off, balance of payments and arms trade, and the conversion question will become partially obsolete in the wake of the RMA and the wider gulf between armed and non-armed countries. Kenneth Arrow had spoken of four possible perspectives on the economics of arms reduction – neo-classical welfare theory, the Keyensian version of macroeconomics, modern growth theory, and the perspective of political choice.[61] Some if it would be not relevant any more, at least for now, for without arms acquiring there would not be much to reduce. Also the supply side explanation about determinants of military expenditure, namely supply of military capability, demand for national security, or function of estimating military expenditure[62] will lose meaning for large areas of the world, deprived of the technology ushering in the RMA. By quarantining war, there will be peace. Or, wars in large parts will be conducted with stones, scythes, swords, and daggers (means of Indian and Rwandan genocides) – much of the time outside the electronic gaze. The history of dialogue on issues of war and peace will remain as skewed as ever.

The other feature in the present situation, as mentioned earlier, is that, while democracy has still to come to terms with the dirty wars of the past, liberal and individualist approach to human rights abuses is now perched on 'humanitarian intervention' as the remedy. Peace politics has to grapple with the unsettled past in an equally unsettling present, and only time can tell if a new politics of justice, dialogue, and reconciliation, whose contours are still hazy, will become capable of making the issue of peace an emphatically critical agenda, and make the economics of arms reduction a meaningful political question. In making sense of the paradoxes of today's peace politics, briefly recounted in the present chapter, we need to know the history of dialogues.

Yet, we cannot have a history of whose history we do not anticipate. The anticipated history of dialogues must lodge itself, therefore, in the interstices of contests – contest between dialogue and authority, conversation and unilateral acts, state and sections of society, war and non-violent means, justice and power, and the established history of war and the silences in that received discourse. This anticipated history consists of discrete moments to be recovered synchronically, as the democratic moment will require. In its recovery and in its material form, dialogue is embedded in contest – contests of various kinds. The moment of dialogue is therefore a moment of contest, in one word, *a democratic moment.*

The world's disorders are bound to leave their mark on visions of peace. Dreamers of pacifism bruise easily. And yet, why have visions of peace attracted so much notice? Because quite simply, peace due to its own contested nature attracts visions, and because of the essentially dialogic nature of peace politics attracts all those who are not reconciled to unilateral acts of authority. It is a puzzle, yet as I propose to show in the pages to follow, the more there is war, carnage, violence, and hordes of specialists to assure humanity that they have the knowledge of ways by which they can provide security from these, the more there is the disbelief in these assurances, and a yearning for more, beyond, that can be captured only in a dialogic inspiration and vision.

In the pages that follow I therefore return again and again to the process by which dialogue constitutes itself, fails, and lives anew, and forms of accommodation and understanding are imagined – not to write a holy book of peace, but to show how the 'theology of peace' exists only as a political question of our time. In doing so, I necessarily allude to themes of contest in the midst of which the agenda of dialogue is shaping up, themes of contest that, though tantalizingly, suggest a possible politics of peace. In South Asia, needless to say in many parts of the world as well, the contest began with the emergence of the post-colonial politics itself in the most grotesque geopolitical form – the politics of the partitioned spaces and times. I therefore begin with the time of the partitioned nations, the time most marked by the failure of dialogue, the time that marks partition, and not dialogue, as the method of resolving difference.

Notes

1 These quotations are from Ajit Bhattacharjea, *Countdown to Partition – The Final Days* (Delhi: HarperCollins Publishers India, 1997), pp. 1-31.

2 On the interplay of the sociologies and psychologies of war and peace, see Colin Creighton and Martin Shaw (eds.), *The Sociology of War and Peace* (London: The McMillan Press, 1987) – particularly three essays, Michael Mann, 'War and Social Theory – Into Battle with Classes, Nations and States' (Ch. 3, pp. 54-72); John Mattausch, 'The Sociology of CND' (Ch. 10, pp. 199-217); Graham Day and David Robbins, 'Activists for Peace – The Social Basis for Peace Movement' (Ch 11, pp. 218-236).

3 Thomas Hobbes, *Leviathan*, reprint, Ed, C.B. Macpherson (Harmondsworth: Penguin Books, 1972), Ch. 13.

4 James Madison, 'Universal Peace' in M. Meyers (ed.), *Sources of the Political Thought of James Madison* (Hanover: Brandeis University Press, 1981), pp. 192-93.

5 Once again we can look in Hobbes for the most apt definition, '…there is no such thing as perpetual Tranquillity of mind, while we live here, because Life itself is but Motion, and can never be without Desire, nor without Fear, no more than without Sense.' *Leviathan*, ch. 6.

6 Cited by Peter Henrici in 'Two Types of Philosophical Approach to the Problem of War and Peace' in Peter Caws (ed.), *The Causes of Quarrel* (Boston: Beacon Press, 1989), p. 156.

7 This is the main argument of Kenneth E. Boulding in his *Stable Peace* (Austin: University of Texas Press, 1978).

8 Ibid. p. 10.

9 Kenneth E. Boulding, *Conflict and Defense – A General Theory* (New York: Harper and Row, 1962), p. 5; in *Stable Peace* he writes, 'conflict is a re-distributional game', and speaks of zero-sum, positive-sum, and negative sum games (pp. 10-13).

10 *Conflict and Defense*, p. 6.

11 Ibid. Chapter 2, pp. 19-40.

12 Ibid. pp. 39-40.

13 Ibid. Chapter 13, pp. 248-276; see also Robert E. Kuenne, 'Disarmament Negotiations as an Exercise in Mature Rivalry' in Walter Isard and Charles H. Anderson (eds.), *Economics of Arms Reduction and the Peace Process* (Amsterdam: Elsevier Science Publishers, 1992), pp. 203-217, where Kuenne admits that the 'game of chicken' may not apply, and that conditions of 'mature rivalry' (developed state of information flow and extensive information-exchange, mutual conviction of the ability of the other to respond effectively and to exploit the weakness of the adversary, and mutual unwillingness to disturb the status quo) may not be pervasive.

14 Ibid. p. 312.

15 Indeed, even when Boulding was writing this, roughly at the same time, impelled by such fundamental wars as Vietnam, theorists and thinkers were considering the peace question beyond conflict. In particular discerning observers were drawing attention to the genre of studies that turned conflicts into games, and civilian speculators (about results of conflicts) into militarists. See for example, the works of I.L. Horowitz, *Games, Strategies and Peace* (Philadelphia: American Friends Service Committee, 1962), and *The War Games – Studies of the New Civilian Militarists* (New York: Ballantine, 1963). In 1967 a massive peace convention was held in Geneva on 28-31 May to discuss the 1963 encyclical of Pope John XXIII, 'Pacem in Terris' (Peace on Earth), where persons like Martin Luther King, William Dauglas, Hans Morgenthau and Abdul Rehman Pazhwak discussed the requirements of peace beyond being just co-

existence in the perspective of intervention in Vietnam, confrontation in Germany, inadequacies of international law, and challenges for the theologians. See, Edward Read (ed.), *Beyond Coexistence – The Requirements of Peace* (New York: Grossman Publishers, 1968). In order to understand how peace studies developed parallel to conflict studies, also significant was the publication in 1965 a collection of the Dag Hammarskjold Memorial Lectures, Andrew W. Cordier and Wilder Foote (eds.), *The Quest for Peace – The Dag Hammarskjold Memorial Lectures* (New York: Columbia University Press, 1965). A quick look into some of the themes covered in the series of Dag Hammarskjold memorial lectures in 1964 would show this other trajectory of the sixties. Julius K. Nyerere spoke on the 'courage of reconciliation' and spoke of the inadequacies of international law (pp. 9-28), Alva Myrdal spoke of the role and responsibility of the UN in achieving disarmament (pp. 149-165), a judge of the International Court of Justice, Philip Jessup spoke of the 'diversity and uniformity in the law of nations' (pp. 293-313), and the then President of the European Human Rights Commission, Sture Petren spoke of the need for 'the promotion of human rights through the Council of Europe' (pp. 331-342).

16 Amitai Etzioni, *The Hard Way to Peace – A New Strategy* (New York: Collier Books, 1962), p. 55.
17 Ibid. p. 42.
18 Ibid. Part 3, Chapter 8.
19 Ibid. p. 111.
20 Ibid. p. 125.
21 *Stable Peace*, p. 25.
22 Ibid. pp. 70-71.
23 Ibid. p. 66.
24 Ibid. pp. 17-18.
25 Anatol Rapaport, 'Conflict Resolution in the Light of Game Theory and Beyond' in Paul G. Swingle (ed.), *The Structure of Conflict* (New York: Academic Press, 1970), Ch. 1, p. 41.
26 Paul G. Swingle, 'Dangerous Games' in *The Structure of Conflict*, Ch. 7, pp.294-295.
27 On this Paul Wehr, *Conflict Regulation* (Boulder, Colorado: Westview Press, 1979).
28 George H. Quester, 'International-Security Criticisms of Peace Research' in George A. Lopez (ed.), *Peace Studies – Past and Future*, special number of *The Annals of the American Academy of Political and Social Science*, Vol. 504, July 1989, pp. 98-105.
29 Ibid. pp. 99-100.
30 Ibid. p. 102.
31 Ibid. pp. 104-05.
32 Walter Isard in his acclaimed mathematical treatment of the science of conflict and peace, *Understanding Conflict and Science of Peace* (Cambridge, MA: Blackwell, 1992) remarks of 'decision-making pathologies' in this way, 'As a participant's set of possible actions is steadily reduced with continuous rounds of actions and reactions in a crisis situation, the behaving unit may find himself in a state of *entrapment*. Having no further time and resources for re-evaluation and search, he is boxed in and perceives one and only possible action, perhaps to carry out his threat to invade. A leader faced with an act of mobilization by an opponent, for example, might see no alternative but to carry out his threat to invade...Once he has acted, he may resort to rationalizations to justify his actions, especially when it has failed or proved inadequate (*post-decisional rationalization*). Note once more that people have limited analytical capabilities and capacity to encode new information when confronted with a tremendous amount of it. Since they cannot absorb all, and since they have no objective criteria for differentiating

among pieces of information thrown at them, it is efficient for them to absorb only that information consistent with the information they have already encoded and to be insensitive to (discard) new information inconsistent with their existing set of perceptions of, and theories about, the world. That is, they find it advantageous to use a *cognitive consistency principle* – a principle used extensively, most often not consciously, but which leads to poor decision making when significant information is discarded.' (Author's italics) p. 71.

33 Chadwick F. Alger, 'Peace Studies at the Crossroads: Where Else?', *Annals*, op.cit, pp. 117-127.

34 Johan Galtung, 'Twenty-five Years of Peace Research – Ten Challenges and Some Responses', *Journal of Peace Research*, 2 (2), 1985, p. 1985.

35 Birgit Brock-Utne, *Educating for Peace – A Feminist Perspective* (New York: Pergamon Press, 1985), pp. 71-76. In this context, the writings of Gene Sharp are significant in their emphasis on non-violent behaviour, for example, *The Politics of Nonviolent Action* (Boston: Porter Sargent, 1973).

36 These citations are from Stanley J. Tambiah, *Leveling Crowds – Ethnonationalist Conflicts and Collective Violence in South Asia* (Delhi: Vistaar Publications, 1997), pp. 3-6.

37 George Lakey, *Powerful Peacemaking – A Strategy for a Living Revolution* (Philadelphia: New Society, 1987).

38 In the context of the almost inevitable discussion of transition and transitional democracy in the current discourse of peace and peace process, the following essays are a pointer to the trend. See, Craig Charney, 'Civil Society, Political Violence, and Democratic Transition and the Peace Process in South Africa, 1990-1994', *Comparative Studies in Society and History*, 41 (1), January 1999, pp. 182-206; John T. Ishiyama, 'Representational Mechanisms and Ethnopolitics – Evidence from Transitional Democracies in Eastern Europe', *East European Quarterly*, 33 (2), Summer 1999, pp. 251-279; Finn Laursen, 'Civil Society and European Integration', *The Annals* (Issue on Civil Society and Democratization), 565, September 1999, pp. 66-78.

39 I do not agree with David P. Forsythe who presents some compelling discussions on *Human Rights and Peace* (Lincoln: University of Nebraska Press, 1993), for ignoring that 'right' itself is a matter of historicity.

40 For example, James Gow, 'A Revolution in International Affairs?', *Security Dialogue*, 31 (3), September 2000, pp. 293-306; for a critical review of such argument, Henrik Syse, 'Ethics, Sovereignty, and Self-Defense – A Rejoinder', *Security Dialogue*, 31 (4), December 2000, pp. 437-442.
One of the most significant books in recent years taking a critical look into the entire issue of intervention in the context of the development of the discourse and ideology of human rights is Michael Ignatieff, *The Rights Revolution* (Toronto: Anansi Publishing, 2000).

41 James Gow, p. 303.

42 On the history of the just war tradition, Richard J. Regan, *Just War* (Washington, DC: Catholic University of America Press, 1986).

43 Special Rapporteur F. Capotorti's definition appears in his *Study on the Rights of Persons Belonging to Ethnic, Religious, and Linguistic Minorities* for the UN Sub-commission on Prevention of Discrimination and Protection of Minorities, UN sales no E 78, XIV.1; Canadian member of the same UN Sub-commission, Jules Deschenes' definition appears in his *Proposal concerning a Definition of the Term 'Minority'* submitted to the Sub-commission and adopted by it 1985 – UN Doc. E/CN.4/Sub.2/1985/SR.13-16. References to the legal texts are from among others Jean-

Pierre Colombey (ed.), *Collection of International Instruments and Other Legal Texts Concerning Refugees and Displaced Persons* (Geneva: Division of International Protection of the Office of the United Nations High Commissioner for Refugees, 1995).

44 One of the most informed discussions on the minority issue in international law is by Patrick Thornberry, *International Law and the Rights of Minorities* (Oxford: Clarendon Press, 1991).

45 See on this, Richard Ashley, 'The Poverty of Neorealism', *International Organisation*, 38, 1984, pp. 225-86.

46 On the concerned literature, see Steven A. Miller, *Debating the Democratic Peace* (Cambridge, MA: The MIT Press, 1996); James L. Richardson, 'Contending Liberalism – Past and Present', *European Journal of International Relations*, 3, 1997; David Long, 'The Harvard School of Liberal International Theory', *Millennium – Journal of International Studies*, 24, winter, 1995.

47 For a formulation of the paradox that democratic institutions are installed by non-democratic means, see Juan Linz and A Stepan, *Problems of Democratic Transition and Consolidation* (Baltimore: Johns Hopkins University Press, 1996).

48 One of the most authentic accounts on this is D. Rueschemeyer, E. Stephens, and J. Stephens, *Capitalist Development and Democracy* (Chicago: University of Chicago, 1992).

49 Lev Grinburg, 'Imagined democracy, Imagined Peace – Reframing Israeli Politics', paper presented in the second Humphrey International Conference on 'Challenges to Democracy – Peripheries as a Vantage Point', Ben Gurion University, Beer Sheva, 19-21 May 2000.

50 Ibid.

51 The Sri Lankan political scientist, Jayadeva Uyangoda calls this as 'democracy and the politics of emancipation as an argument for democracy's victims' and argues that 'democracy as a political project needs to be democratised'. Unpublished paper, 'Democracy and the Politics of Emancipation', presented at 'The Dialogue on Democracy', Jawaharlal Nehru University, 18-20 August 2001, New Delhi. In another paper in the same dialogue, another Sri Lankan social scientist, Darini Rajasingham, speaks of, as the title of submission suggests, 'The Fetishism of Democratic Institutions' and speaks wryly in the context of democracy and civil war in Sri Lanka, 'Perhaps it is a sign of the learning curve that has taken place in the development industry that the World Bank is in the process of formulating a 'Reconstruction, Reconciliation, and Rehabilitation Framework' for the Government of Sri Lanka with the assistance of UN agencies.' (p. 1).

52 I address this problem specifically in 'State in the Revision of Space and History Today' in Ranabir Samaddar (ed.), *Space, Territory and the State* (Hyderabad: Orient Longman, 2002), Chapter 8, pp. 166-184.

53 The dilemma of the liberal internationalism in enforcing peace is evident in many current writings on bringing peace through force and intervention. See for example, Antonio Franceschet, 'The Ethical Foundations of Liberal Internationalism', *International Journal*, LIV (3), Summer 1999, pp. 443-481; Andrew McGregor, 'Quagmire in West Africa – Nigerian Peacekeeping in Sierra Leone, 1997-1998', *International Journal*, LIV (3), Summer 1999, pp. 482-301.

54 Obviously I borrow the word from Michel Foucault. But my intention is here is different. Foucault links the capillary existence of power to the organization of political power and the transformed conditions of sovereignty. My argument has been that conditions of globalization make constituencies and technologies of dialogue and democracy widespread, distributed across several layers of political and civil society. And, this happens not because the state has almost collapsed and power resides in a

capillary mode, but because of emergence of new technologies of dialogue and peace; and this, notwithstanding the most centralized existence of power in today's world of *pax Americana*.

55 On the controversy see, Linda Weiss, *The Myth of the Powerless State* (Cambridge: Polity Press, 1998).

56 For a comprehensive survey on this, Klaus Dodds, 'Political Geography I – Globalization of World Politics'.

57 Paul Virilio, *Open Sky* (London: Verso, 1997).

58 Klaus Dodds, 'Political Geography I'.

59 Naomi Roht-Arriaza at the end of her detailed survey of 'Institutions of International Justice', *Journal of International Affairs*, 52 (2), spring 1999, pp. 473-491, comments, 'These results are not bad, for the first 50 years. But the existence of these multiple mechanisms cannot hide the fact that the political will these mechanisms, even in the most egregious cases, is still usually sorely lacking'.

60 Carlos Nino, *Radical Evil on Trial* (New Haven: Yale University Press, 1996).

61 Kenneth J. Arrow, 'The Basic Economics of Arms Reduction' in *Economics of Arms Reduction and the Peace Process*, pp. 57-67.

62 Carlos Seiglie, 'Determinants of Military Expenditure' in *Economics of Arms Reduction and the Peace Process*, pp. 183-202.

Chapter 3

In the Time of the Partitioned Nations

Two Declarations – of Independence and Partition

Is it not curious, for we all respect the craft of history, that historians writing of independence end up in writing of partition, and writing of partition end up by writing of messy independence, and in trying to escape the bind damn almost all other historians, and end writing neither of independence nor of partition?[1] If one adds up all these histories one cannot fail to be struck by the impossibility of the theme, impossibility of our social sciences to grasp the problem posed by the history of our own being today, our thought and memory. A history of partition is a history of how independence came, featured in our minds, and still arrives; and a history of independence is a history of the impossibility of retaining the innocence of the dream of independence. When you are curiously watching this game, you suddenly realize that writing on partition is not about writing of partition. It is about the particular ways that 'contain partition' in politics, wherever there is politics of the nation – in territory, citizenship, administrative system, group identity, law, literature, language, and now in history. Singularly incapable of writing of partition, of events marked by a situation that Edward Said termed as 'pessoptimism' in *The Politics of Dispossession*,[2] history does not leave the theme and let it go, nor the event of partition leaves the comfortable nest of history. Tossed everywhere, to every wind, memory only keeps on returning partition, depositing it so to say, back to the nest, for the violence that the de-historicisation of the event causes is too much for everyone. The last hurrah of the Westminster model of de-colonization does not fade away. In this model, nation assumes statehood by claiming the universality of its own specific form demanding that all voices submit to it; a country's gaining of independence means in actuality transfer of power; transfer of power boils down to power being transferred to warring groups; and, civil war and partition become, in place of dialogue, the form of settling differences.

I must pause here and make clear what I am attempting in this chapter. In opening the chronicles of the politics and practices of dialogue, I want to show through this note on the politics of its opposite, that is partition, how partition has become the universal geopolitical sign of our time in which dialogic politics has to make out its place. This is not thus as I have hinted already, one more attempt to write a history, or a part of the history, of partition. It is not even a comment on the histories, for that history is impossible, at least today. We live in partitioned times; it is in our post-colonial being, sufferings, reunion, pessimism, and strivings. That history of the history, which partition created, and therefore to which partition belongs, will have to recognize this presence to even fathom what made the event momentous; in other words it will have to deny its craft, and shamelessly delve into the politics of today. Also, what history is left there to write on partition? One can of course write, when writing of partition, of its prelude, or of the imperial process of divide and quit, or its residue, or the trauma, the violence, orgy, the human sufferings, and the catharsis. We can quarrel on the method, on the object of writing, but as the dusk falls and the owl of Minerva flies out, even the owl is lost. It is a history lost in the quagmire of the present, hence incomplete. Inscribing itself into the current process of state-formation, partition relives every hour of the nationalist life. The more distant it is from its time of occurrence, the more appears its contemporary nature, and more powerful appears its oracular role. For all our insecurity and illness, we go back to the origin, and for all our solace, security and identity, our shelter is the same. At once an event and a sign, a history and a currency, partition lives in the post-colonial times to the extent that we should begin now using the phrase 'partitioned times' instead of the term more commonly used, 'post-colonial time'.

The mystery of a sign that began as an event is endless. We are told that a sign derives its meaning only from the existence of other signs. Bereft of any autonomous relation with what it signifies, the sign derives its validity not only from the context, but also from the virtual expanse spread out with other signs. The signs also owe their lives to the origin; therefore they are prophetic. You may scrutinize them, as indeed we do at the appearance of each sign of partition in our political life, and we cannot put our trust in any explanation as to why it happened, and that it will not happen again. It is impossible for us to distance ourselves from it. The last century that still continues is deservedly the century of partitions.

Suffice it to mention that the very process of state formation in the 'new areas' in world politics was shaped in a substantive way by the great power strategy of partition to realize spheres of influence. Thus Germany was partitioned, so was Korea, then Vietnam; Ireland, India, Yemen, Palestine, Cyprus all these lands were subjected to partition by the great powers. Radha Kumar's researches on Bosnia or Ambassador Holbrooke's graphic account of the modulation of US policy in the Yugoslav crisis to partition the country, has shown how for quite some time partition has been a managerial strategy of stabilizing a situation that seemingly is totally out of control, and in the process becomes the usher of a new round of instability.[3] The extent to which the states formed in these areas in the post-war period have been influenced by that mode of origin has not been historically measured and appreciated yet, not even in the case of Germany.[4] In short, even though how the Second World War ended has a centralizing narrative, possibly on justified grounds, here too we have several histories of war, end of the war, and then peace, partition, independence, and new states. South Asia was decolonized through the strategy of partition; as a key element in state-formation, partition was not however unique to the sub-continent. As already mentioned, Rwanda, Ethiopia, Korea, Cyprus, Yugoslavia, Iraq, Afghanistan, Germany, and many other countries have been either partitioned, or virtually partitioned, or have been on the verge of partition as a part of the post war settlement, new state-formation and the creation of a new states-system in cold war and post cold war times. Because this was never seen in terms of a global politics of the current times, and was seen as 'events of madness', partitions became cases (of what? insanity? barbarity? independence?) for groups of area studies specialists, and remained historically inadequately studied in their significance in terms of 'globalization and fragmentation'.[5] Yet, as we know, partition has been a common political practice. Robert Schaeffer reminds us in his global review of partitions that, since 1920 thirteen countries have been divided resulting in thirty-seven successor states, wars between these successor states in Korea, Vietnam, India, Palestine, Yugoslavia, and the Soviet Union, have failed to provide durable political solutions, and reminds us of the words of the Irish diplomat, Conor Cruise O'Brien, that partition has been 'the expedient of the tired statesman'.[6]

The image of fatigue, however, may be misleading. For it conveys the picture of the parent state trying utmost to stop partition, whereas in reality the parent state with each of its political steps during the prelude to partition acted to make the 'inevitable happen'. Thus, in policies of divide and quit, cold war partitions, balancing acts, 'divisions for democratization' ('decolonization by default'), and 'internal ethnic solution', we have the form of a political practice appearing again and again. Crucial in such recurrence are not only features of the form, but several histories that the practice of partition creates in its wake, ranging from the history of continuing wars and massacres of stranded minorities, to the histories of desires to re-unite and of reunions, evolution of forms of shared forms of sovereignty, hybrid political arrangements to perpetual low level peace, new ethnicities, coexisting times of past and present in the politics of the present, never-ending politics of borders and boundaries, fragmented identities, and newer forms of accommodation, co-existence, and dialogue all these predicated by continuing waves of population flow with their impact on state-formation and subsequent state-development.[7]

Again crucial in all these is how the post-partition state forms. Political historians are now speaking of periodic 'resizing of states' – expansion and contraction as bounded entities, which States as power configurations seldom accept with resignation or dignity.[8] Indeed as Gurharpal Singh shows with reference to Punjab, how modern state relies heavily on a myth of uninterrupted continuity of its size; the myth becomes a self-evident truth; and the rhetoric devours ultimately the entire political class of the land.[9] Thus 'resizing' (of external boundaries) and 'reshaping' (of internal boundaries) happen at great cost. Leaders and statesmen never come to grip with the mystery of how resizing leads to reshaping, and reshaping leads to resizing.[10] The reshaping of internal boundaries in colonial India led to her resizing; as we show with an instance in the next chapter, the resizing still continues to reshape the state. In this reproductive presence of partition, we have the mystery and the clue, namely the organic relation of geopolitics to modernity size with the soul.

The emergence of modern India was through one of the biggest population transfers in human history. Between 1946 and 1951, nearly 9 million Hindus and Sikhs came to India, and about 6 million Muslims went over to Pakistan. Of the said 9 million, 5 million came from what became West Pakistan and 4 million came from East Pakistan. Immigration into India, particularly eastern and northeast India has continued under the shadow of the exodus/influx of that time with of course new patterns but with substantive continuities also.[11] Its impact on state and politics at national, regional, and local levels, in terms of institutionalized and non-institutionalized domains, has been critical, but again unfortunately inadequately studied. The irony is clear, clearer when we remember the circumstances of South Asia. The post-colonial nation form was to be the ultimate form of political institution. For the realization of this form, fundamentally through a constitutional process, it required the production of non-state people from time to time. Thus the entire process of decolonization including partitions, wars, and the deliberate acts of enforcing territoriality produced non-state persons in the country. But this process of production is so violent that it gives rise to anxiety, is this form going to be the ultimate form? Will not the massive flows of population give rise to other structures that rupture the fusion of the state and the nation? Indeed we find that the influx of people makes the process of nationalization ever delayed, ever incomplete, for every community produced by the process of state-making is substantially imaginary; therefore there is always the possibility of newer communities being created out of the national process of creating an appropriate ethnic core.

Think of the theme of inadequacy that partition brings to politics of the nation, particularly in South Asia and India. Partition was to make the nation complete so that the subject could now become the citizen and the nation could now fly into the full panoply of state. This desperate effort to be adequate was to weigh heavily on the nation's subsequent political conduct. Nationalist strategies became revisionist. They were to be the testimonies of a perennially inadequate nation trying to be adequate through ethnic revisions by means of redrawing communal and ethnic boundaries,[12] territorial re-divisions within the country, revision of demographic composition in structures of rule, and an uncertain style of political management that expressed itself in a variety of forms – diplomacy, federality, tools of war and accords, and deliberate attempts to change the geopolitical imagination of the nation along lines of some mythical space expressed through legends of various homelands.[13]

The impact of this revisionist present has not been negligible.[14] It has created a surge of interest in the partition of the sub continent, though historical scholarship has continued to argue along predictable lines. Nationalists say that this was the ultimate betrayal to the ideal of composite secularism that had been all along the main basis of the anti-colonial movement. To some historians, partition was the most damning testimony on a nation that had failed in coming to be, and to them a nationalist history of a such a nation cannot bring out the history of the violence, sufferings and trauma still retained in the memory of the masses. The left has argued (ironically few had put their faith in the argument of the Left whose position became less convincing in the sub-continent following the collapse of the Soviet Union, but its argument has received recently a renewed interest) that partition was a particular form of de-colonization of the region, the retreat of a great power, a way of ensuring the residuary existence of the colonial mode of power, and the final ascendancy of the bourgeoisie in the region defeating the upsurge of the anti colonial radical masses in the period 1945-47. But in this last argument also, probably carrying greater credibility than others, we have more of political explanations and less of history. The radical explanatory argument consequently becomes a history with big spaces and gaps. And interestingly, all these arguments, which link the destiny of the nationalist movement with the movement for a separate identity and the arrival of independence, have to rely as necessity, as matter of imperative, on *history* therefore the question of political particularity escapes all of them, that particularity revolving around the issue of a post-colonial nation's capacity to dialogue, and the fatal attraction of a decolonized polity to partition as method of settling differences. This indeed remains at the bottom of the intriguing and thus the interesting conjecture, the arousal of interest in partition and the stranglehold of (received) history over the event with its consequent impact on our capacity to dialogue.

Writing on partition therefore means understanding the birth of a non-dialogic polity through the prism of the present. Anything less than full engagement with this specificity is unlikely to 'explain' the event *the transition*. Since the event is linked to a transition transition from colonialism to independence/post-colonialism, nationalist secularism to separatism and communalism, and a united entity to a vivisected region, we have here all the peculiarities of a transition debate, namely the characteristics of tautology.[15] It is this dissatisfaction with the story of birth that remains at the root of the arousal of memory, the strivings for different possibilities of the same history, the deliberate construction of a region called South Asia, and what Anthony Giddens had noted, the heavily counter factual nature of future-oriented thought an essential element in the reflexivity of modernity,[16] more significant in the present context of discussion, the roots of dialogic thought. The received history of partition is clearly inadequate in understanding the characteristics of the post-colonial dialogic situation. Even when all details of partition are excavated, dissatisfaction in terms of *explanation* will be incessantly produced. Such dissatisfaction is because of the transformations produced by partition: from a colonial state to a welfare state; from subject to citizen; nation as a locus of desire to one of depravity; tropes of respectability to those of malign; from 'our nationalism' to 'their sectarianism'; and finally from the language of commonness to one of grievance. Partitioned independence has been the imaginative trigger it has triggered claims to permanently ascribed communal and secular discourses. The structures and rhetoric in the separation of one from the other discourse have underpinned the distinctions in the sub continent. Partition thus will be never a settled fact in the sense of settling something, settling the birth of a nation, settling disorder; in producing the differences it will remain unsettled and unsettling.[17]

A modern India today is patently dissatisfied with its origin in 1947: three wars (and many more near wars) with her neighbour, her own insecurities and insecurity syndrome, the permanent division of the nation into majorities and minorities, the failure of a state induced capitalist developmental agenda, insubordination of the permanently underprivileged classes, and a refusal to accept the modernist consummation from all sides, right, centre, the left, all demanding the redefining of the agenda of modern nation fuel dissatisfaction with the origin, and mark the dialogic possibilities of current time. And all this repeatedly reopens the partition question. So long the official position had been that communal violence was *madness*, partition was the only *rational* act of that time; now as genealogical heresy spreads, partition's legitimacy no longer remains unaffected and can no longer escape the gaze. Its rationality now appears as strictly contingent (thus disputing the inevitability argument); the contingency in turn reactivates memory, creates mythic spaces, and provokes alternative discourses of nationhood. Not unnaturally therefore we find the rationalist explanation of partition being interrogated, reason's externalization, that is partition, being investigated by the genealogical heretics. And increasingly, a possible politics of dialogue is now placed in opposition to the politics of partition

Event Not Yet Over, Not Yet an Event

Thus, in dealing with partition that triggers always the imagination of a possible politics of dialogue, what we are grappling with is not a historical event subject to death, but an event signed into the present, therefore an event not yet over, not yet an event. To write of its history we shall have to invent a language that will not stop at the past but will take into account our present that has made the past of partition irrevocable. This kind of symmetrical convergence of the event and the present requires a different language, capable of producing many commentaries. Reflexive, fabulous, journeying into the origin and the actual event, and then zooming back onto the present, historicizing, almost at the same time critiquing the language that will have these attributes, or at least will be capable of serving these needs, waits to be developed in the interest of understanding the fate of the partitioned lands such as India, and a possible politics of dialogue.

If we can speak of a law of partition, briefly it is that, partition not only de-colonizes a land, it re-colonizes, ethnicizes, and releases a blinding array of formations which to sustain themselves reinvent the tool. In short the law of the existence of partition is its reproduction through the enforcement of ethnic decrees. The institution of partition attracts the light of nationalism in which it hides itself. The lightness of the instrument hides the sharpness of the effect, the cleanliness of the act hides the un-cleanliness of the result, and the short duration of the enactment does not anticipate the long duration of its life. This is of course not to say that there is no long partition. Empires such as the Austria-Hungary and Ottoman Turkey have gone through 'long partition', while partitions such as the Indian partition has been abrupt. It took just four months in 1947 for the gentlemen of a commission to neatly finish their assignment of boundary redrawing; it has taken the international political managers and warlords just four years in the last decade to finish off a land.[18] Some have been partitions of *ancient regimes*, some of modern states. Equally important in studying the ubiquitous presence of partition is the issue of its genealogy. Some have occurred through referendum, some through imperial decision. States have divided themselves and have agreed to international re-division in a region, while some states have partitioned as a result of de-colonization. In all these cases, the history of partition began indeed not merely in its origins, but more in what it left for its victims – the local inhabitants.

It is important therefore to see in a study of the form of partition at what stage of its career, the idea of mass expulsion became normal.[19] Also important is the inquiry as to how the event of partition reduces the historically available dialogic options for the people whom it is going to devour. The law of partition thus resides in the *method.* This method enables all symptoms of political convulsions to converge in the irreducible form of partition, and yet this method successfully hides its essentials to the extent that historians and other social scientists think that its anonymity can be battled against only in the form of fragmentary discourse. The method remains unexamined. That is why it is essential to examine the form and the method, and this can be done in at least two ways.

First, we have to show the difference between earlier acts of separation and dismemberment the non-national and the pre-national imperial forms and the de-colonizing re-colonizing forms of today.

Second, we shall have to show how current politics cannot exist apart from existing in partitioned times.

In order to go ahead with this question of the form and the method by which partition is put into effect, and due to which partition attains a never-ending life reproducing itself, we must spend a moment on what is immanent to the method, in other words what lends to the form its life, larger than life. And here let us briefly recall what was suggested earlier in this chapter.

Partition leaves in its wake independence, a dream of sovereignty fulfilled. It is combined with optimism. At the same time it ensures only half-fulfilment of dreams, it spreads around dejection and pessimism; it creates the impossible desire to get the pre-partitioned days and the pre-partitioned land. It realizes the dream of homeland, but at the price of creating what Salman Rushdie calls, 'imaginary homelands'.[20] All kinds of footloose people emerge at the end of the tunnel refugees, migrants, immigrants, non-state persons, vagabond widows, exiles, and aliens who will forever live in these imaginary homelands. Never happy to transact their identity of escapees of violence with a certification of citizenship, they impart a continuous instability to the state that is born through the partition and to the state-system that is born from such de-colonized, re-colonized states. In this 'pessoptimism', there is all-pervasive excess always excess. In grief, hope, remembrance, amnesia, in quarrel and gesticulations of making up, the excess shows up. Repetition and excess become the way of the spiritual life of a nation, and of all other political communities aspiring for nationhood. People cannot settle down. They become almost willing targets of fascist propaganda. Partitions create not only imaginary homelands, they create ethnic decrees to pursue those homelands; they unleash memories that become the site of fascist politics.

Invariably such memories are fragmentary, memory becomes the passion of mind, and victim-hood becomes the soul of the nation. Everyone becomes a victim of the victim. The nation is victim of the imperial policy of divide and quit. The minority becomes the victim of the nation. The individual refugee seeking shelter in the other half of the partitioned land becomes a victim of the alien state that s/he had considered his/her own, and which s/he now finds the unwilling host, for the state too claims to be the victim of conspiracies. In this situation of excess, in which partition had first appeared (we have to only think of the condition of

India in the forties of the last century), it becomes again and again the tempting tool for a *final solution* of a footloose situation, of the inherent and impossible tensions that the political history of colonialism, domination, dismemberment, migration, displacement, democracy and self-determination throw up. Hence, it is not enough to point out that partition leaves in its wake powerful memories that no history can explain or anticipate. The chronicler of the Rwandan genocide of 1994 has commented,

> We may agree that genocidal violence cannot be understood as rational; yet we need to understand it as thinkable. Rather than run away from it, we need to realize that it is the 'popularity' of the genocide that is its uniquely troubling aspect. In its social aspect, Hutu/Tutsi violence in the Rwandan genocide invites comparison with Hindu/Muslim violence at the time of the partition of colonial India. Neither can be explained as simply a state project. One shudders to put the words 'popular' and 'genocide' together, therefore I put 'popularity' in quotation marks. And yet, one needs to explain the large-scale civilian involvement in the genocide...to understand the logic of its development...to make the popular agency in the Rwandan genocide thinkable...to problematize the relationship between the historical legacy of colonialism and postcolonial politics...[21]

And further,

> I argue that the Rwandan genocide needs to be thought through within the logic of colonialism. The horrors of colonialism led to two types of genocidal impulses. The first was the genocide of the native by the settler. It became a reality where the violence of colonial pacification took on extreme proportions. The second was the native impulse to eliminate the settler...When the native killed the settler, it was violence by yesterday's victims...this violence of the victims-turned perpetrators always provoked a greater moral ambiguity than did the settlers' violence.[22]

Thus comes natives' genocide,

> In the annals of colonial history, the natives' genocide never became a historical reality. Yet it always hovered on the horizon as a historical possibility...What distinguished native violence from the violence of the settler, its saving grace, was that it was the violence of yesterday's victims who have turned around and decided to cast aside their victim-hood and become masters of their own lives...'The colonized man finds his freedom in and through violence.'...The great crime of colonialism went beyond expropriating the native, the name it gave to the indigenous population. *The greater crime was to politicise indigeneity in the first place*: First negatively, as a settler libel of the native; but then positively, as a native response...To understand the logic of genocide, I argue, it is necessary to think through the political world that colonialism set into motion. This was the world of the settler and the native, a world organized around a binary preoccupation that was as compelling as it was confining. It is in this context that Tutsi, a group with a privileged relationship to power before colonialism, got constructed as a privileged alien settler presence, first by the great nativist revolution of 1959, and then by Hutu Power propaganda after 1990.[23] (Quotation from Franz Fanon and italics author's)

Memory played a big role in making victim-hood and the alien/settler dichotomy permanent features of the post-colonial life. We require an understanding of the structure of that memory. We require to understand the politics of that memory, an understanding that can cut through the ambiguity of de-colonization, currently known in even more an ambiguous term, 'post-coloniality', and wake us to the fact that contrary to the commonly held belief, political modernity needs memorial exercises to an increasing degree. The reason is not only that, the modern is the most hegemonic way of exercising power, so that it does not rest till memory also is colonized,[24] but that, in order to reinforce the modern, memory has to be invoked again and again. With memory at the service of state formation and state consolidation, the passage of time smoothes out the wrinkles. Memory's truth defeats the political truths of the time. Partition creates 'pessoptimists' who are good subjects for literature, photographs, cinema, and poignant plays, but singularly incapable of re-democratizing the politics of the land. In this milieu of a sleepwalking nation where instability is the sign of the day, the only stable institution appears to be in form a cartographic act that will solidify all that was melting into air. The act of re-drawing the boundaries is therefore only the beginning, its life or the larger than life, begins thereafter.

We cannot afford to make a mistake here. In creating a space where every element seems to obey the sole principle of territorial identity, political representation born through partition turns democracy inside out. Allowing popular representation to rule in a space of territorialism, a space that is only a surface, no more than a polished stone, the incisions draw figures much below, at the core of the self, where identity emerges in 'person' that we know by the name of ethnicity in place of democracy, where there is no place for any other politics besides populism in ethnic form. Now we can understand why an act of partition can never be smooth. *Popular representation may constitute politics of the post-partition nation, but the politics of the self constitutes popular representation*. We have to only think of the small and big instances of massive genocide created by acts of cleansing in this country and elsewhere in this region, so that representation can become most popular, purest, and most authentic. Representational politics, as the events of 1946-47 in this subcontinent had shown, and as again after almost fifty years the destruction of the Babri mosque and following communal frenzy proved, partition makes the convergence of democracy and identity impossible. This dissociation is central. The result is an insurmountable fissure, an unending series of phantasms that in the spatial-temporal duality can only invoke the original act in a desperate attempt for a final solution. One more act, one more effort to finally solve, and we know this will be never final.[25] An endless deferring of *final solution*, for that solution has meaning only in the field of signs and acts that prompt it, and therefore cannot escape the bind.

In such nations born through partition, re-colonization takes the form of independence, ethnic purity takes the place of self-determination, formal sovereignty takes the place of freedom, dreams dry up, demagogues take the place of democrats, and pimps that of the representatives of the people.

Politics of Time and Space

Partition in this way creates *majoritarian* states, otherwise called *ethnocracies*.[26] In this development five factors merit our attention – the irreversibility of the time of partition, the specific character of violence with which partition is effected, the substituting logic of majority in place of plurality, the geo-political imagination lodging itself at the heart of post-colonial politics, and finally the mystery around the history of partition – the problem of the knowable and unknowable. I began with the last aspect. Let me discuss briefly the other four not in an order, but in their relations.

Let us begin with the issue of geopolitical imagination of the nation that is essential for it to turn into an ethnocracy. When in any case beyond the intellectual arguments about the required history of the birth of a nation, lies before us the partitioned region of South Asia, de-colonized and arranged into a states system, it will not be irrelevant to remind ourselves that globalization is mediated here through the order ushered in by partition. With partition, we have today South Asia, what was earlier a vast stretch of land known as the sub-continent, whose margin vanished imperceptibly into other regions the Turco-Persian part of the continent, and what is known today as Southeast Asia.[27] That un-definability is now gone for ever; the home where the nation was born is lost; we are now inhabitants of a region constructed on the reality of vivisection. This vivisection has produced a variety of homelands that symbolize a home that is gone, and is now able to relive only in the form of reflexive images, to be more accurate in form of reflexive nationalisms, known as 'homelands'.

Now that partition and the succeeding fifty years have ruled out easy or programmatic solutions of either complete reunification or total dispersion, the implication is both *universality* and *difference*, for old pre-partition times containing many values were never enough to ensure respect for each other. Europe is increasingly talking of reunion. Southeast Asia is talking of union in form of association. People here are learning from them and want South Asia to 'return in its history and geography like one who is returning home'. But returning where – to a common cultural space? But culture would mean, first of all, identity, that is, difference with others, and then difference with (in) itself. How then to accommodate the differences, or cope with the permanent process of creating others? Or, will it return to a common economic space? But the truth is that, marked by multiple colonial linkages and the exclusion of the majority from the 'economy' it had never been a common economic space at all; this common economic space can arguably be created but we cannot return to it. Or, will it return to a common historic space? But this historic space includes partition, and today we cannot return to it excluding that epoch. Like every history, this history of partition, that is the history that created it and the history that it created, can be summarized. Then how can we include this summary in this other history of commonness, whose summary will have other movement, memory, promise, and identity? All this simply means we cannot *return*; we can only *turn* on an acknowledgement of differences, the differences of a unity, in unity, with other unity.[28] The implication of this metamorphosis of commonness into difference for a dialogic politics is significant, because with the possibilities of a return to commonness being predicated by this irreversible metamorphosis, today's dialogic politics cannot simply be based on a programme of returning to the old ways, but must find out new ways of negotiating differences, and situating reconciliation at the heart of a politics of justice. It also implies that dialoguing in such circumstances can be only a contentious act. I take up this theme in details in a later chapter on friendship and reconciliation.

However, at least for now it is necessary to remember, that today's dialogic politics is featured by the specifics of the geopolitical aftermath of partition that includes the creation of a region called South Asia.[29] In this process there is discovery of old fault-lines. Orthodox and fanatic Islam, backward sub regions, non patriotic frontier tribes, small neighbours prone to blackmailing the stable and strong nation, historic connections with other regions, natural area of influence – all these are being rediscovered, much in the same way fault-lines are being rediscovered in Eastern Europe in the process of Europe's regionalization today.[30] The irony, therefore, is briefly this: in regionalization of South Asia, unity is the theme – that is how the nation is to return home. But in forging a policy of uniting, it rediscovers the old fault-lines – in fact the nation has to mark these out, redefine them in order to erect the new political-economic and security arrangements, and relate on those bases. Therefore designations of such as, underdevelopment, permanent fundamentalist threats to liberal values and civilized existence, backwardness, are found out and they become the reinforcing material for a neo-modernization theory which can map out the cultural region of South Asia – the home of this nation. That is how the core of the region is built. In the resultant marginality of certain elements to build the core of a region, partition still remains the reference point. The agenda of finding back the home seems to have an inviolable validity, so too remains the validity of partition – the most stringent event or theme contradicting the former exercise.

In short, these two declarations of independence (partition and freedom – the latter associated with the advent of a new region) mark out as if in a hypothetical theme the possible politics of dialogue in South Asia. The janus faced independence gave South Asia political power, democracy and an assurance that the dream of ethnic liberation is worth pursuing; it also created communities of the afflicted and aliens in their homelands – both.[31] It structured in a deep way the politics of dialogue in the region. If, as seems clear by now, the minority categories (religion, language, territory, caste, tribe) in this region, called the 'home of the nation', are historically formed and continuously revised in the process of formation, and acquire specific geo-political forms equally prone to periodic revisions, then there cannot be an easy recourse to an administrative or state-centric management of the dialogic agenda, assuming that such categories are fixed points of descriptive reference. On the other hand, the contentious politics of dialogue marked by features of partition has to take into account the more complex accounts of their historical location, differential access, internal cleavages, and potentialities of transformations.

At a more fundamental level, dialogic politics faces the question of a modern representative politics in post-colonial conditions of South Asia born out of partition. The question classically stated would be like this: representative political system is based on representation of necessity, identity and the self, in other words, interest. Where interest has been articulated around mostly group or a collective, representational problematic will be, how is the polity to organize the hierarchies of representations? On one hand, representational politics validates the nation, which is the highest group or community, on the other hand it gives space to other interests, hierarchically, for by according supremacy to the nation it has admitted the logic of hierarchy. Therefore the genealogy of a modern dialogic politics in South Asia is caught in the representational bind. If membership of a representative polity in South Asia has been thus a transactional identity only, based on the necessity of exchange of unequally endowed individuals, and of unequal agencies, it has three implications: first, political dialogue most of the time is structured by hierarchies in the representational system, and finds very difficult to cross the limits posed by the rules of transaction in order to address the contentious issues; second, the dynamics of representation is ruled by a spatial order, hence dialogue cannot ignore the geopolitical realities of a post-partition representational political system; third, in this perspective of a stalemate brought about by 'equally' valid claims to draw on history, only *justice* can be the basis on which reconciliation – the functional mode of dialogue – can work itself out. This is where we can anticipate the possible politics of dialogue – the possibility of transcending the limits of the reality of partition.

Dialogue, in short, stands in an intriguing relation with time in as much with space, conditioned as these two are in a fundamental way by an absolute politics of separation, known as partition.[32] Without the past of cleavages and partitions, there is simply no dialogue. But without the past being represented in a different way, that is, without the utopia which is also a received past, there is no possibility of dialogue transcending the limits of the present. Jurgen Habermas has insisted on the singularity of the past in referring to the Auschwitz experience, and has seen in the attempt at its banalization the danger of a drive towards a *sonderweg* – a separate path. We too are faced with the banalization of our past, the past of partition to an extent, where we fail to see the role of partition in the desire of the post-colonial state to develop its 'own' geopolitics.[33] The meticulously researched essay by Joya Chatterjee on boundary delimitation and demarcation in eastern India by the Radcliffe Commission[34] leaves no one in doubt of the explosive result of a combination of geopolitical imagination, territorial politics, and nationalist democracy. The geopolitical ambitions and imaginations linger and increase, they numb our revulsion to a rogue past; they create what Habermas termed as banal past. But to this we must add, that the traces of this past and the access to this past are differential, not only making the banalization of past difficult, but also making possible the dialogic politics of today.[35] One can give an instance. The geopolitical imagination of homelands and nationalisms that lay at the heart of partition has led to a theology of the border. Borders convey sanctity; yet born of partition and therefore daily negotiated by hundreds and thousands of people, borders have become naturalized. For, the truth is that, just as borders are products of partition, and the mental borders conjured up by geopolitical imaginations of the nations-in-the-making have given birth to partition, these borders have in time become the site of the practices of accommodation, interaction, and negotiation, drawing on historical flows and conversations among peoples, communities, and solidarities.

It is necessary, therefore, to understand the issue of territorialism as the product of partition in a more concrete way, predicated by time and space. By cutting the territory, also by making the task of guarding the territory defined by the colonial rulers as India (we have to only think of the British frontier policy in the sub continent) an imperative, partition has accentuated the territorial consciousness in the region. The logic of territory, namely standardization, formalization, centralization and configuration, euphemistically called in the region as *territorial integrity*, has now turned into a state-strategy to rationalize, affect, influence and control. Originating from an 'unclean partition', the territorial consciousness of the nation in the sub continent will eternally disturb the neatly segmented universe of the standard political map of this region of South Asia, the home of this nation. In this geopolitical universe inhabited by nations, states, and nations-in-the-making – all of them drawing from a partitioned past – the politics of dialogue confronts its possibilities and impossibilities.

Democracy, Ethnicity and Partition

Space, as we have seen, thus plays crucial role in the emergence of states, known as 'majoritarian' states because they are marked by the production of majorities and minorities, and the polities reinforce themselves with politics of the majority. Such majoritarian states are criss-crossed with lines separating majorities and minorities, and homelands from the aliens. Also we have indicated that in this emergence of ethnic space extremely significant is the sense of the irreversibility of time. Because partition appears as the final act, perhaps undesirable but unavoidable, perhaps bloody but necessary to avoid more bloodshed, perhaps an unhappy separation but a separation to stop genocide, the irreversibility of partition overwhelms the nation, its leaders, and publics. In this situation the public mind becomes an ethnic mind, for the only thing on which the nationalist public can discover its self is in the comic soon to turn farcical form of identity – that can assume only an ethnic form. To the extent to which democracy becomes a victim of the nationalist strategy of partition to achieve statehood, ethnicity wins. As a law, if one were to say of that, partition makes the relation between nationalism and ethnicity on one hand and democracy on the other inverse. The irreversibility of the act makes such an inversion possible. The nationalist leaders could be only either gullible or misleading, when they would say, as the prime minister of undivided India presiding over the partition of the country to become the prime minister of the post-partition India had said within one year of the act,

> Partition came and we accepted it because we thought that perhaps that way, however painful it was, we might have some peace to work along our own lines. Perhaps we acted wrongly. It is difficult to judge now. And yet the consequences of that partition so terrible that one is inclined to think that anything else would

have been preferable. That partition has come, and has brought in its train other vast changes. There is no going back now to India as it was before the partition.[36]

After sacrificing conversation and dialogue, and embracing partition, the nation-form can thus assume only an ethnic form. When democracy cannot be sufficient for providing a collective name, anti-colonialism and anti-imperialism prove similarly inadequate, and the conversation becomes so contentious that a democratic resolution of the 'nation question within the nation' (minority, nationality) proves impossible, it is time for the final, irreversible step towards ethnic (majoritarian) statehood. Yet, and this is the strangest part, no one takes this step, it just presents itself, and everyone bows to the inevitable. Thus, we have the run of assertions: the British did not divide the country; they had no other way left but to bow to the wishes of the subjects. The Muslim League did not cause it; it was forced to seek a homeland due to the obduracy of the Hindu Congress in sharing power with the Muslims. And the Congress did not want it, but to avoid genocide it had to welcome a divided India. Partition appears as the axe of God, and the nation must go forward to attain statehood. Time is therefore as important as space in giving birth to the post-colonial politics of state and nation. Ironically, only the de-colonized understands how the ethnic space builds on ethnic time, what the time of colonialism and the time after mean in terms of self.

Besides building on the irreversibility of time, the ethnicization of politics takes place on the irreversible seizure of the nation by patriarchal political power. The non-democratic aspects of a nationalist politics that had wanted to negate at every opportunity the autonomy of various political formations growing within the anti-colonial space now reaches their apogee. The sacrifice of countless women by the newly independent state to implement partition and secure the post-colonial order represented the final assault of a nation that was fast turning into an ethnic formation. Surprisingly, excepting very few unrepentant left intellectuals, few historians connected the violence on women and violence on the peasants during the transfer of power, and found out how an extreme and pervasive violence marked the birth of the nation-state. On the countless books on partition, if we have therefore few works on violence on women, we get fewer works on violence on the peasants. At least, the studies by Urbashi Butalia, and Ritu Menon and Kamala Bhasin[37] and others get some attention, though lamentably they are still read only as specialized feminist histories, and not the history of a counter-revolution, though it is true, that only through a feminist reading of that past we can get at times an idea of the general counter-revolution. But studies such as Amalendu Sen Gupta's *Uttal Challish – Asamapta Biplob* (*The Surge of the Forties – The Incomplete Revolution*) are lost in the discourse of partition, in which no one refers to the almost simultaneous peasant upsurge and the history thereof. That is another history; a vernacular history; pushed out of centralized narrative of partition. Amalendu Sen Gupta has written in Bengali, others have written in Hindi, Assamese, still others in Malayalam and Urdu; in the centralizing narrative of partition and state-formation those histories are lost. In a desperate appeal for a federal or con-federal history, that can answer the question, *can the history of the genocide ever be known?*,[38] there are recourses to oral histories, and arguments in defence of fragments. But the truth remains that all these are parts of a passive revolution without which the nation could not have come into being, that is to say into statehood. Quarantining all the deviants – the women, the rebellious peasants, the obdurate political sects, and the indigenous peoples – partition was implemented and the post-colonial state was formed. Built on the irreversibility of the decline of all other fortunes excepting the ethnic destiny, the nation marched into statehood.[39] One has to look only little closely into the deliberations of the Indian Constituent Assembly to see how the nation went back on the promise of a new deal for minorities, on accords with various unwilling political formations, how it became one of the key instruments of passive revolution. So severe was the discipline of nationalist violence on society and so complete was the hegemony that even the limited practice of the Constituent Assembly to continue dialogue with the deviant sections was restrained, put a stop to, and those who wanted the Constituent Assembly to recognize these other voices besides the nationalist one, needed explicit sanction of the nationalist political oligarchy to put in a few lines in the constitutional dialogues and debates. In the ongoing constitutional negotiations and games, provisions for minorities were not issues of dialogue, but, in the language of one of the most prominent voices of the State, became a gift of the majority to the minorities, a result of bargaining between the dominant majority claiming to represent the state and the disparate minority groups who were not sure individually and collectively how to address their powerlessness in face of a strong majority-centric political class determined to maintain the power-structure transferred to it by the colonial rule through partition. India's Home Minister Sardar Patel drove home the point by commenting, just before the Constitution was to be to given the final shape, on 26 May 1949,

> It is not our intention to commit the minorities to a particular position in a hurry. If they really have come honestly to the conclusion that in the changed conditions of the country that it is in the interest of all to lay down real and genuine foundations of a secular state, then nothing is better for the minorities than to trust the good sense and sense of fairness of the majority, and to place confidence in them.[40]

And then he said,

> What do the minorities desire? Do they want to have any share in the Government of the country and in its administration? I tell you, you cannot have a genuine seat in the Cabinet if you segregate your self from the rest of the community, for the cabinet can only act as a team in a harmonious manner and unless every member of the Cabinet is answerable to a common electorate the Cabinet cannot function in a fruitful manner. Are you prepared to give up your right of representation in the Government...There cannot be any divided loyalty. All loyalties must be exclusively centered round the State. If in a democracy you create rival loyalties...then democracy is doomed.[41]

Clearly, the nationalist-republican space was re-built this way in the aftermath of partition. Partition of the colonial space, which was in reality of a complex of majorities, minorities, castes, tribes, classes, territories, different categories of subjects and citizens, a dominating power structure and a variety of dominated entities trying to unify in a nation, defined the way in which the conversation on minority rights would proceed in the Assembly. Jinnah was the evil whose ghost was to be exorcized in order to make the nation and purify the national space. Shibban lal Saksena, a member from the United Provinces objecting to treating the scheduled castes as a minority said,

> I would like to draw the attention of the Assembly to one important declaration. It is this. It will be remembered that Mr. Jinnah has often tried to include the Scheduled Castes in the minorities; and on June 26, 1946, in a letter from Maulana Abul Kalam Azad to Lord Wavell, and the latter's reply thereto, Lord Wavell is reported to have said, '...if any vacancy occurs among the seats allotted to the minorities, I shall naturally consult both the main parties before filling it.' Mr. Jinnah has thus included the Scheduled Castes among the minorities. But so far as we are considered, we consider the Scheduled Castes as belonging to Hindus, they are not a minority; they have always formed part of us.[42]

So was also the matter about Sikhs, as Munshi said, 'In view of the special situation in Eastern Punjab the whole question will be considered later.' But the fallacy of disaggregating minorities, and dismembering them was quickly pointed out when another member Rizwan Allah spoke,

> Sir, I beg to raise a point of order on this amendment. This is a report of the Minorities Committee. Different provisions have been laid down in this report about various minorities. So far as the Sikhs are considered, no decision has been arrived at in the Minorities Committee Report about them. It is stated in this Report that the matter about Sikhs will be decided later on. Now an amendment ahs been tabled to replace a Province instead of Sikhs and thus in place of a minority an issue about territory is brought in. This is a report for the minorities and has nothing to do with any province...[43]

In this interplay of ethno-politics and geo-politics, which partition had brought about and whose child had been the partition, the strategy of the state was to define democracy by a strong majoritarian logic that would combine both reasoning of the ethnic and the territorial, and thereby relegate the practice of dialogue on issues of rights and justice to the domain of 'protection', and not make it integral to democratic nationhood. The massacre of a section of citizens had created the people – the source of sovereignty of the republic. And now this *people*, as distinct from the *citizen*, and defined through partition in no uncertain terms, defined who were to be in it, under what terms, and who belonged outside.[44] Thus, H. J. Khondekar, an Assembly member, referring to Ambedkar's politics of the preceding twenty years of demanding separate representation of the scheduled castes and opposing the inclusion of the scheduled castes in the minorities, said,

> The first name is that of Dr. Ambedkar, and you all know, that from the time of the second Round table Conference till the Minority Sub-Committee of the Advisory Committee assembled, he relinquished the demand for joint electorates. On the question of this demand his message to all Harijans of his country, who

> belonged to his party, went to the extent that they were not even Hindus that they wished to have a colony separate from the Hindus, that they were not within the fold of the Hindu religion, and it was for this reason that they desired separate electorates. This thing has been going on the country for the last fifteen years with the result that a sort of discord has been created between caste Hindus and Harijans of Dr. Ambedkar's party...I feel happy to state that when this matter relating to joint and separate electorates came up before the Minority Sub-Committee, Dr. Ambedkar did not press the claim further but withdrew it on the ground that he had no argument in support of the principle...It was because of Lord Morley Minto that Muslims got separate electorates and the result was that our country was divided into two...If this is accepted either for Harijans or for our Muslim brothers, then it would mean fulfillment of what my friend Mr. Jinnah has always said, 'Muslims of India and Muslims of Pakistan' – which means the preparation for Pakistan within India. Much suffering has caused already. India has been divided into two. Brother Muslims have got what they wanted, and was for their benefit. Having got that, they should be good enough not to try to create Pakistan within India and should not bring an amendment of this sort in this House.[45]

If this was not enough to mark out in a disciplinary note that political dialogues must not continue impinging on the nationalist control of the emerging republic, Sardar Patel was further blunt. Warning all those, who were speaking of the need for dialogue on issues such as rights of the minorities, and separate electorates, Patel said in barely concealed language of venom,

> ...But now the separation of the country is complete and you say let us introduce it again and have another separation...If the process that was adopted, which resulted in the separation of the country, is to be repeated, then I say: Those who want that kind of thing have a place in Pakistan, not here. Here, we are building a nation and we are laying the foundations of One Nation, and those who choose to divide again and sow the seeds of disruption will have no place, no quarter, and I must say that plainly enough...Why go on saying, Oh Muslims were not heard; Muslim amendment was not carried...If that is going to pay you, you are much mistaken, and I know how it cost me to protect the Muslim minorities here under the present condition and in the present atmosphere...You have got a separate State and remember, you are the people who were responsible for it, and not those who remain in Pakistan...To the Scheduled Caste friends, I also appeal: Let us forget what Dr. Ambedkar or his group have done. Let us forget what you did. You have very nearly escaped the partition of the country again on your lines...[46]

Accounts of independence are thus at the same time the prison guards of other accounts, in particular of contested conversations. There is now enough documentation on the way in which dialogues had gone on among various sections of the political society in the anti-colonial political space, particularly in the preceding twenty years of partition and independence, on the geo-political and ethno-political future of India. Those accounts are an embarrassment to all, because a dialogic past may not be appropriate to the ethnicized present, or to our ethnic futures. But clearly they marked the dreams and strivings for making a different, dialogic India, and partition marked their deaths, at least their drying up, till time and reconfiguration of the political space would allow the dreams to re-emerge though in different forms.

What happens when dreams dry up? I shall suggest four results and quickly go through them in order to suggest that the greatest function of partition is disciplinary. While historians are busy looking at the dark side of the independence of the new nation, namely violence, homelessness and hatred as the cardinal principle of politics, for once let us draw our selves away from the narratives of insanity and see what they mean in terms of democratic politics. First, constitutionalism (whether succeeds or not) in the new nations naturalizes the dark side, and promising a return to normalcy marginalizes the dissenters and heretics. Second, this normalcy based on an enactment of violence and hatred kills the spirit of reconciliation. The polity can speak of only then *retributive* justice and not *restorative* justice. The promised restoration of dignity after the hour of retribution never arrives. Third, democracy becomes organic, in other words intolerant; people so defined by democracy do not allow other peoples not so defined by democracy to co-exist, and push the latter out.[47] Finally the issue of self-determination that had caused so many sleepless nights of the political managers resolves its own dilemmas by a bizarre technique. It kills all others selves so much so that there need not be any other self but only the geopolitical self of the aspirant. We shall need several essays to work out the history of the discipline by which the modern state arrives; Charles Tilly calls it contentious history. But at least we can say this much that a theory of the passive revolution, which inevitably accompanies state-formation, must investigate these aspects to find out why partition has become the universal act in the age of liberal democracy – the institutionalized form of our domestic and geopolitical histories. To grapple with the dilemma that democracy places before us by appearing with its stepbrother that carries the twin names of *geo-politics* and *ethno-politics*, my argument in this book will be to emphasize the need to rethink different forms of justice and dialogue – victor's justice and dialogue and survivor's justice and dialogue – in the context of democracy, so that we can acknowledge the play of co-existing identities and coexisting futures.

Partitions have not worked as solution to ethnic conflicts, if it has had any role, that role has been to make ethnicity the all-pervasive truth of life, conditioning relations, dialogues, and norms of justice, and then it has transformed ethnic divisions into racial divisions. In lieu of providing resolutions it has restructured conflicts on the colonial model of settler/native and inner/outer distinctions. At best it has been a clumsy strategy of exit for the hitherto ruling power, at worst a bloody tactic of containment. As the Dayton Peace Agreement shows, partition can be only a temporary separation of forces agreement, which can never take off towards plural dialogue for accommodation and peace, because as I show in the next chapter, partition can have only a downward spiral to ever more inner levels. Strewn before us are the historical examples of how each time partition has been countered with proposals for alternative solutions, each time these solutions have been made to appear as weak, idealistic, and unrealistic, because the situation has appeared as making an irreversible turn from the dialogic to exterminatory, and more important, a turn from the chamber politics of few members of a select political class to democratic, populist/popular. It is all the more necessary therefore to recognize the weaknesses and strengths of these alternative possibilities of dialogue, the only way to resist the organized power of genocide.[48]

The light of partition has held sufficiently long even when the sun of post-colonial development became unbearable, and multiculturalism, secularism, toleration, anti-poverty programmes, and national integration became the elect-ideologies, to warrant for us a scrutiny of the time of wars and peace we have gone through. In the ethnic futures of the post-colonial nation buttressed by these elect ideologies, the form of partition may feign simplicity. But this form wears its instincts on its sleeves; like Dostoyevsky's Father Karamazov, that corrupt and deep old man, the act of partition appears as simple, because it manages to appear as inevitable. The geopolitical history of the region much of which is made up by these ethnic futures with which nations are growing holds the clue to the charm and irony of an arbitrary act. Pressed and interrogated closely, the simplicity wears thin. The line between friendship and enmity, rights and accommodation, sovereignty and care is drawn with blood, and those who take too much notice of that, will always appear as naive or preposterous to the rational actors of international politics who know that deepest feelings have the iron necessity to be disciplined into the grey dignity of modern nations and calmly preach that war and peace as inevitable phases of a pre-destined life.

It will be necessary, therefore, to look more closely into some of these geopolitical histories, often appearing as ethnic futures, to find out how cartographic acts have become symbols of the twin phenomena of trans-valuation and focalization, whereby a geopolitical history transgresses all other histories, overwhelms them, and becomes the focal point of the political histories of our time. I shall speak here of one such history – the history of the line of control in Kashmir, not to go over an account hundred times told in bits and pieces in diplomatic enclaves and demarches, and in agitated tones in the map rooms of army headquarters, but to keep in mind the stark reality of a geopolitical logic invoked again and again in deciding our unities and differences – the enmities and friendships of this post-colonial age.

Notes

1 Two instances, written from two almost opposite angles, would make out the point. Sucheta Mahajan, *Independence and Partition – The Erosion of Colonial Power in India* (New Delhi: Sage Publications, 2000). She takes to task the works produced out of 'generously funded fellowships...from Oxbridge, Chicago, or even lowly Canberra', 'subaltern studies (with) deep complicity with imperialist scholarship', 'Cambridge-subaltern pedigree', studies by Sugata Bose, 'obtuse logic' of Jaya Chatterjee, 'rejection of secularism of Partha Chatterjee', and 'most existing works on the Congress and the communal problem in the 1940s' – pp. 17-37; less damning, but equally adversarial response to other historians' writings, is Gyan Pandey, *Remembering Partition – Violence, Nationalism and History in India* (Cambridge: Cambridge University Press, 2001).

2 Said uses the term following Emile Habibi's *The Secret Life of Saeed, the Ill-Fated Pessopitimist – A Palestinian Who Became a Citizen of Israel* (Columbia, 1989), in Edward Said, *The Politics of Dispossession – The Struggle for Palestinian Self-Determination, 1969-1994* (New York: Vintage Books, 1995), n. 23, p. 422.

3 Radha Kumar, *Divide and Fall? Bosnia in the Annals of Partition* (London: Verso, 1997); see also the review essay on this book by Arvind N. Das, 'The End of Geography', Biblio, March-April, 1998, pp. 8-9; Richard Holbrooke, *To End a War* (New York: The Modern Library, 1998), particularly pp. 159-166. On this theme further, Clive J. Christie, 'Partition, Separatism and National Identity – A Reassessment', *Political Quarterly*, 63 (1), January-March 1992, pp. 68-72; the most detailed work till date, Robert K. Schaeffer, *Severed States – Dilemmas of Democracy in a Divided World* (New York: Rowman & Littlefield Publishers, 1999).

4 Indeed only with the re-appearance of 'one' Europe, we now see what the partition of Europe did to the continent. The least that we can say is that with old rivalries reappearing and the states system in Central and Eastern Europe being redrawn, Europe often reminds us of an older stage in state formation. The impact of the post-1945 arrangement of spheres of influence in East Europe in particular does not always seem very profound. For a relevant discussion, see Geza Jeszensky, 'More Bosnias? National and Ethnic Tensions in the Post-Communist World', *East European Quarterly*, 31 (3), September 1997; and Sarah A. Kent, 'Writing the Yugoslav Wars – English Language Books on Bosnia (1992-96) and the Challenges of Analysing Contemporary History', *American Historical Review*, 102 (4), October 1997.

5 Clive J. Christie's argument in 'Partition, Separatism and National Identity – A Reassessment' about the need to separate two kinds of partition, ideological and identity generated, can be accepted only in a broad format of decolonization where the simultaneous processes of partition, decolonization and new state-formation were ushered in by war and the decline of old imperial tradition.

6 *Severed States – Dilemmas of Democracy in a Divided World*, 'Introduction', p. 1.

7 Some of these features are discussed in details by Tai Yong Tan and Gyanesh Kudasia, *The Aftermath of Partition in South Asia* (London: Routledge, 2000).

8 Gurharpal Singh speaks of 'The Partition of India as State Contraction – Some Unspoken Assumptions' (Ch. 4) in his *Ethnic Conflict in India – A Case-Study of Punjab* (Basingstoke: Macmillan Press, 2000), pp. 56-72. He reverts to the theme of expansion/contraction again in 'Resizing and Reshaping the Indian State – the Punjab Problem in a Comparative Perspective' (Ch. 12), pp. 194-218. Gurharpal Singh builds on the idea of partition as a mode of state contraction and resizing on the basis of I.S. Lustick's work, *Unsettled States, Disputed Lands – Britain and Ireland, France and Algeria, Israel and West Bank-Gaza* (London: Cornell University Press, 1993).

9 *Ethnic Conflict in India – A Case-Study of Punjab*, p. 196.

10 On this relation, Oren Yiftachel, '"Rights-Sizing" or "Right-Shaping"? Politics, Ethnicity, and Territory in Plural States' in Brendan O'Leary, Ian S. Lustick, and Thomas Callaghy (eds.), *Right-sizing the State – The Politics of Moving Borders* (Oxford: Oxford University Press, 2001, Ch. 12, pp. 358-387.

11 I have discussed these in my book, *The Marginal Nation – Transborder Migration from Bangladesh to West Bengal* (Delhi: Sage Pub, 1999); on this see also the essays in Ranabir Samaddar (ed.), *Reflection on Partition in the East* (Delhi and Calcutta: Vikas Publications and Calcutta Research Group, 1997).

12 On ethnic boundaries and frontiers, I am grateful to Oren Yiftachel of the University of Ben Gurion for his writings and deep understanding. See, for example, Oren Yiftachel and Avinoam Meir (eds.), *Ethnic Frontiers and Peripheries – Landscapes of Development and Inequality in Israel* (Boulder, Colorado: Westview Press, 1998), particularly 'introduction', pp. 1-16.

13 Even in the colonial time, the demand for homeland was raised by several communities, most notably Muslims and the Nagas. The Muslim demand for homeland is probably the more researched theme in sub-continental history. On the less-known Naga demand, Asoso Yonou, *The Rising Nagas* (Delhi: Vikas, 1974); Udayon Misra, *North East India – Quest for Identity* (Guwahati: Omsons, 1988); M. Horam, *Naga Insurgency: The Last Thirty Years* (New Delhi: Cosmo, 1988), and Charles Chasie, *The Naga Imbroglio* (Kohima: Standard Publishers, 1999).

14 I deal extensively with the revising form of a post-partition nation in my *A Biography of the Indian Nation 1947-97* (Delhi: Sage Publications, 2001); I draw upon the ideas expressed there in Chapter 1.

15 This has been noted by one of the perceptive commentators on the 'transition debate'. See, Ellen Meiksins Wood, 'Capitalism, Merchants and bourgeois Revolution – Reflections on the Brenner Debate and Its Sequel', *International Review of Social History*, Vol. 41, part 2, August 1996, pp. 209-32.

16 Anthony Giddens, *The Consequences of Modernity* (Cambridge: Polity Press, 1988), p. 154.

17 This feature of the situation will be clear to anyone who goes through the massive compilations by Mushirul Hasan (ed.), *India's Partition – Process, Strategy, and Mobilization* (Delhi: Oxford University Press, 1993), (ed.), *India Partitioned – The Other Face of Freedom*, 2 Volumes (Delhi: Roli Books, 1995), and his own work, *Legacy of a Divided Nation – India's Muslims since Independence* (Delhi: Oxford University Press, 1998).

18 One is reminded of the infamous 51-49 formulae of the Western powers for solving the Bosnian crisis (51 per cent of the territory of Bosnia-Herzegovina under Federation, that is Muslim-Croat, control and 49 per cent under Bosnian Serb control) that led to absurd drawing and re-drawing of control lines on maps during the Dayton negotiations. As partitions show, while on one hand political discussions during partitions are tentative efforts to build a political framework for (some kind of) a joint future, on the other hand discussions on drawing of boundaries are in effect a continuation of the strife and wars that have led to partition. The same thing happened in the Indian sub-continent in 1946-47, to be repeated fifty years later in the Balkans. On the 51-49 formula, see *To End a War*, pp. 294-297; also, Oya Akgonenc, 'A Precarious Peace in Bosnia-Herzegovina – The Dayton Accord and its Prospect for Success' in Stefano Bianchini and Robert Craig Nation (eds.), *The Yugoslav Conflict and its Implications for International Relations* (Ravenna: A. Longo Editore, 1998), pp. 61-70.

19 I am indebted to Eric Hobsbawm for raising before me some of these issues during a discussion on partition.

20 Salman Rushdie, 'Imaginary Homelands' in Salman Rushdie, *Imaginary Homelands – Essays and Criticism, 1981-91* (London: Granta Books, 1992), pp. 9-21.

21 Mahmood Mamdani, *When Victims Become Killers – Colonialism, Nativism, and the Genocide in Rwanda* (Princeton, NJ: Princeton University Press, 2001), introduction, p. 5.

22 Ibid. p. 6.

23 Ibid. p. 9.

24 In *The Marginal Nation* I show how violence and memory of violence lead to forced migration and then renewed violence on the immigrant.

25 In this connection it is significant to note that writings on partition have meaningfully turned from writing of the event to what partition has meant in its afterlife, and have noted the reproduction of the form. Thus, Suvir Kaul in his introduction to the Kaul (ed.) *The Partitions of Memory – The Afterlife of the Division of India* (Delhi: Permanent Black, 2001) notes, 'We thus need to ask not simply how Partition created nation-states of the subcontinent, but also how these nations require partition, or its more local iterations, in order to justify state authority. It is clear, for instance, just how much the state needs the fear of Partition, and of future Partitions, to maintain its investment in externally and internally repressive agencies'. (p. 9).

26 Oren Yiftachel uses the concept in his essay, 'Ethnocracy and Its Discontents: Minorities, Protest and the Israeli Polity', *Critical Inquiry*, 26 (4), Summer, 2000.

27 On the construction of a region and its relation to nation, I have written earlier, *Whose Asia Is It Anyway – Region and the Nation in South Asia* (Calcutta: M. A. K. Azad Institute of Asian Studies/Pearl Publishers, 1996), chapter 3.

28 On a similar question concerning Europe, Jacques Derrida, *The Other Heading – Reflections on Today's Europe*, trans. by Pascale-Anne Brault and Michael B. Naas (Bloomington and Indianapolis: Indiana University Press, 1992), pp. 4-20.

29 On the following argument I have written at length in Ranabir Samaddar, *Whose Asia is it An way? Nation and the Region in South Asia.*

30 The best instance of such rediscovery is of course S.P. Huntington, 'The Clash of Civilizations?', *Foreign Affairs*, 72 (3), 1993, pp. 22-49; Huntington, 'If Not Civilizations, What? Paradigms of the Post Cold War World', *Foreign Affairs*, 72(5), 1993, pp. 186-194. On an analysis of such paradigms in regionalizing Europe, Elizabeth H. Prodromou, 'Paradigms, Power and Identity: Rediscovering Orthodoxy and Regionalizing Europe', *European Journal of Political Research*, 30 (2), September 1996, pp. 125-154.

31 I am indebted for this idea of a janus faced independence to Michael Rogin's essay, 'The Two Declarations of American Independence', *Representations*, 55, Summer 1996, pp. 13-30.

32 I have written in details on how time plays a decisive role in shaping nationalism – a factor ignored by some prominent theorists of nationalism, is significant. See Ranabir Samaddar, *Paradoxes of the Nationalist Time – Political Essays on Bangladesh* (Dhaka: University Press Limited, 2002).

33 Even perceptive observers miss the role of 1947 in the development of Indian geopolitical thinking – an event that marked the shift of the nation from marking out a cultural region/area to a geopolitical area as its 'natural habitat'. See for example a RAND study by George K. Tanham, *Indian Strategic Thought – An Interpretative Essay* (Santa Monica, CA: National Defense Research Institute, 1992).

34 Joya Chatterjee, 'The Fashioning of a Frontier – The Radcliffe Line and Bengal's Border Landscape, 1947-52', *Modern Asian Studies*, 33 (1), 1999.

35 See in this context, Navnita Chadha Behera, 'Forging New Solidarities – Nonofficial Dialogues' in Monique Mekenkamp, Paul van Tongeren, and Hans van de Veen (eds.),

Searching for Peace in Central and South Asia (Boulder, Colorado: Lynne Rienner Publishers, 2002), pp. 210-236.

36 Cited in Mushirul Hasan, *Legacy of A Divided Nation – India's Muslims Since Independence*, p. 100.

37 Urbshi Butalia, *The Other Side of Silence – Voices from the Partition of India* (New Delhi: Viking, 1998); Ritu Menon and Kamla Bhasin, *Borders & Boundaries – Women in India's Partition* (Delhi: Kali for Women, 1998).

38 Urbashi Butalia in fact refers to the question citing James E. Young, *Writing and Rewriting the Holocaust – Narrative and the Consequences of Interpretation* (Bloomington: Indiana University Press, 1990).

39 See the admirable survey of ethnic cleansing by Andrew Bell-Fialkoff, *Ethnic Cleansing* (New York: St. Martin's Griffin, 1999), particularly chapter 3, 'Cleansing as a Metonym of Collective Identity', pp. 57-115.

40 *Constituent Assembly Debates* (hereafter CAD), Volume VII, p. 272; Patel's remarks brushed aside the weak voices of the Muslim representatives who admitted that time had changed, that since they had showed their integrity to the nation, separate electorate for the benefit of the minorities should be maintained. It is clear that the truth games about minority problematic were being played out in the shadow of partition, Kashmir war, and the memory of massacres, and continuing migration of thousands and thousands. Speech of Md. Ismail, Ibid. p. 277; also p. 283. This opinion however met with objections of some other Muslim members. For views identical to those of Patel, see the lecture by another Constituent Assembly member, B. Pocker Sahib Bahadur, who had exclaimed in exasperation, 'The majority is a majority, and the minorities are minorities,' and further, '...the idea of getting representation from religious groups is simply ridiculous...The minority must remain a minority. Now before a minority there is only one alternative: It is to be loyal to the majority, and cooperate and gain the confidence of the majority.' – *CAD*, Volume VII, p. 212 & 218.

41 Ibid. pp. 223-224.

42 Discussion on the Schedule, *CAD*, Volume 7, p. 234.

43 Ibid. p. 240.

44 Max Jean Zins calls this the 'citizen-massacres' in his essay ' The 1947 Vivisection of India – The Political Usage of a Carnage in the Era of Citizen-massacres.' In Mushirul Hasan and Nariaki Nakazato (eds.), *The Unfinished Agenda – Nation-Building in South Asia* (Delhi: Manohar, 2001), pp. 49-78.

45 *CAD*, Volume 7, p. 266.

46 Ibid. pp. 271-272.

47 Michael Mann, 'The Dark Side of Democracy – The Modern Tradition of Ethnic and Political Cleansing', *New Left Review*, 235, May-June 1999.

48 Radha Kumar mentions some of these in 'Lessons Learnt – The Options Ahead', *Transeuropeannes*, 19-20, 2001. I shall however suggest, in addition to the lessons that Radha Kumar draws, that we pay attention to the way trust plummets down in contingencies of partition, therefore to the ways in which trust stops potential partitions that must have been lurking at us at this time throughout the world. In this context we must go back repeatedly to the moment of 'partition', the moment of 'decisive break' when trust gives way to killing separation. Our histories and new histories have been as yet insufficient in dissecting the architecture of that moment. For a revelatory account of one such moment in the history of the sub-continent, see Sudhir Ghosh, 'To Trust or Not to Trust' (Ch. 7) in his autobiographical account, *Gandhi's Emissary* (Boston: Houghton Mifflin Company, 1967), pp. 133-176.

Chapter 4

The Ineluctable Logic of Geopolitics

Enmeshed Histories

I narrate here the strange case of a line that has refused to freeze. The line of control (LOC) in Kashmir has not only not stabilized it has subsumed within its geopolitical history all other dimensions of the situation of the land, virtually partitioned by this line, including its internal ethno-linguistic variety. Deeply lodged in the quagmire of the geopolitics of post-colonialism, the history of the line of control is a history that has pursued other histories of Kashmir as the unrelenting shadow, particularly the histories of the ethno-lines of which Sumantra Bose has written so vividly in a recent essay.[1] Sumantra Bose's essay makes compelling reading because he draws our attention to two Kashmirs one of guns, bombs, suicidal attacks, *jehadi*, security forces, and the other, the lived experiences of the people. He notes that while the first has dominated public consciousness outside Kashmir and has monopolized the discourse on Kashmir due to all kinds of reasons relating to authoritarian security, which I shall summarize as *reasons of state*, the analysts have ignored the second, which he thinks is crucial for finding out the path to peace. He also shows how a study of the second Kashmir, that is the lived experiences of the people, leads one to a multi-layered conception, with nuances, of the land and the problematic, necessary for such an understanding. In this context he points out an absolute retardation of democratic institutionalization in Kashmir as a result of the disjuncture between the pluralism in the politics and society of Jammu and Kashmir and what he terms the 'anti-plural institutional space'. In short, Bose is suggesting here the need for a turn in security thinking. But in order to appreciate the need for such a turn in thinking, it is necessary to probe in the first place, how plurality is eaten up and a partitioning line overwhelms all other realities. Then we shall realize that partitioning strategies generated from reasons of state and authoritarian security cannot free us from the geopolitical compulsions of conflict and enable us to switch from revenge to reconciliation, monolithic peace process to plural peace process, dialogue to plural conversation, and thus, retributive justice to restorative justice. In presenting the two narratives – partition, and plurality and democracy in the single narrative of a cartographic line, I am trying to address three fundamental questions relevant to a politics of dialogue and peace.

The first question to ponder over is, why and how did the ethno-lingual lines in Kashmir and the layers of the society there collapse into one and the internal lines reflecting the plurality of the land submerge in a very monochromatic conflict scenario? Every conflict situation has within it, as I have already hinted at the end of the last chapter, the dual process of 'trans-valuation and focalization', whereby all features of a situation except one get lost and only one aspect of a conflict-situation becomes the focal point with harmful consequences for its solution. I try to show in this connection, how contrary to the common idea that political entities make borders, borders sometimes make and solidify a state, in the process erasing many of the 'internal' varieties related to accommodation and co-existence.

The second issue calling our attention is that, this process of 'trans-valuation and focalization' helps a regime of *control and expanse* to strike roots in conditions of globalization. Borders being fundamentally linear forms of controls, soon generate a power of their own. And as a modular form they multiply so that they can become a model. In this way they attain the power to defy at times their creators to become masters of their own career.

Third, in the resultant shrinkage of space for plurality, the issue is, can democracy resist the fatal attraction of geopolitics, foster respect for accommodation, and overcome its own deficit?

In short, partitioning lines often acquire a kind of autonomy. In view of the repeated wars and border skirmishes between India and Pakistan, it has been asked, how and why have the arrangements to keep controls not succeeded? Why did agreements after giving the impression that they were going to last fail, why did an understanding after being achieved fade away? One expert has asked, 'Would the CSBMs (confidence and building measures) have made a difference'? His own answer is, 'CSBMs long worked in the European theatre between the nations of NATO and the Warsaw Pact, because these countries had no un-resolved territorial disputes that would have resulted in war. But what of other areas of the world, where borders are still disputed and where CSBMs have not been instituted?'[2] And then he observes, the complexity of the border problem between India and Pakistan in Kashmir sets 'distinct political limits' to the pursuit of such measures between the two countries.[3] The precise question is, what are those complexities, such as are often ignored in administrative-juridical decisions on borders, and what are the dynamics in the history of the LOC, which while allowing it to appear as amenable to political wills and decisions of its masters, make it in reality autonomous to an extent where it becomes able to engulf other histories? If Sir Cyril Radcliffe had ignored this fact in the great boundary-demarcation operation in 1947, the leaders of the two states, India and Pakistan along with the international community, also had not anticipated the intractability of the problems that a clumsy line in Kashmir agreed upon by the two states would create.[4] Consequently, the line of control that was arrived at the conclusion of cease-fire in the first Kashmir war has time and again invited revision. Both states have attempted to give the line the nature of a permanent marker India trying to lend it the nature of an international border, Pakistan using the line as a marker of its unaccomplished mission to change the border. The wars of 1965 and 1971, the Shimla agreement, the dispute in the Siachen area and the undeclared war in the snowfields, the persistent intrusion/infiltration/homecoming, and then the 1999 war on the heights in Kargil, show the brittleness of the agenda of settling politics though partitioning lines.

Yet if the agenda is brittle, why is the persistence with the agenda? Here is the importance of the three issues mentioned earlier. As persistently lines produce instability, the states also keep on trying to stabilize them. In the time of globalization there are few better ways to have dividends from conflicts than to have *local* wars over borders that make the milieu *global*, and therefore make the political returns for the state high. Borders and quarrels over lines fetch rich dividends in politics of the region. Borders as lines have the capacity to keep a war local, in fact they can keep it unannounced, yet they make the politics of lines vocal and global. The history of border conflicts and wars in Africa, Balkans, Caucasus, West Asia, shows that as a political technique, a border war is one of the best methods to consolidate the state. Its risks are low; returns are high. The truth is that even after the lines have been institutionalized and given the form of a border, lines remain lines. The prehistory of the border is the line; and the history of the line looms large on the later day history of borders, war and peace.[5] Even dialogues over borders seemingly to stabilize borders unwittingly become part of the *strategy of control and expanse* in the current age of global politics.

Chronicle of a Geopolitical Destiny

In October 1947, a force of invading Pathan tribesmen, with some help from Pakistani army crossed from the Northwest frontier and invaded the Kashmir valley. The Hindu Maharaja of this Muslim majority state signed a deed of accession with India to gain India's help in repelling the invaders. Before accepting this accession Pandit Nehru, the Prime Minister of India, sought permission from the leader of the National Conference, which was the strongest political party of the state. The next day Indian troops landed in Srinagar and within a short time this escalated into open warfare between India and Pakistan. The problem reached the international community when India decided to take the question of Kashmir to the United Nations on 1 January 1948. A cease-fire line (CFL) was drawn under the auspices of the UN on the same date next year. The line left Pakistan in control of Gilgit, Baltistan and a narrow strip of the western part of an area that consisted of the Vale of Kashmir, Poonch and Jammu that along with Ladakh remained in India's possession remained Ladakh. The cease-fire line instead of being only a line of cessation of hostilities, demarcated areas of possession, put the stamp of acceptability and finality of mutual areas of control, and led to four wars and more than two near-war situations between the two countries of India and Pakistan. Pakistan continued to consider the line as provisional; India desired to treat it as permanent. The flexibility initially exhibited by both countries was soon over.[6] The two states solidified their respective positions on Kashmir, and the line signified that the core of the two nationalisms has hardened.

In the first few years, when the cease-fire line was evolving that would in effect divide Jammu and Kashmir and leave the two countries in possession of two parts, Pandit Nehru consistently upheld the temporary nature of this line. He had reiterated on a number of occasions his desire to refer the issue to the people of Kashmir. On October 30, 1947, he sent a telegram to the Pakistani Prime Minister

Liaquat Ali Khan stating, 'Our assurance that we shall withdraw our troops from Kashmir as soon as peace and order are restored and leave the decision about the future of the State to the people of the State is not merely a pledge to your Government but also to the people of Kashmir and to the world.[7] Then through a radio broadcast on November 2, same year, he declared that the people of Kashmir would decide their political future through a plebiscite, which could be held under the auspices of the United Nations.[8] We need not go here along the tired path of arguments and counter-arguments over whether India by reneging on its commitment to hold plebiscite was right or wrong. But in the context of our discussion, considerations about the earlier flexibility deserve attention. The earlier flexibility had given rise to the hope about the ability of the cease-fire line to hold and about its capacity to lead to necessary following steps for a solution of the issue. This resulted in greater flexibility also towards the question of mediation regarding the cease-fire line. The flexibility was reflected in India's attitude towards third party mediation as well. India showed willingness to discuss the issue with representatives of UNCIP (United Nations Commission for India and Pakistan), be it McNaughton, Dixon or Graham. It was Frank Graham who while initiating direct negotiations between India and Pakistan had declared that,

> Instead of the United Nations' representatives continuing to report differences to the Security Council, may the leadership of over 400,000,000 people, with the goodwill and assistance of the United Nations, join in negotiating and reporting an agreement on Kashmir and thereby light a torch along the difficult path of the people's pilgrimage to peace.[9]

The feature of the CFL was that none of the three involved actors, India, Pakistan and the United Nations, envisaged a continued international policing of the line as the instrument to find a solution of the Kashmir problem and through it a settlement of the issue of a permanent border between the two countries. The UNCIP was most moderate; India oscillated between the two positions, sometimes upholding both, at other times stressing the need to find a solution or saying that since the accession was final there was no problem left; and Pakistan treated it *ad tempore*. As a line that signified essentially a bilateral issue, whether in the pledge to hold plebiscite, or in prospecting possibilities of mediation, the CFL remained true to its history of a line of partition. We can speak here of three such features to make clear the ironies and paradoxes inherent in the situation.

First, the line had been established with the intention of keeping the two contending states apart and thereby providing the space to find a solution of the Kashmir problem through political means. But the line itself provided impetus to both parties for adding military strength. Therefore it should not surprise us to find that both states in direct contravention of the resolution of the Security Council (21 April 1948, S/726) treated the CFL as the new element in further strategizing. The resolution had enjoined upon Pakistan (vide A/1/a) to 'secure withdrawal from the State of Jammu and Kashmir of tribesmen and Pakistani nationals not normally therein who have entered the State for the purpose of fighting, and to prevent any intrusion into the State of such elements and any furnishing of material aid to those fighting the State'. Similarly it enjoined upon India (vide A/2/a) to 'put into operation in consultation with the Commission a plan for withdrawing their own forces from Jammu and Kashmir and reducing them progressively to the minimum strength required for the support of the civil power in the maintenance of law and order' when the tribesmen had withdrawn and cessation of fighting had become effective, to ensure (vide A/2/c/i) that 'the presence of troops should not afford any intimidation to the inhabitants of the State' and to initiate steps towards establishing a 'plebiscite administration' to which 'such powers as the latter considers necessary for holding a fair and impartial plebiscite' (vide B/8). Clearly there was a vision of a political solution accepted by all the actors, and the CFL was not to be part of that solution. Yet in face of developments brought to life by the line, it proved to be only a liberal dream in a situation marked by post-colonial identities and collisions.

Second, the CFL was not to become a commentary on the dispute. The UNICP resolution of 13 August 1948 had enjoined (vide Part 3) that the 'Government of India and the Government of Pakistan reaffirm their wish that the future status of the State of Jammu and Kashmir shall be determined in accordance with the will of the people and to that end, upon acceptance of the truce agreement, both Governments agree to enter into consultations with the Commission to determine fair and equitable conditions whereby such free expression will be assured'. The subsequent resolution of the UNCIP of 5 January 1949 also made the CFL *unambiguously* only the beginning of a policy whose chief instrument was intended to be something else. In this resolution (vide 6/a, 7/a, b, c, d, & e) the UNCIP envisaged a liberal solution based on restoration of civil liberties, protection of minorities, demilitarization, in short, *condition of* peace. But the line itself was the *agent provocateur*. The line had signified *control*, and the political situation in the valley itself marked by rising nationalism followed by a rethinking among the nationalist leadership in Kashmir, the increasing hardening of the Indian government's stand, the dismissal of the Sheikh Abdullah ministry, and the reneging of the political commitment of the Indian State to the people of Jammu and Kashmir resulted in making the UNCIP resolutions stand on their heads. Instead of the CFL being the subservient to a political arrangement, the line now dictated what the subsequent arrangements would turn out to be in later periods.

Third, the CFL was meant to be a *local line*, but with the onset of the cold war it became a part of a *global line*. Thus notwithstanding the apparent softness of Nehru's policies and the occasional reasonableness, global alignments prevented this 'softness' and 'reasonableness' being meaningful. When Pakistan accepted the role of one of the most trusted allies of the United States by signing the Baghdad Pact, and becoming a member of SEATO, the prime minister of India made an offer to the Pakistani leaders in 1955. He asked that international boundary be stabilized at the cease-fire line and stated that India had no desire to take 'the part of Kashmir which is under you' (that is Pakistan), by fighting.[10] Indian offer came in amidst fears of the possibility of Pakistan flexing its newly acquired political-military muscle to change the course of the divide and of China turning into a political rival in Asia following the Bandung conference. Indeed, it was claimed 'Zhou En Lai openly gave his imprimatur to the Pakistani decision to join SEATO and CENTO in 1955 by declaring at the Bandung Conference in 1955 that the two countries had achieved a mutual understanding on the question.'[11] India began to increasingly refuse a change in the status quo in Kashmir following the establishment and the consolidation of the CFL, and this was highlighted by Noon when he complained to the Security Council in 1957 that a number of proposals (about eleven in number) were made by United Nations to solve the Kashmir impasse but India had accepted almost none of these. India had agreed by the offer of Pandit Nehru to Noon in 1958 to refer certain disputes in the eastern border to outside arbitration. But when Noon refused to accept the validity of arbitration on only one particular issue and pressed for similar arbitration over Kashmir, India refused. With the change of government in Pakistan and the usurpation of power by military dictatorship, the line in Kashmir became an even more protagonist of global entanglements. Ayub Khan signed a treaty with the United States, which agreed to provide Pakistan with non-conventional weapons and construct launching sites for missile. Nehru's official biographer is of the opinion that such a treaty completely changed Indian posture toward Pakistan. What he does not say, however, is that this was also the time of the emergence of China as a great power. The global politics of bi-polarity had pitted India against China. India's self-avowed leadership of the Afro-Asian nations came into direct collision course with China's position. This was also the time of growing border dispute with China and the exacerbation of dispute towards the end of fifties and the beginning of sixties. This shift in Indian policies reflected in the hardening of the Indian stand not just regarding the Indo-Pakistan border but also regarding the Sino-Indian border, particularly in the climate of reverses against China in the border war of 1962. In such a situation, defence and security again came to occupy a priority higher to that of political solution. In the words of Y.B. Chavan, then the Indian Minister of Defence, 'In the current climate of hostility and tension, however, we have...to take necessary measures for defence of our territorial integrity against any aggressive threat, the more so because of our experience last year...The first programme of defence preparedness is, one of expansion of our Armed Forces'. And, Krishna Menon, the Indian representative to the United Nations, a few months later said, 'I think we have come to a position today that any attempt, any giving away under these circumstances will be used against us. We cannot subscribe to any resolution in the Security Council...'[12] Even the efforts of the Kashmiri nationalists for another Indo-Pakistani dialogue failed in such atmosphere. The global milieu in which these two countries had been de-colonized and had been left to sort out the political issues of sovereignty among themselves had within ten years come home to roost.

As a line, the CFL thus not only evolved into one that effectively divided Jammu and Kashmir and left the two states in possession of two parts, it has remained till date true to its history of erasing all other aspects of the dispute including grievances of the people into a territorial one, giving birth in the process to the doctrine of bilateralism. The war of 1965 was to bring out the precarious nature of the line more clearly. Contrary to the idea that the Nehru epoch was one of a pursuit of a liberal policy and political solution, the war of 1965 was a not a *discontinuity*, but a *logical outcome* of the preceding fifteen years.

The war in 1965 between India and Pakistan over Kashmir lasted for twenty-two days. It started with incursions across the cease-fire line from the Pakistan side and went on to becoming a full-fledged war. A cease-fire was arranged once again under the auspices of the United Nations. Once again a war had started with incursions by tribesmen from the Pakistan side of the CFL and Pakistan was soon involved in a full-scale war, whether it had wanted it or not. But it did not succeed in a change of the status quo and the Indian strategy on Kashmir to maintain the CFL as strongly as possible, make it the kingpin of its strategy, and devise other policies appropriately was clear for the first time and reinforced now. If Pakistan had so long criticized India of reneging on her commitment on plebiscite, India could now argue with equal vehemence that Pakistan was violating the international arrangement for maintaining peace in the disputed area. The war had closed the circle. According to Chester Bowles, the American Ambassador to India, the Indian government came away from the conflict with a new sense of confidence in its military and political power. He said,

> As a result of all these factors, the GOI is in no mood to present Pakistan, by negotiation, with a position in Kashmir, which Pakistan failed to achieve by fighting. Nor does GOI agree that Pakistan should be rewarded by the UN for a last-minute willingness to water down a relationship with an expansionist China which Pakistan should not have entered into in the first place.[13]

Like being attached to the ghostly other, internationalization has always brought back particularisms, in this case bilateralism. The more global factors cast their shadow, the more could a country argue that bilateralism was the only way for a sane solution to the imbroglio. The war of 1965 and the following pact in Tashkent established the dialectic of globalization and bilateralism that continues to haunt the fate of the borderline and along with that the fate of the disputed land and her people. The nationalist agenda that the state of Pakistan could not concretize in the first decade of its existence could now be firmly put in place. India could no longer hope to be a world power but could assert herself as a regional giant. Pakistan could no longer hope to establish the *Medinian* age on earth, but it could say now, that India, and India alone, stood in the way of fulfilling her national destiny charted out in 1940-47. The CFL gained a new sanctity in such a situation. Even international observers sensed the development. The UN Secretary General U Thant, in a report to the Security Council, described the Indo-Pakistan cease-fire as 'precarious' because of numerous incursions of the cease-fire line.[14] The policy of treating the line as a political solution became even more confirmed as the line itself went on to become more unsettled.

The discussions at Tashkent resulted in the Tashkent Declaration of 10 January 1966. It was agreed that the two powers would withdraw their forces to pre 5 August 1965 position. Thus status quo was maintained. The joint communique issued at the end of talks noted that, 'the Pakistan side pointed out the special importance of reaching a settlement of the Jammu and Kashmir dispute'. It signified that Pakistan had given notice that Kashmir still remained a disputed territory notwithstanding the Tashkent pact, while India would continue to maintain that its sovereignty over Kashmir was non-negotiable, and that this sovereignty was at most predicated by the conditions achieved through the establishment of the CFL.[15] The Tashkent accord enjoined upon both the countries to withdraw to positions they held prior to 5 August 1965 (Art. 2), to observe cease-fire terms (Art. 2), to follow the principle of 'non-interference in the internal affairs of each other' (Art. 3), to 'continue meetings at highest and other levels on matters of direct concern to both countries' (Art. 9) and to 'set up joint Indo-Pakistan bodies which will report to their governments in order to decide what further steps should be taken' (Art. 9). Even at that time one could see that 1948 was only being re-enacted. What constituted 'internal affairs', and did that include Kashmir? What was joint effort, and if that meant bilateralism as the declaration seemed to suggest, did the withdrawal signify only a *status quo ante bellum*, or a lead to new effort to conflict resolution? Finally, and this follows from the earlier two posers, was the declaration a ratification by the two countries only of the cease-fire arranged by the United Nations (vide Security Council resolution of 20 September 1995), or a declaration on the Kashmir dispute? All these indicate that the declaration was a continuity of the muddled times that was characterized by the liberal hope of a solution and a simultaneous militarization of the issue. The clearest evidence of the muddled thinking was in Art. 1 of the declaration, which said that, 'in the background' of the interests of peace in the region and the interests of the people of India and Pakistan, 'Jammu and Kashmir was discussed, and each of the sides put forth its respective position'. Was then there a Kashmir dispute? Or, was it a border (that is to say a territorial) dispute? India felt that its stand had been vindicated. The Pakistani leadership was no less Nehruvian. Its President Ayub Khan in a statement of 14 January 1966 defended the declaration much in the same way as Bhutto defended the Shimla pact five years later, with the words, 'this Declaration will open new avenues for the settlement of the Kashmir issue'.[16] Bhutto later on made much too of the issue by criticizing the former President that whereas in the Tashkent declaration the phrase 'in the background' had been used, in the Shimla statement the words were, 'pending the final settlement'.[17] But such substitution proved of not much significance as the geopolitics of borders had proved insurmountable in the path of a political solution in both 1965 and 1972. And the hope that this could be done is what we term as the muddled hope of the times, the historical destiny of post-colonial liberal vision of building a full sovereign statehood without having to face the harsh options posed by the geopolitics of the age. The existence of such measures as a demilitarized zone (DMZ) along the CFL or hotline between the director-generals of military operations (DGMO) also would not have meant much because this was not a case of border dispute, or a case of maintaining a line to separate two armed countries or camps (we have to only think of the DMZ in Korea, or the wall in Berlin), but one where the *real* was always indicating an *other*, the line was indicating a territory, and the territory was indicating the line.[18] We have to only think in this context the terms of the agreement of 25 January 1966, on withdrawal of troops from occupied areas. The agreement not only entailed withdrawal of troops, but also virtually tried to lay down the dynamics of the beginning of a demilitarized zone. It spoke of removal of mines (Art. 5/a), use of unarmed military personnel for lifting of mines and dismantling of all other defence works and complete ban on use of civilian or civil labour for this purpose (Art. 7), continuing the agreement affecting restriction on flights of aircraft (Art. 10), and other measures for civilianizing the occupied areas. It may be argued that this was intended to facilitate the withdrawal process only, and that therefore nothing more should be read into this. But the question is precisely there, why after thinking of such details for the reduction of tension in Rajasthan/Sind, Amritsar/Lahore, Jammu/Sialkot, and Akhnur/Chhamb (from River Chenub NW 8061 to Mawa Wali Khad NW 7770) sectors, could not the agreement proceed towards the obvious, that is the creation of a demilitarized zone? Therefore contrary to the poser whether such a measure would have worked, the question is, why did not the instability along the CFL lead to such a measure? We are indeed confronting here the dilemma of Holmes that the point is not what would have happened if the dog had barked, but why the dog had not barked at all.

We know that the war in 1971 had started on a different issue, but as surely as ever it again ended with the efforts to stabilize the divide in Kashmir. India had supported the independence movement of the Bengalis in East Pakistan. Such support came in the wake of inauguration of the ancient silk routes between China and Pakistan, also large-scale atrocities of the Pakistan army on the people of East Pakistan, and the massive flow of refugees in the country across the border. The Indian policy of finishing the 'eastern question' was so successful that Pakistan declared war on India. At the end of the war Bangladesh was liberated and according to the Ministry of Defence India held 479.96 square miles of territory across the 1948 cease-fire line in Jammu and Kashmir. The Shimla agreement of 1972 put in even more sharper focus the by now set course of the strategy of the parties in the dispute the respective parts of the disputed territory would be recognized as being areas under control and possession of the two states, in one case (India) it implied reinforcing sovereignty, in the other (Pakistan) reinforcing control. The most far-reaching part of the Shimla declaration was that the two countries had 'resolved to settle their differences by peaceful means through bilateral negotiations or by any other peaceful means mutually agreed upon between them'. This, of course, immediately became a subject of widely differing interpretations. On the basis of this statement the Indian government contended that the issue should be settled bilaterally and not through third party mediation. Such a strategy had considerable advantage for a regional power that was wary of the internationalization of the issue due to various reasons such as, the China factor, cold war considerations, the unfinished businesses of state consolidation in the region, as well as internal complications in the valley itself. But where the Indian state failed was in *converting the divide in Kashmir into a border*. About the divide the agreement stated:

> In Jammu and Kashmir, the line of control, resulting from the cease-fire of December 17, 1971, shall be respected by both sides without prejudice to the recognized position of either side. Neither side shall seek to alter it unilaterally, irrespective of mutual differences and legal interpretations. Both sides further undertake to refrain from the threat or the use of force in violation of this line. (Para IV/Art. 1)

India's possession of some 354 square miles north and west of the cease-fire line to the line of control was thus confirmed by Pakistan. This included strategically important points in the Tithwal region and at Kargil, which were to be severely contested later. The dispute over the divide was recognized as a bilateral issue as a result of which Pakistan was robbed of its ability to pressurize India through its alliances, but India also failed to stabilize the line though it was recognized as 'a line of control'. It remained merely *a line of control* functioning as a border, yet lacking its sanctity. The difference from an international border was made more apparent because in the text the clause on the line of control (LOC) followed immediately that on the international border (Para IV/Art.1). Therefore, *control* remained the accepted justification for the divide in Kashmir. Theoretically then at least, if a country failed to control a line, the line could be altered as it was in 1972. Although there was the clause against unilateral alteration of the line by either of the countries, this did not reflect in any way on the likely results in case of a revolt within the territory in control of one of the two countries. Shimla could thus at best buy a fifteen year peace whence the logic of the line would reappear with greater intensity, for the next war would be not on Kashmir but on the line. In this context, Bhutto's formal position of leaving it to the people of Kashmir to work out their own destiny acquires special significance.[19] No wonder that after the Shimla agreement that upheld bilateralism, and reinforced the cease-fire line by turning it into a line of control (LOC), the Pakistani President Zulfikar Ali Bhutto defended the principle of bilateralism against his domestic critics with these words,

> We said, 'pending the final settlement' and only after that did we mention respect for territorial sovereignty...Then we had said that Indian and Pakistani forces would be withdrawn to their sides of the international border first. Next 'in Jammu and Kashmir the line of control, resulting from the ceasefire of 17 December 1971 shall be respected by both sides without prejudice to the recognized position of either side. Neither side shall seek to alter it unilaterally irrespective of mutual differences and legal interpretations. Both sides further undertake to refrain from the threat or use of force in violation of this line.' Now here again you will see that because we believe the dispute of Jammu and Kashmir exists, we did not discuss it only in the background of a peaceful settlement between India and Pakistan. We have de-linked the international border from the Jammu and Kashmir dispute...This is important...Let me now come to the next point. It has been said that the doors of our United Nations have been closed on us and that we are going to settle our problems by bilateral means. As a matter of fact, we have been pleading for bilateralism for a long time. It is not that India compelled us to accept bilateral procedures. Bilateralism has its own benefits...It certainly does not draw a curtain on the rest of the world...I had, in 1966, written a book, not as well known as a book of a former President, but nevertheless a book in which I had clearly stated that it is in our best interest of Pakistan to have a foreign policy based on bilateralism and that we must give bilateralism a chance as far as relations between India and Pakistan are concerned...With due respect to the United Nations, I should say that it has not yet been able to deal with the Kashmir problem effectively. On the contrary, with the passage of time the United Nations' position on this dispute has weakened...There is nothing in the Shimla document, which says that we shall not go to the United Nations...We have to go step-by-step

to tackle each and every problem gradually, seeing the results consolidated and then moving to the next step. This step-by-step approach always advocated by Pakistan was accepted by India.[20]

The ramifications of the Shimla agreement, though not immediately apparent, were therefore far-flung. Efforts to stabilize the LOC contained within it roots of further instability. Henceforth, there would not be a Kashmir war, but constant incursions across the line of control. Also there would be undeclared war wherever the line would be found un-demarcated and vanishing in the deep snow of the Himalayas. Shimla ushered in the age of low intensity conflicts over the LOC. Both states would henceforth continue with the policy of integration with India taking the most pronounced step with an accord with Sheikh Abdullah. Shimla agreement thus symbolized the *territorialization* of the issue. The CFL, transformed into LOC, signalled that the line was the best instrument for the unfinished businesses of state consolidation in the region. The fact that the LOC could not become the border showed that the line remained merely a line of control functioning as a border, yet lacking its sanctity. Therefore, control remained the accepted justification for the divide in Kashmir. On the legitimacy of control depended India's ability to tackle popular protests and militancy and present itself as the moral power. Although there was a clause against unilateral alteration of the line by either of the countries, this did not reflect in any way on the likely results in case of a revolt within the territory in control of one of the two countries. Shimla could thus at best buy an eighteen-year peace.[21]

Moreover, as one observer of Indian foreign policy has pointed out, the policy of treating the line as permanent was reconfirmed in the context of a policy that accorded highest priority to borders.[22] It began in late 1960s in an atmosphere marked by the new emphasis on 'security, territory and prestige', and an increased concern for acquiring military power. The desire to become militarily strong by buying sophisticated instruments from abroad led to keeping the nuclear option open while conducting the first Pokhran test in May 1974 and launching in space orbit an indigenously designed and built 35 kilogram satellite. Acts like the Armed Forces Special Powers Act (1956) got its final shape in 1972 when it was amended. This Act was imposed on so-called disturbed areas, which were largely the bordering states.[23] Then came the National Security Act (1980) and the Terrorist Affected Area Ordinance (1984). The national security lobby took a definite shape and came to influence external affairs of the country in a decisive manner.[24] This was the milieu in which concern over line overrode all other concerns, most significantly concern for democracy. At different international gatherings the two countries continued to express concern for peace, but the border policies of the two post-colonial states reflected the fatal attraction of geopolitics.

The failure to stabilize the line became revealing with the continuing inability to reach an agreement on Siachen. Siachen has been the world's most strategically absurd high-altitude war over an non-demarcated border beyond a point known as NJ-9842, fought at elevations exceeding 6,000 meters, costing India 3 men and Rs. 5 crores a day,[25] thousands of soldiers being exposed to frostbite, hypoxia and severe mental stress. Siachen contrary to the starry-eyed journalists and defence reporters became a proof not of anyone's bravery, but of the expendability and low value of Indian and Pakistani lives towards defending a cartographic act. India claimed that Pakistan had planned to occupy the area. Pakistan had in turn claimed the LOC ran east to the Karakoram pass and declared that it had been in occupation of the area. This decision to extend the LOC was unilateral. So was unilateral India's action to airlift soldiers and thus militarize the dispute. The bilateralism of the Shimla Agreement proved of no value notwithstanding the five rounds of talks between 1986 and 1989. And when after the fifth round, agreement seemed round the corner, the understanding proved fruitless in the tense atmosphere around the Babri Masjid. Thus secretaries, military experts and advisers could not firmly propose any measure. Nor could the political leadership of the two countries resolve on sensible steps like withdrawal of unilateral extension of the LOC, establishment of a de-militarized zone, joint delimitation of a line from NJ-9842 northwards to the border with China, formulation of ground rules to govern future operations and finally redeployment of forces to mutually agreed positions. Instead, only a cease-fire, prior to talks on other 'modalities', was proposed. To Pakistan, this meant, in effect, that the Indian government was not ready to leave the advantage of having its troops at elevations 2,000 feet above the Pakistani positions on the Saltoro Ridge, no matter what the cost and how little the strategic advantage. Everything now firmly anchored to 'security', and in this case security around a glacier far from the Karakoram highway with no direct access to it through a major road from any direction, to the two negotiating states therefore it seemed, nothing constituted 'unacceptable damage' and so there could be no question of thinking how to avoid it. The armies stood, rather crouched in snow-covered bunkers in most of the area, opposite to each other along a 70 km. Line of Control (LOC), 110 km. Actual Ground Position Line (AGPL), 10 km. International Boundary (IB), and a 70 km. Unnamed Boundary (UB) opposite the Shaksgam Valley. The Siachen glacier, 76 km. long, served as a crucial wedge between Pakistan and China, otherwise linked by the Shaksgan valley of 5,160 km. that was ceded to China by Pakistan in 1963.[26] The impasse showed that the line had the capacity to create self-locked positions, and in such situation the language of conflict-resolution remained fragile. Thus nothing came out of the Siachen negotiations. As Nehru had laid down the line years ago, no one was to expect anything, 'substantial...out of a discussion of Kashmir issue on official level. The only possibility (was) noting (down of) the various lines of approach without commitment.'[27] Thus Siachen negotiations that began in January 1986 about twenty months after fighting began on the heights of the glacier were discontinued in November 1992 without any agreement reached. The parties could not agree on the major points. Pakistan did not agree to a cessation of 'cartographic expansion' of the LOC from the agreed terminus point of NJ 9842 to the Karakoram pass on the border with China. India did not agree to the immediate establishment of a DMZ (de-militarized zone) at the glacier. The two parties could not exchange authenticated maps showing current military ground dispositions, as that would have tied them to the notified level of deployment of forces. The delimitation of boundary northward to the border with China proved obviously difficult in such total lack of trust. Ground rules could not be formulated to govern and/or prevent future military operations in that area, a fact that was demonstrated with Pakistan's enactment of a Siachen on India in the form of Kargil war. With all these failures, an agreed redeployment of forces was out of question.[28]

Added to this policy of reasserting lines was the milieu in the eighties of the last century, when the drive for power and resources transformed the tango to a three-some game of India-Pakistan and China.[29] The situation was a pointer for the future because it made obvious how an assertive policy regarding the markers of territoriality could trap a country's policies into the morass of territorialism and territorial politics.[30] But from the point of democracy, the result could be only harmful. A pronounced policy towards the borders and militarization of areas around the borders led to a sense of alienation among the people living in the area. In 1989 Kashmir witnessed the strongest outbreak of nationalist movement clearly articulated in secessionist terms. It shattered the illusion that by demarcating the line of control, Kashmir problem had been solved. Pakistan began to provide diplomatic, moral, and other forms of support to the Kashmiri movement to which the line was nothing or meant 'Indian control', and that Pakistan might provide covert material support to the Kashmiri secessionists was apparent from the time of the Shimla agreement. In its signifying function, the line now signified the blurring of the distinction of the proxy war, the Kashmiri war of nationalism, and the India's war against secessionists in Kashmir and against Pakistan providing support to the Kashmiris. Hitherto, the divide in Kashmir represented a sort of an ethnic divide. When Pakistan had tried to foment ethnic tensions earlier in 1965, it had failed and *Kashmiriyat* seemed as strong as ever. But in 1989 *Kashmiriyat* could not save the day for India, and therefore by strengthening the LOC India could not arrest destabilization in Kashmir. Enough has been written already on the civil war in Kashmir. We need not repeat the accounts here, but can only ask, what else fate could visit a politics that had relapsed into geopolitics fully and unconditionally?

In the wake of the civil war in Kashmir the line became not only the border between the two countries, but it had developed into an impregnable line in order to control and suppress the rebellion in the valley. India wanted to fence the international border in the Akhnoor sector in 1996 by raising a barbed wire. Pakistani Rangers raised objections to such fencing of what they called a 'working boundary.'[31] Apparently in a protest note in 1993, Pakistan had referred to the same area as the international border. It was also said that the Pakistani protests were due to 'fears about a possible Indian nuclear test or the deployment of India's short-range Prithvi missiles on the Pakistan border'.[32] The area, popularly known as the chicken's neck, was the one of fierce contest between the forces of the two countries in the 1971 war. With the Siachen war continuing, it was a sign that armed conflict around the line would soon recommence. The Kargil war was thus within the logic of this highly illogical situation. India had got part of the area when the cease-fire line had been stabilized into the line of control in 1971. It was therefore not ironic that incursions would occur here. But, at the same time, the border war in 1999 with the line's past fifty years of entangled history was to show that beneath the ineluctable logic of geopolitics, there was the irony of representation. Line of control, control of line this was all a matter of representation, the cartographic representation of the many issues of post-colonial state formation.

When the undeclared war broke out in Kargil in 1999, Pakistani officials defended their actions by saying that the LOC was 'not intended to become a permanent line of division in opposition to the wishes of the people of J & K'.[33] Pakistani Foreign Minister Mr. Sartaj Aziz repeated his country's traditional stand that the line of control was not acceptable to his country as an international border. It was merely a temporary line in a 'disputed territory'. He further said that the divide was not acceptable also to the people of Kashmir. A member of the Indian National Security Council Advisory Board in a plaintive tone added that an attack across the LOC 'would not constitute an attack on the sovereignty of either nation'.[34] Pakistan complained that India had earlier violated the LOC thrice in 1984, 1986, and 1988.[35] Indian reactions ranged from acrimonies blaming 'the malign character of a hostile neighbour which defines its very identity in anti-Indian terms', to more sentimental theories of stab on the back.[36] The Indian Prime Minister stated in a press conference that the 'LOC is defined and identified in the maps of the two countries which are sealed and bear signatures of representative from both the sides. In this light, any change in the LOC is not negotiable.'[37] He also stated that India was always open to talks but if the purpose of these was to alter the LOC then 'the proposed talks would end before they begun...'[38] In other words, there could be negotiation only on and by maintaining the status quo. One retired foreign service official however in an exasperated tone declared, that Pakistan should not be allowed any position near the Srinagar-Leh road, for 'line of control border(ed) on the impossible'.[39] But, ironically within less than a year, the army was telling that the line was not settled, telling that the campaign in the media that the army had a lost a peak in Kargil heights to Pakistan was a 'smear campaign' because Point 5353 was never in Indian possession. One newspaper not unnaturally screamed exactly one year after the border war had ended, 'Post-conflict, LOC Course Changes, Jockeying for Peaks On'.[40] In spite of the claim that the LOC was settled, the fact became clear that the LOC in that area followed an imaginary line connecting high points such as points 5070, 5108, 5245, 5353, and 5608, the numerals being the altitude of the high points in the mountain range depicted in meters. Who would like to control what part of this imaginary line depended on the inter-visibility exercises by the two armies on their detailed contour maps to have a command of the range and the roads and the valley below.

The Kargil border war put the final seal on the conversion of the Kashmir problem from a political one into a security one. Even as late as early 1990s there was an understanding among the Indians that Kashmir was a political problem. According to one commentator 'until 1994-95 emphasis was on alienation of the people (erosion of internal autonomy, corruption, rigged elections, repression) and Pakistan's moral and material support for militancy was read within this context...Today alienation, popular aspirations, and need for democratic solution find no reference in any official pronouncement.'[41] It was pointed out that rigging in the 1987 elections led to disenchantment with the electoral process among the Kashmiris and triggered armed militancy. In most newspaper reports of 1995, for example, it was argued that the problem in Kashmir was a problem of reinstating democracy as 'Article 370 has been battered beyond repair...Electoral fraud for four decades largely accounts for the people's alienation and suggests the need for an accord which will render such fraud, violation's of Kashmir's autonomy and suppression of its people's aspirations impossible.'[42]

The transformation of a political problem to a security problem that Kargil border war symbolized accompanied huge increase in defence expenditure. Between 1989 and 1999, military expenditure in India rose by 27 per cent, i.e., from $7.8 billion to $9.8 billion.[43] This was apart from the money spent on the Border Security Forces, who fall under the jurisdiction of the Ministry of Home Affairs and on building and maintenance of roads in border areas. Besides the region was now characterized by deployment of nuclear or near nuclear weapons. Although it is difficult to compute costs of going nuclear, the Human Development Report for 1999 noted that India and Pakistan would spend over $15 billion on nuclear arms over a ten-year period.[44] Such increased arms expenditure led to a concomitant increase in the number of security personnel in Kashmir one soldier for every 14 Kashmiris, and all this while Kashmiris went without such basic necessities such as electricity and food. In Kashmir throughout the last decade electricity was unavailable for seven to nine hours even in winter months.[45] As for food it is impossible to grow anything in the region during winter. Particularly the Kargil district spread over 14,036 kms is generally devoid of any natural vegetation and has barren rocky mountain all around.

The fields around the LOC became the killing fields. Some of the typical media reports in 1999 were such as the following. On June 17 two women were killed and another suffered serious wounds in an unprovoked and ceaseless firing by Indian troops from across the Line of Control in different villages of Neelum Valley and sources claimed that the Indian troops were targeting almost all the villages in the upper and lower belts of the Neelum Valley.[46] On June 30, 'Indian troops recaptured two major heights on Jubar Hills, some 2 km from the Line of Control in the Dras sub sector...These two heights and the recapture of Point 4700 and Black Rock yesterday cost the Indian Army heavily. As many as 26 army personnel, including three officers killed in the operations. The Indian toll in the Operation Vijay has mounted to 201 killed, 384 wounded and nine missing.'[47] On June 2 'Information on the number of intruders killed was being gathered but it was estimated 400 has been accounted for.'[48] On June 20 'Two persons were killed and four injured in Pakistani mortar shelling and firing along the Line of Control (LOC) in Poonch and Jammu sectors...official sources said here today.'[49] In Surankote, Poonch, on June 29, 'In a gruesome incident, militants massacred 15 members of three families including three women and three minor girls and set ablaze their houses at village Morcha Bachchai.'[50] But this was about who died. What happened to the living? Again many news reports appeared such as the one that in Kargil on June 3, 'Like the people in Dras and Kargil, almost all the residents of Batalik have moved out. Civilian movement has been banned in the area.'[51]. Or, on June 15, 'Families were moved from about 300 villages along a 72 – kilometre stretch of the border after the evacuation was announced over mosque loudspeakers.'[52]

This paradoxical history of a line – its reassertion and fragility poses a challenge to the conventional practice of peace building. In a context wherein the territorial line's fate was mixed up with those of two more lines the line psychologically separating Kashmir and India, and the various demarcating lines within Kashmir – how could a monolithic peace policy hope to succeed? A *de jure* bifurcation fifty years ago risking ethnic cleansing, *grant of azadi* through a roundtable, intervention by other countries, turning the CFL into a permanent border, imposition of a sizeable multinational army to police the CFL and preside over a DMZ, a permanent Kashmir commission whichever way one chooses to look at the likely scenario, it is extremely difficult to disentangle the issue of priority – the line, or, the political future of the contested land whose people also from time to time have gone on contesting the position that others have the right to involve in a contest over their own land? Or, eliminating the rebels by killing them one by one, so that it is one less rebel to confront, or ethnic cleansing, or ethnic dilution – the bizarre solutions that the Indian state with the rebels copying Indian strategy seem to follow at times? All solutions from the time when Sir Robert Owen as the UN mediator in September 1950 had mooted his scheme of solution have been based on the idea of partitioning Kashmir (limited or extensive). Proposals for territorial adjustments or exchanges, multi-featured autonomy package for the valley, the maintenance of soft borders (customs-free) for travel, trade, and in a more radical version bi-national employment have as their core territorial partition. And such proposals have now led the Indian state (ironically the rebels also) to find out *who are the ethnic Kashmiris*? That search is as much of a quicksand as is the answer, namely that the overwhelming Kashmiri-speaking valley-Muslims are the ethnic Kashmiris. In such context, any Trieste-type solution would involve re-partition of the region inside. In the 1954 agreement on Trieste between Italy and Yugoslavia, it was agreed that the city would have access to both sides of the partition line, thus doing away with most of the disabilities of partition while laying the foundation stone for future cooperation. The inner lines of Kashmir are against such solution here. Bilateral, multilateral, con-federal, co-sovereignty based – all proposals are based on one line or the other, while Kashmir remains bound with several lines with three competing realities clamouring for highest attention – the dividing line between India and Pakistan, that is the LOC, the line dividing the two Kashmirs, and the *Intifada* in Kashmir (that has re-emphasized other lines in Kashmir).

The central tenet of the paradigm of peace building is the assumption that it is a step by step strategy, a phased policy by which the capacity to hold peace increases, and a multilateral custody of the process ensures the supply of the political substance of the restoration exercise. Thus restoration of peace is linked to restoration of polity, state, and *politics*. The history of the line of control in Kashmir shows the limits of such strategy; for, where the conflict is built on political linkages, how does restoration of *politics* sidestepping the *politics of plurality that linkages are all about* help?[53] Indeed a manual for early warning will tell us that watching how a conflict develops is a question of policy, of politics. Such a manual tells that first step is understanding the context (factors, location, history, actors, framework), then identifying conflict indicators (political, economic, socio-cultural, institutional; also indicators such as structural, triggers, accelerators), then identifying opportunities of peace (initiating events, mediators, facilitators, options, agenda items, peace building capacity), and steps for response development (instruments for conflict prevention, analysing potential peace actors and institutions).[54] The strangest part of the history of the conflict in Kashmir is that there was never a system installed for early warning and early response, hence no attempt to disentangle the issue, consequently we find a closed circle of determination of line, conflict, cease-fire, reinforcement of the line, conflict, and cease-fire. The gradualism and the multilateral strategy inherent in the policy of peace building, as exemplified in Kashmir through strengthening security of the line, is only an edition of supply side economics, for here too the approach is to supply the inputs of peace, to concentrate on the problems of supply.

Geopolitics and the Democratic Deficit

The Kargil border war over the LOC signalled that the line's greatest function has been to symbolize a double metamorphosis. We have already spoken of the first metamorphosis, the transition of a political problem into a security problem. The second has been also equally critical.

It showed the symbolic role of the border in the collapse of all internal variations in the society and politics of Jammu and Kashmir in an un-variegated single whole either the polity came to be known as Jammu and Kashmir on the Indian side or Azad Kashmir on the Pakistani side. Therefore obvious factors such as that 50 per cent of the population of Jammu and Kashmir live in Jammu and about 3.5 million residents of Azad Kashmir and the Gilgit region have little in common with the people of the valley were ignored. Or, was ignored the fact that about 80 per cent of the 2 million people in Azad Kashmir hailed from traditional Rajput ancestry and formed part of the cultural, linguistic and ethnic constituency of the erstwhile Jammu province. More fundamentally, the ethno-linguistic variations within the entire region were glossed over in the perennial search for ways to stabilize the line and thereby the region. The line in that sense built upon a surreal existence of two homogenous blocks (Indian Kashmir and Pakistani Kashmir) that suppressed the reality of an amazing variety of existences within the erstwhile region known as the state of Jammu and Kashmir.

After all it is well to remember the most often forgotten fact that the region consists of three distinct sub-regions. Divided into six districts (Baramulla and Kupwara in the north, Srinagar and Badgam in the central part, and Anantnag and Pulwama in the south), the Valley with about 4.5 million population, is overwhelmingly Muslim, primarily Sunni Muslim, with a sizeable Shia minority, and is overwhelming Kashmiri-speaking. Jammu lying south of the valley is considerably larger and socially much more heterogeneous. It has also six districts – in the northern and western parts Poonch, Rajouri and Doda, and in line east and south, Jammu, Udhampur and Kathua. Doda district has a Kashmiri-speaking Muslim majority and Hindus constitute 40 per cent of Doda's population. Similarly Muslim-majority populations are in Rajouri (65 per cent) and Poonch (85 per cent), but they are Gujjars, Bakerwals, Rajputs and other small communities with the main language being Pahadi, a dialect of Punjabi. The line that separates Jammu from Pakistan is both *de jure* (with Pakistan Punjab) and *de facto* (with Azad Kashmir) borders. Rajouri and Poonch formed part of a common linguistic-cultural and economic zone with Mirpur and Muzaffarabad on the western side, the western Punjab districts of Rawalpindi, Jhelum and Mianwali, and the North-West Frontier Province districts of Abbottabad and Mansehra. Within these two districts the *towns* of Rajouri and Poonch are centres of administration, trade and commerce, and education with the majority being Hindus and Sikhs. The Gujjars are a nomadic people, herdsmen living in mountain areas. Among the Hindus of Jammu also who are the majority of its population (65 per cent), there are varieties of caste, location, language and to complicate matters more three of its six Jammu districts are Muslim majority areas with significant minorities. Conversely, one of the Jammu region's three Hindu-dominated districts, Udhampur has a similar sizeable Muslim minority of about 30 per cent, including both Kashmiris and Gujjars. It is like peeling an onion.

The LOC, as we see it now, began not only a line of cease fire but also a partitioning line, and like all other partitions has now the tendency to reproduce itself. In a situation as the present time, partition-centric proposals are always an easy temptation for rulers who are far removed from the democratic imperative of dialogue to deal with social diversity, and plurality. Such temptations are now increasingly in evidence. Some want to trifurcate the region into three separate units, Hindu-majority Jammu, Muslim-majority Kashmir and Buddhist Ladakh, though as was shown in the preceding lines that there is neither a pure Hindu Jammu, nor a pure Buddhist Ladakh, nor even a pure Sunni Muslim Kashmir. Some therefore have a counter-proposal to trifurcate the Jammu region instead, effectively separating its Muslim-majority areas from the Hindu-majority areas. Some again are proposing that the valley with some more areas (from the other side and from Jammu) be considered as de facto autonomous (sovereign) with powers of high protection remaining with the Indian state. We may not be far off from the time when we shall again witness bloodbath, a spiral of reciprocal sectarian violence, expulsions of people belonging to 'other' communities from numerous locales in purifying homelands, and hordes of 'stranded' minorities. We must remember in this context also that the line impacted on Pakistan as well. From a state born with the declared purpose of providing the Muslims of the sub-continent their 'nation', it became an Islamic state. From a liberal notion of Sunni Islam the spirit of the nation turned to an unending edition of orthodox Islam that even the founder of the *Jamat-i-Islami*, Maulana Maududi, were he to be alive now would not have approved of today. The theology of *Jihad* re-emerged in a controversial form to break the sanctity of the line, for this seemed to be the principal obstacle to liberate the land. Schools to train religious warriors proliferated to supply the insurgent groups with human resources, while trans-national Islamic foundations and the state of Pakistan took care of the financial resources. As the infatuation with the line grew, and it became the focal point of the political class of Pakistan including its security establishment, the social, political fabric, even its economy, arrived at times at point of virtual disintegration.

In such a convoluted milieu on both sides of the line, naturally the patience of the rulers and the elite is wearing thin. 'Some sort of settlement through some sort of partition that will give rise to some sort of new entities that will bury the hatchet' is now the dominant thinking. And when that comes through partitions in the guise, it will be the final gift of the line to the sub-continent. Indeed as the origin of the Dixon plan shows, proposals of internal partitions always went hand in hand with the strategy of stabilizing the Kashmir dispute with a line of control. With the fictive internal homogeneity of the two entities gone, the rationale for the line would be no more.

All this discussion leads to one fundamental issue that we can anticipate in this chapter, but to which we have to return again and again. The issue is how much has democracy in the way it has been received in South Asia been relevant to the process of a 'democratic negotiation of the conflicts'? In the context of the shrinking space for plurality, the issue is, why could not Indian democracy resist the fatal attraction of geopolitics, foster respect for plurality, and help the process of a democratic negotiation of the Kashmir question? By posing this question we not only reflect on the liberal theories of international relations, but on the more fundamental problematic of *democratic deficit*.

Parliamentary democracy in India was intent upon building a *nation* that could not have been built without building a fictive ethnic core based on a certain geopolitical imagination. This explains why the democratic nation in India adopted a dual, apparently paradoxical, strategy. On one hand, the nation in India claimed that retention of Kashmir, made safe by the CFL/LOC, was needed to distinguish India from ethnocracies, to show that there was no dividing line between the minorities and the 'homogenous' Hindu majority; on the other hand this very democracy needed the CFL/LOC as a dividing line between secular India and Muslim Pakistan. Thus on one hand went on the policy of integrating Kashmir with India (culminating with the accord with Sheikh Abdullah), on the other hand went on imagining a Kashmir which had a similar fictive core, *the valley*, to the extent of neglecting the plurality of Kashmir society and politics. On this basis appeared a monolithic peace process – India and the Sheikh, just as earlier it had been India and the Raja Hari Singh whose accession was considered to have laid the ground for unity of Kashmir with democratic India. Democracy reduced to such geopolitical containing of 'masses' cannot but become organic.

Thus what began as a ceasefire line quickly became a partitioning line, and like all other partitions has now the tendency to reproduce itself. In a situation as the present time, as Sumantra Bose has rightly pointed out in the essay referred to at the outset, partitionist proposals are always an easy temptation for rulers who are far removed from the democratic imperative of dialogue to deal with social diversity, and plurality. Such temptations are now increasingly in evidence. Some want to trifurcate the region Hindu-majority Jammu, Muslim-majority Kashmir and Buddhist Ladakh, though there is neither a pure Hindu Jammu, nor a pure Buddhist Ladakh, nor even a pure Sunni Muslim Kashmir. Some therefore have a counter-proposal to trifurcate the Jammu region instead, effectively separating its Muslim-majority areas from the Hindu-majority areas. Some again are proposing that the valley with some more areas (from the other side and from Jammu) be considered as de facto autonomous (sovereign) with powers of high protection remaining with the Indian state. We may not be far off from the time when we shall again witness bloodbath, a spiral of reciprocal sectarian violence, expulsions of people belonging to other communities from numerous purified homelands, and hordes of stranded minorities. We must remember in this context also that the line impacted on Pakistan as well. From a state born with the declared purpose of providing the Muslims of the sub-continent their 'nation', it became an Islamic state. But more important from the point of the politics of democracy is that riding high on a republican spirit, democracy does not tolerate pluralism. As Michael Mann puts it, there is a dark side of democracy, and liberal democracy has not been able to keep the dark side hidden.[55] Democracy often operates in the name of the people, periodically constitutes the *others* outside the people, and makes conflict with 'them' the mainstay of democracy.[56] The effort to make democracy 'organic' makes it incapable of accommodating and hence intolerant of various political forms of existence. Democracy becomes increasingly ethnic. If this is true of India, it is so regarding the 'democratically elected' political leadership of the valley also, namely Farooq Abdullah and his wagon. Parliamentary democracy does not necessarily appreciate social, political and cultural demands of plurality, and the fact that expansion of political forms and flexibility in the notion of citizenship is essential for a renewal of democracy, and that such expansion and flexibility implies not only more political and juridical individual rights, but also a capacity to address collective demands through imagining newer political forms including forms of shared sovereignty.[57]

Indian democracy forgot that that these demands when ignored rapidly resulted in violence. This failure has also carried a lesson that expansion of political forms and flexibility in the notion of citizenship was essential for a renewal of democracy, and that such expansion and flexibility implied not only more political and juridical individual rights, but also a capacity to address collective demands.[58] As against such a development, democracies have tended to be organic. With the birth of organicism, mass murders started. It began with cleansing. Democracies have shown the capacity that this can end in genocide. The question about the supposed efficacy of democracy to reconcile with others is bound to arise as experiences of democracies converge with those of non-democracies of various types. The Calvinistic urge of democracies to bring others to submission has proved to be extreme, which is why today there is a revival of interest in the first ten-fifteen years in the life of the line, in a review of the performance of political democracy in handling the dispute, in short interest in a genealogical exercise.[59]

One reason of this incapacity is because democracy relates basically to the question of representation of will. How is the will of the people of Kashmir to be represented? How much? How durably? Will it be represented through a constitution? But as Article 370 in the Indian constitution showed that a provision embodying an accord can forget the plurality that was the beginning of the article itself. An election? But it leaves out so many and legitimates 'organicism'. Or, is the will to be represented by fresh accords? But as most of the peace accords in India and elsewhere in South Asia shows, as the spirit of reconciliation is soon overwhelmed by constitutionalism that reinforces older forms of power. Therefore in the task of redefining democracy through de-linking it from its nation-form, and enabling it procedurally to foster reconciliation, the fundamental challenge comes from conservative geopolitics.

The incapacity of an organic democracy to undertake the procedural culture of dialogue comes not only from deep-rooted geopolitical reasoning, almost a ‘geopolitical common sense’, it comes from the way in which democracy acquires an organic character. Take, for example, how we came to know of the last border war in Kashmir, in other words how the war was represented to us. With satellites beaming onto the readers and viewers, reports of war were made of battle reports, profiles of families of dead soldiers, political reviews, military sketches, geopolitical analysis, diplomatic evaluations of the implications of the war, and also the definitions of patriotism. With so much to feed the media, there was a continuous selection of what to feed and what to leave aside or suppress. The media called for holy war, consensus for waging the war, fixing those who refused to fall in line, and focalized the issue that the war over line of control was the essence of all conflicts for the past half century in the country. With velvet and now with heavy hand, the democracy knew how to deal with others.[60] So we were not only being informed about the war, or rather you were being very little informed of the actual war, but of political issues that were fed as war reports and comments. Thus, the media formulated what was the line, and why was the war over control of the line. It spoke of the exodus and why in spite of repeated shelling from both sides and the massive displacement of people on the border, the line was a line of control and not a line marking out a killing field; why the war had to be publicized in all its glory; why the war had to be waged both hot and cold, that is to say with guns and letters, gestures, meetings, and visits. We were also told what was the definition of patriotism, and therefore who were patriots and who were the scoundrels. We were told besides, how a particular religion was susceptible to fanaticism, how political parties should behave, how the financial impact of the war would not be significant, and even if it were, the nation had to bear it, and finally, how this was a local war but the country stood to gain globally from this. In brief, the war over the line represented all the political issues that this democracy was facing in the most congealed form. As we say, *for life unto death*. Without democracy, without the presence of interest groups such as security lobby or the media lobby this was inconceivable. Without media, a global armament industry, new espionage techniques, rapid communications, and quick political mobilization, the war could not be fought, and then peace could not be attained. With political democracy mobilized for war in this way, dialogues could not but become part of the battles over control. The border war of 1999 stood not on the ruins of earlier accords, but on their succession.[61]

To understand in this backdrop, where all issues have been congealed in the form of a line and have brought a great tragedy in its wake, we must probe a little more into how geopolitical reasoning has become organic to democracies and transformed democratic thinking into almost what we can call a ‘geopolitical common sense’.

To understand in the backdrop of focalization, in the form of a line, of all issues on Kashmir, and the tragedy this brought in its wake, we must a probe little more into how geopolitical reasoning becomes organic to democracies and thus transforms into almost a 'geopolitical common sense'. Sankaran Krishna in discussing post-colonial insecurities speaks of 'hegemony as spectacle' and refers to Clifford Geertz in showing the spectacular aspects of statecraft in South Asia.[62] Geertz had written of the theatre state of Bali, where the state, while condensing within it three layers of meaning of its existence as state status, pomp and governance – was a 'theatre state' in which rulers were impresarios, priests the directors, the rest including the peasantry the supporting cast, crew, and the audience. He wrote of the stupendous ceremonies, sacrifices, dedications, mobilizations, and amassment and display of wealth that were not a means to rule, but the essence of rule, 'not means to political ends: they were the ends to themselves, they were what the state was for'. In this theatre state, 'power served pomp, not pomp power'.[63] It is wrong to think, Sankaran Krishna reminds us in the vein of Geertz that, while modern state politics seems to effect a recession of the spectacular aspects and an elevation of the aspect of governance, this recession is at the level of appearance only. The combination of diminished attention to pomp and status and the continuing importance that nations give to 'performance' is a sign of the modern.[64] In Kashmir, the post-colonial state got all the opportunities of display of its pomp and authority. That Kashmir needed on the part of the Indian state efficient governance, development, honesty, and above all trust in the accord with independence-minded politics there, humility, and continuing efforts to reconcile was not difficult to realize, and certainly did not need the hordes of policy advisers that the Indian state mobilized from time to time. That it needed a rapprochement with Pakistan, diminution of state power and sovereignty there, was also certainly within comprehension there.

Instead, what happened was, and continues to be, an unending series of spectacles from the famous horse rides of the first Prime Minister of India, Jawaharlal Nehru, in Gulmarg to the regalia and the vast cantonments of armed forces in the valley. A democratically elected parliament could not but think of Kashmir in geopolitical terms not because it had rationally decided, at least there is no evidence to that, to 'keep the valley' at all costs on considerations of political and territorial security, but ceremonial imperatives of a big state demanded that India would and could follow nothing else. This explains, why there was never any Kashmir policy as such, that it evolved as it responded to its spectacular self and to similar acts of its feuding blood brother, Pakistan, and why territory, religion, the land, all helped make geopolitics a part of democratic common sense. After all keeping Kashmir the way it was conceived was crucial for keeping India the way it was conceived when the nation was born, that is to say, a liberal democratic constitutional state protected by a huge army and paramilitary forces and buoyed by big power ambitions, co-existing with poverty, exploitation, continuing western and neo-colonial domination, big business led industrialization with the state acting as the midwife, and selected measures of devolution and decentralization of power and administration. Boundaries in this context made the space of post-colonial democracies sacred, and elevated the truth of territory to divinity, while at the same time marginalizing other truths as fable. The failure to keep Kashmir was inconceivable to this sense that had guided the national polity from its beginning. This also explains besides, the incapacity of democracies, particularly post-colonial democracies, to resolve issues of geopolitics.[65]

The issue of procedural capacity of a democracy to dialogue assumes significance when states think that by declaring ceasefires militancy can be contained and borders protected. India while declaring that it would not talk with dictators, declared a unilateral ceasefire during the month of Ramadan on November 19, 2000. This was hailed by many as the first positive step in the direction of peace after the collapse of the August 2000 peace initiative of the Hizbul Mujahideen, which had also declared a unilateral ceasefire, by ordering its armed forces on the LOC to exercise 'maximum restraint' to 'strengthen and stabilize' the ceasefire in 'occupied Kashmir'. Pakistan reciprocated Indian unilateral move. However, the failures of the earlier attempts show that a *Camp David* style peace process cannot work in South Asia. A formalist exercise in conflict resolution that does not go into the basics of the status quo is bound to be stillborn. The hallmark of the American Administration's policy of 'waging peace' in West Asia has been not to disturb the status quo (for example, ignoring the fundamental issues of Palestinian displacement and Jewish settlements). The fact that intent of the Lahore declaration could not be translated into policy is significant in itself, for the peace process remained singular and refused to take into account the plural character of the issue and the process. The terms of the unilateral ceasefire indicate that India is still not willing to talk to Pakistan. A refusal to accept the plurality of the process of peace will not lead ceasefires towards a larger peace. South Asia will remain bogged down in a Camp David-like process where, ostrich-like negotiators bury their heads in the sand. A ceasefire is generally achieved through negotiations. Steps before and after lead ceasefires to peace. Pakistan and India, particularly India, can turn the current ceasefire into an opportunity for opening up for a fresh peace process by immediately initiating steps to restore rule of law in the valley, initiating de-militarization, ordering its forces to respect rights of the people, and beginning dialogue at various levels. India can simultaneously lift all restrictions against peaceful public activities there. This will enable the people to discuss their social and political problems and their future in a democratic manner. As Kashmiri civil society recaptures its lost space, the vexing question, '*Whom do we talk to in Kashmir*?' will lose its relevance. A vibrant Kashmiri civil society will provide the answers.

Not that the political class in Indian democracy did not know that even after the Jammu Kashmir Liberation Front (JKLF) had rested arms, the security forces pursued its cadres and most of its political cadres were killed. Or, the Hejbul Mujahideen wanted dialogue.[66] Or, the Hurriyat notwithstanding its notorious vacillation welcomed talks. Or, voices have been present against the proposal to trifurcate Jammu and Kashmir along communal lines. Or, plurality that liberal thinkers celebrate is being misused at this very time to foment ethnic enclave making, exactly as was done in India's northeast to such harmful consequences. Or, that the Sikhs in Kashmir even after the Chattesinghpura massacre have refused to leave Kashmir. Or, there has been a rebirth of civil society organizations such as

Mothers of the Disappeared, the Jammu Kashmir Forum of Civil Society Organizations, and the initiatives of the Pakistan-India People's Forum for Peace and Democracy. Hence is the required two-fold inquiry, will nationalist democracy organic as it is, be capable of generating an outlook that will encourage restoration of dignity, compromise and reconciliation on the basis of *minimal justice*? And, can the democracy we inhered as gift of de-colonization procedurally promote dialogue?

Unfortunately, a policy of peace based on conservative geopolitical reasoning has shown itself often as part of the politics of war. Such geopolitical policy of peace has been attempted to wear out the foe, to create a stalemate, to exasperate him, to choke the agenda of reconciliation and peace with which the adversary has come forward due to fatigue or desire for minimal justice. Geopolitical settlement de-linked from the agenda of a 'democratic peace', that is, the participatory route of wide-range dialogue, is typical of a policy of authoritarian security where the state thinks that by talking to the leader of the adversarial camp, a tête-à-tête between two leaders, peace is achieved. Such thinking divorces the condition and feelings of people, and their desire for justice and activism for peace from the process of peace. To embark on a path of plural dialogue and plural peace process, we must first understand, that the pursuit of a monolithic peace process is not congruent with flexible arrangements, new forms of political accommodation, and an imaginative geopolitics. The challenge before democracy is if it will be able to recover from the clutches of the older forms of geopolitics with which the Westminster model of de-colonization had triumphed in South Asia.

The line, that divides Pakistan and India in that region, also divides the two parts of Kashmir, indeed in several parts. That line has been replicated already within the country, dividing how many times, how many entities, and in how many forms, historians will find out someday.[67] Visible and invisible frontiers have been created. The feature of these *nouvelle frontiers* is that they are being produced internally; they are not vertical lines separating two spaces, but concentric circles continuously dividing and then locating these to rejoin them in the universe of the nation. Law, citizenship, rights, obligation, and morality – all are caught in this universe of concentric circles. In a situation where geopolitics has imprisoned democracy, the agenda for democracy begins with a realization that the reproduction of frontiers as a disciplining tool is not a condition of normalcy. Taking that as normal means a lack of acknowledgement of plurality, critical to note here, plurality of peace constituencies. After all, any policy of reconciliation in Kashmir will have to begin with a recognition of that specific history of *Kashmiriyat* which tells us of its internal forms of accommodation (between tenantry and the middle classes, between several sub-regions, between communities, between Kashmir and rest of the sub-continent, between religion and tolerance) as it tells of similar internal fault-lines. Democratic politics will have to begin with a refusal to accept a revision of this past, if it is to free itself from the lure of imperial forms of nation and geopolitics, and think of reconciliation, justice and accommodation. The dividing lines in Kashmir still the time of the nation, at the same time lend it a form, indeed many forms – of nationalism, post-colonial politics, neighbourhood, enmity, identity, and finally the institutional form of political discipline. Power unless given spatial form remains in limbo, as if the national power is grievously weakened because it lacks focus, it has no source. Geopolitics not merely supplies the nation-form that focus it also creates what Foucault calls the *heterotopy* – a site of more than one social and political existence, several spatial settings combined in one – a multiplicity of locative spaces in one site, in this case the line.

The significance of the border – the dividing line – has only increased in this way in the wake of conflicts, rather that diminishing. In a perceptive essay on the 'fence/border problematic' in Israel-Palestine peace question, an Israeli peace researcher Tamar Hermann has argued, both the historical legacy and the present political preferences of Israel's leadership and public make an implementation of the separation option far more likely than the integrative option, while ruling out the co-existence option. A recently published book, she says, dedicated from cover to cover to promoting the idea of separation, eruditely summarizes the 'objective' need for separation. Integration, it maintains, implies six dangers to the State of Israel: if Israel fails to separate from the Palestinians, its political and moral responsibility for the well-being of the latter's millions will overburden its welfare system to the point of collapse; only strict separation can prevent the influx of millions of Arab immigrants into Israel – a process which will result in far-reaching demographic transformations, to a point where the Arab vote will determine the features of Israel's political map, jeopardizing Israel's economic achievements and modernization prospects; immediate relations between the more affluent Israeli Jewish society and the impoverished Arab societies are bound to lead to a destabilization and 'relative deprivation' in the latter, thereby adding a class struggle aspect to the already grave national conflict; if indeed the borders are opened, the influx of Arab immigrants will lead to the collapse of Israel's substructural systems such as transportation, water, sewage, etc. Moreover, if the barriers are lifted, a sharp increase in crime and corruption inside Israel is anticipated, in light of the profound normative and behavioural differences between the Jewish and the Arab societies, and their respective control systems; last but not least – integration will expose Israel's 'soft underbelly' to Arab insurgency and terror attacks.[68] Tamar further argues that this suggests that the pre-Oslo combination of the zero-sum definition of the conflict with the greater Israel vision, based on security or religious-national grounds or both, seems paradoxically to have served well the colonialist mode of relations as it sustained the integrative spatial policy imposed by Israel on the Palestinians between 1967 and 1992. During these 25 years of occupation, the Green Line, which separated the West Bank and Gaza from Israel before the 1967 war, was practically erased, making de facto the entire Land of Israel/Palestine west of the Jordan river into one entity, albeit internally torn apart and discriminatory. This practical spatial integration seems incapable of surviving the post-Oslo combination of the still predominant definition of Israeli-Palestinian relations as fundamentally antagonistic with the Israeli sober assessment that withdrawal from the territories is a sine qua non for management of the conflict, and now provides fertile ground to nourish the idea of separation. The post Oslo conceptual combination now accords the Israelis full self-legitimacy to confine 'them' [the Palestinians] to the territories on the one side of the fortified border, while 'we' (Israelis) alone reside on the other.

The similarity does not end here. A second Palestinian uprising is on, involving a great deal violence on both sides, Israeli chauvinism is now at its hardened most, reminding one of the situation of sixties and seventies, causing setbacks in almost every sphere of joint activity as well as a resurgence of all the old, negative stereotypes. There are plans for all kinds of unilateral solution, without seeking or obtaining Palestinian consent thereto. The 'fence policy' has again become tempting. The issue of future boundaries has not become irrelevant; on the contrary – it is being dealt with incessantly, both formally and informally, even while the talks are suspended, and is continuously debated by experts, crisis managers, and political figures in Israel. And with this the theoretical and practical problems of a politics of reconciliation are now before a politics of peace. An acknowledgement of this, Tamar says, does not mean questioning the relevance of the current discussions on reconciliation that have contributed considerably to our understanding of the causes, components, and dynamics of various sorts of conflict, as well as the conditions for their resolution. But it implies certainly the need to take into account specific issues, problems, and practices of negotiating borders and boundaries leading to conflict and reproduced on an extended scale by the conflict.[69]

In the context of the geopolitical realities that have thrust themselves upon the politics of Kashmir, the counter politics of dialogue has to probably turn them on their head. Thus while there has to be a recognition that the conversion of LOC into an international border cannot be the solution, there has to be also a recognition that the line cannot be abolished overnight and it will remain in place and will be respected at least temporarily. At the same time, the line would be transcended by the institution of 'soft borders' for facilitating flow of goods, services and people towards restoration of a shared life. Accompanying this process there has to be dialogue on making Kashmir on both sides 'autonomous' again through de-militarization, withdrawal of all intervening powers, dismantling of the ruthless summary power structure, initiation of a new constitution-making in Kashmir, which will be observed by a dialogic group of interlocutors, and guaranteed by both countries who among themselves can work out a joint defence council buttressed by a series of agreements and treaties. Similarly the lines that divides Jammu and Kashmir from within have to be similarly turned on their heads by making Jammu, Ladakh, and the Northern Areas 'mandated territories' whose autonomy will be guaranteed not only by the two countries, but the autonomous body governing Kashmir also, and which will be administered by autonomous councils of these areas. And all these dialogues and guarantees will be predicated by a supervening guarantee of protection of minorities and their rights that will have a place from the beginning in any plan of restoration of peace and freedom in Kashmir. Autonomy, to summarize, does not have a once for all form of existence; it deepens and must be allowed to assume thereby variegated and simultaneous forms. I do not intend hereby to suggest solutions. But these ideas and their possibilities broached and appreciated in various forms show that the connecting links between geopolitics and democracy cannot be wished away, rather they have to be appreciated. These links also suggest, that dialogue has to be plural, because issues, constituencies, histories, disputes, and unities are also plural.

Here is then the agenda for democracy, as one philosopher has asked, 'How to assure that this non- democratic institution (frontier) *per excellence*, which is never collectively mentioned, but serves to monitor individuals and populations, becomes a stake in politics whose modalities of use and transformations can be negotiated?'[70] In such a question we have the beginning of a realization of how the political class in a democracy becomes incapable of coping with the torment of the *inside/outside* – democracy that appears as an *internal* problem and geopolitics that appears as a problem related with the *outside*.

In order to do so we shall have to move further along that path of understanding the problematic of democratic deficit – the deficit produced by democracy internally relating to its capacity for dialogue, accommodation, tolerance, understanding, and providing autonomy. For that we shall bring into discussion some of the legal-political accounts in this region on the issue of autonomy as a form of accommodation but not to reflect once more on the old question of whether minority protection has been adequate in post-colonial political formations, but on how the contradictory demands of republicanism incessantly producing ethnic frontiers, and the claims for autonomy have marked democracy – how these polities have ensured the transformation of the question of rights and accommodation into an 'ethnic question' to render the existing arrangements of autonomy as forms of accommodation and products of dialogic provisions inadequate – to render the nation's morning of tolerance grey.

Notes

1 Sumantra Bose, 'Kashmir at the Crossroads Problems and Possibilities', *Security Dialogue*, 32 (1), March 2001.
2 Sumit Ganguly, 'Past Attempts at Mediation and Conflict Prevention in South Asia: Would CSBMs Have Made A Difference?' in Sumit Ganguly and Ted Green (eds.), *Mending Fences: Confidence and Security Building Measures in South Asia* (Delhi, Manohar Publications, 1997), p. 73.
3 Ibid. p. 85.
4 There are several accounts on the indication of future complications in the Gurudaspur award, that made land-link between Kashmir and rest of India possible. Of them see, Nicholas Mansergh and Penderlal Moon (eds.), *The Transfer of Power, 1942-47*, Vol. X (London, 1981); H.V. Hodson, *The Great Divide* (London, 1970); Lord Birdwood, *Two Nations and Kashmir* (London, 1956); M.J. Akbar, *Nehru* (London, 1988); Latif Ahmed Sherwani, *The Partition of India and Mountbatten* (Karachi, 1986).On the clumsy partition in the east, the best till date is Joya Chatterjee, 'The Fashioning of a Frontier – The Radcliffe Line and Bengal's Border Landscape, 1947-52', *Modern Asian Studies*, 33 (1), 1999, pp. 183-242.
5 Discussions on links between internal and interstate conflicts sometimes fail to take into account the catalytic role of borders in forging the links. See in this context, David Carment and Patrick James, 'Internal Constraints and Internal Ethnic Conflict: Towards A Crisis-based Assessment of Irredentism', *The Journal of Conflict Resolution*, 39 (1), March 1995, pp. 82-109.
6 On the flexibility of the early Indian stand we write below. On a similar flexibility in early Pakistan policy, see Mehrunnisa Ali, 'Jinnah's Perception of Pakistan's Foreign Policy and the Kashmir Issue', *Pakistan Horizon*, 43 (2), 1990, pp. 57-70.
7 Quoted in K. Sarwar Hasan (ed.), *Documents on the Foreign Relations of Pakistan: The Kashmir Question* (Karachi, 1966) p. 71.
8 Ibid. p. 75.
9 S/2967, paragraph 57, *UN Security Council Official Record 8*, Supplement No. 1, 1953, p. 14, cited in Paula Banerjee, 'The Tenuous Control over the Line of Control' in Helmut Reifeld and Ranabir Samaddar, *Peace as Value* (Delhi: Manohar, 2001), p. 303.
10 Gopal, *Jawaharlal Nehru: A Biography*, Vol. 3, (Delhi: Oxford University Press, 1989), p. 86.
11 *Rajya Sabha Debates*, Vol. XXII, 26 April 1958, pp. 1047-55.
12 Lok Shabha debates, cited in Paula Banerjee, 'Tenuous Control over the Line of Control'.
13 Chester Bowles, EMB Delhi, 760 – Part I, India-Pakistan Working Group, Situation Report, 24 September 1965, *LBJ Papers*, NSF-India Memos (2 of 2) Vol. V, 6/65-9/65, cited in Banerjee.
14 'A Chronology of Significant Events,' Research Memorandum, June 24, 1966, *LBJ Papers*, NSC History, Indo-Pak War, Vol. III, State Department History, cited in Banerjee.
15 *Kashmir Dispute*, p. 31.
16 President Ayub Khan's Statement 14 January 1966; reprinted in *Pakistan Horizon*, op. cit, pp. 120-124.
17 Zulfikar Ali Bhutto's speech, *Pakistan Horizon*, p. 138.
18 Also noted by Sumit Ganguly in *Mending Fences*, p. 77.
19 Bhutto, *Pakistan Times*, 19 July 1972, p. 1.
20 Address delivered at the Pakistan Institute of International Affairs, Karachi, on 31 July 1972, published in *Pakistan Horizon*, op. cit.

21 Years later after Z. Bhutto had made the statement, the Pakistan President, General Pervez Musharraf, said on the eve of his visit to India for the Agra summit on July 14-16, 2001, 'I will be flexible on Kashmir LOC is the problem, not the solution', *Outlook*, 23 July 2001; also reported in *The Telegraph* in a report by Drees Bakhtiar on 7 July 2001, 'Pervez Rules out LOC Solution' of saying, 'LOC was at the core of the (Kashmir) problem and could not be seen as a solution to the dispute'.

22 'The Tenuous Control over the Line of Control'.

23 On the history of this infamous act, the single most insistent research has been done by Gautam Navlakha; from his many writings, see for example, 'Armed Forces Special Powers Act', *Economic and Political Weekly*, 32 (6), February 8, 1997.

24 According to one observer, 'the growing power of the Indian military-bureaucratic complex has also greatly contributed to the expansion of India's military clout. India's intervention in East Pakistan in 1971, compounded with other military interventions...has given the armed forces the appropriate bases to constitute themselves into a powerful lobby.' Harsh Kapur, *India's Foreign Policy 1947-92: Shadows and Substance* (New Delhi/Thousand Oaks/London: Sage Publications, 1994) p. 99.

25 This figure of financial cost is old, a more recent figure (that too 3 years old) is US $ I million a day. See Patwant Singh, 'Too Many Chinks in the Armour', *Asian Age*, 29 January 2001, p. 6.

26 'Battle in Siachen is of Survival', *The Times of India*, 15 January 2001, p. 4.

27 Nehru to the cabinet secretary, 6 April 1953, cited by Noorani, 'Kashmir's Woe – I,' *The Statesman*, February 13, 1995.

28 On Siachen negotiations, see Robert G. Wirsing, *India, Pakistan, and the Kashmir Dispute* (New York: St. Martin's Press, 1994), pp. 195-216.

29 Jean Luc Racine, 'The Uncertain Triangle: India, China and Pakistan – The Regional and International Dimensions', unpublished essay. I am grateful to Racine for allowing me access to this essay.

30 On this period see, Kanti Bajpai, 'CBMs: Contexts, Achievements, Functions' in D. Banerjee (ed.), *Confidence Building Measures in South Asia* (Colombo: Regional Centre for Strategic Studies, 1999) ; on the 'China factor' see Gary Klintworth, 'Chinese Perspectives on India as a Great Power,' in R. Babbage and Sandy Gordon (eds.), *India's Strategic Future: Regional State or Global Power* (New York, 1992) ; on Operation Brasstacks, see K. Bajpai *et al.*, *Brasstacks and Beyond: Perception and Management of Crisis* (New Delhi: Manohar Publications, 1995). On the compulsions to look at the Kashmir question from a geopolitical perspective, see Robert W. Bradnock, 'Regional Geopolitics in a Globalising World – Kashmir in a Geopolitical Perspective', *Geopolitics*, 3 (2), Autumn 1998, pp. 1-29.

31 Pakistani foreign minister, Aseef Ahmed Ali's complaint to the UN General Assembly on 5 October, 1994, reported in *The Statesman*, 6 October 1994.

32 *The Statesman*, 2 February 1996.

33 Ashraf Jehangir Qazi, Pakistan ambassador to India, in an interview to PTI, and quoted in *The Hindu*, 9 August 1999.

34 Bharat Karnad to *The Telegraph*, 20 June 1999.

35 *The Nation*, 8 July 1999.

36 *The Telegraph*, 17 June 1999.

37 *The Indian Express*, 12 June 1999.

38 *The Times of India*, 8 June 1999.

39 J.N. Dixit quoted in 'Line of Control: Bordering on the Impossible', *The Outlook*, 12 July 1999.

40 *Indian Express*, 6 September 2000, p. 5.

41 Gautum Navlakha, 'Kashmir: An Idea Which Cannot Be Suppressed,' *Economic and Political Weekly*, January 29, 2000, pp. 253-255.
42 AG Noorani, 'Poll in J & K – I: A Record of Fraud,' *The Statesman*, April 13, 1995; see also Sisir Gupta's report, *The Statesman*, January 21-23, 1995; Noorani, 'Kashmir's Woe – I,' *The Statesman*, February 13, 1995; M.L. Chibber, 'Indo-Pak Ties: Need to Create Public Opinion,' *The Statesman*, April 24, 1995.
43 PR Chari, 'India's Defence Expenditure: Can It Be Reduced,' *Defence Expenditure in South Asia: India and Pakistan*, RCSS Policy Studies 12, June 2000.
44 Report released on August 17, 1999, www.meadev.gov.in/govt//indnucld.htm, pp. 18-19.
45 Gautum Navlakha, 'Kashmir: An Idea Which Cannot Be Suppressed,' p. 253.
46 'Two Women Killed in Indian Firing', *Dawn*, June 18, 1999, p. 1.
47 'Two Peaks on Jubar Hills Recaptured,' *Hindustan Times*, June 30, 1999.
48 'IAF Carries Out Fresh Strikes in Kargil, *UNI Report*, June2, 1999.
49 '2 Killed, 4 Injured in Pak Shelling in Jammu Region,' *UNI Report*, June 21, 1999.
50 'Army Called Out in Surankote Following Protests' *Daily Excelsior*, June 30, 1999; www.dailyexcelsior.com/99june30/news.html.
51 'An Uphill Task Against Intruders,' *The Hindu*, June 4, 1999.
52 'Villagers Shifting from Areas near Border,' *Dawn,* June 16, 1999. The estimated number of displaced people was put as nearly 1,50,000; the newspapers reported that some of these were displaced for more six months, and even the war would be over, due to snow-covered road at that altitude, people would not be able to return to their homes in Drass and other areas. For details of displacement, see the special issue of *Refugee Watch* on displacements in Kargil war, No 7-8, 1999.
53 See in this context Roland Paris, 'Peace building and the Limits of Liberal Internationalism', *International Security*, 22 (2), fall, 1997, pp. 54-89.
54 See for example the *Manual for Early Warning and Early Response*, issued by FEWER (Forum for Early Warning and Early Response), London, 1998.
55 Michael Mann, 'The Dark Side of Democracy: The Modern Tradition of Ethnic and Political Cleansing', *New Left Review*, 235, May-June 1999.
56 Muzamil Jaleel, the correspondent of *The Indian Express* reported how most Kashmiris felt alienated from the Kargil war, though the war was fought in their name. In his words, 'the war exposed both India and Pakistan in the valley'. Muzamil Jaleel, 'It Was Not Our War' in Sankarshan Thakur and others, *Guns and Yellow Roses: Essays on the Kargil War* (New Delhi: Viking, 1999), pp. 63-94.
57 See for a pertinent discussion, Etienne Balibar, 'Is European Citizenship Possible', *Public Culture*, 8 (2), winter 1996, pp. 355-376, later on published in James Holston (ed.), *Cities and Citizenship* (Durham, 1999), pp. 195-214.
58 Balibar, 'Is European Citizenship Possible?'
59 In this exercise, in recent years the interest has been on to finding out how democracy integrated Kashmir with India. On this, apart from the books and articles mentioned above, see, Prem Sankar Jha, *Kashmir 1947: Rival Versions of History* (Delhi: Oxford University Press, 1996); Vernon Hewitt, *Reclaiming the Past: The Search for Political and Cultural Unity in Contemporary Jammu and Kashmir* (London, 1995); Stanley Wolpert, *Zulfi Bhutto of Pakistan: His Life and Times* (Delhi: Oxford University Press, 1993); R Grover (ed.), *The Story of Kashmir: Yesterday and Today* (Delhi: South Asia Books, 1995); Sumit Ganguly, *Origins of War in South Asia: Indo-Pakistani Conflicts since 1947* (Boulder: Westview, 1986), Sumit Ganguly, *The Crisis in Kashmir: Portents of War, Hopes of Peace* (Cambridge: Cambridge University Press, 1999); Sumantra Bose, and Victoria Schofield, Kashmir in Conflict.

60 For the role of media in the war, see the detailed dossier prepared by the South Asia Forum for Human Rights, *The Ten Week War in Kargil: A Dossier*, SAFHR Papers 6, Kathmandu, 2000; also Geeta Seshu, 'Media and Kargil: Information Blitz with Dummy Missiles', *Economic and Political Weekly*, 9 October 1999, pp. 2917-19; also the collection of essays by some notable journalists, Sankarshan Thakur *et al*, *Guns and Yellow Roses*, op.cit.; on a general level, Tim Allen and Jean Seaton (eds.), *The Media of Conflict: War Reporting and Representations of Ethnic Violence* (London: Zed Books, 1999); Jean Seaton, 'Why Do We Think the Serbs Do It? The New Ethnic Wars and the Media', *Political Quarterly*, Summer 1999, pp. 254-270.

61 Readers interested in this must read the collection of media reports in the *The Ten Week War in Kargil: A Dossier*.

62 Sankaran Krishna, *Postcolonial Insecurities – India, Sri Lanka, and the Question of Nationhood* (Delhi: Oxford University Press, 2000), chapter 5, 'Hegemony as Spectacle – The Theater of Postcolonial Politics', pp. 129-165.

63 Clifford Geertz, *Negara – The Theater State in Nineteenth-Century Bali* (Princeton, NJ: Princeton University Press, 1980), p. 13.

64 *Postcolonial Insecurities*, p. 130.

65 In this context, one may see Richard K. Ashley, 'The Geopolitics of Geopolitical Space – Towards a Critical Theory of International Politics', *Alternatives*, 12, 1987, pp. 403-434.

66 On this see the revealing interview by the Hezbul chief, 'What's the Use of Hurriyat Visit to Pakistan, Asks Hizb Chief', *The Times of India*, 16 January 2001.

67 Peter Radan refers to the mimetic situation in a significant essay 'Yugoslavia's Internal Borders as International Borders – A question of Appropriateness', *East European Quarterly*, 33 (2), Summer 1999, pp. 137-155; see also P. Diehl and G. Goertz, 'Territorial Changes and Militarized Conflict', *Journal of Conflict Resolution*, 32, March 1998, pp. 103-122.

68 Dan Schueftan, *Disengagement: Israel and the Palestinian Entity* (Tel Aviv: Schoken, 1999, pp. 23-24, [Hebrew]), cited by Tamar Hermann, 'Reconstituting Boundaries as a means of Peacemaking: The Israeli-Palestinian Case', unpublished paper, February 2001; I am grateful to Tamar for allowing me access to her work on the Middle-east peace process.

69 Tamar Hermann, 'Reconstituting Boundaries as a Means of Peacemaking', unpublished paper; I am grateful to Tamar for allowing me access to this essay.

70 Etienne Balibar, 'Algeria, France: One Nation or Two?' in Joan Copjec and Michael Sorkin (eds.), *Giving Ground – The Politics of Propinquity* (London, 1999), p. 171. See also his 'The Borders of Europe' in Pheng Chea and Bruce Robbins (eds.), *Cosmopolitics – Thinking and Feeling beyond the Nation* (Minneapolis, 1998).

Chapter 5

Autonomy and the Requirements of Minimal Justice

Claims for Recognition and the Need for a New Dialogic Order

South Asia's democracy, as many other post-colonial democracies, meets its ordeal when it has to prove its capacity to promote tolerance, ways and forms of accommodation, and prevent the state from becoming runaway with a majority-centric structure and orientation. Reinforced by globalization, the politics of the majority and minority more than any thing else characterizes democracy that is raising its guards against all sorts of aliens, which include immigrants, refugees, minorities of various types, and all those decreed as outsiders to the world of wealth, pomp and circumstance. All protests in such condition are 'ethnic', 'anti-democratic', and by a stretch of a yard more, 'terrorist'. For a dialogic politics the question posed in such milieu is, can it imagine out of the present political practices newer forms of political understanding marked by accommodation? Can rights become more than marks of claims, and become the marks of a dialogic order, where in face of the designs of the political class, politics can develop new ways of co-living beyond the current geo-political and ethno-political realities, and assume newer spatial-spiritual forms or dimensions of a dialogic polity?

In this chapter, my inquiry is in that direction, and proceeds in three ways. First, I examine the political background of the failure of existing liberal and communitarian ideas against which the demands for newer arrangements for autonomy at every level arise. Second, I examine in the mirror of the imperatives of a dialogic order the limited success and failure stories of existing arrangements for autonomy in South Asia and in other parts of the world through a broad and rapid survey. Third, through these two inquiries I try to show how a new dialogic order can propose fresh forms of accommodation on the basis of the principle/s of minimal justice. Since, investigation in requirements of justice and autonomy calls for an examination of the current discourse of rights connected with existing institutional politics, discussion of minority rights occupies a substantive part of my inquiry towards the spatial-spiritual forms of a new dialogic politics.

Forms of autonomy are judged by requirements of *minimal justice*. Such an examination will offer us insights into the minority problematic, the limits of the conventional solution, and the possibilities of transcending what liberal politics could only pose as a (minority) 'question'. I shall also argue that these requirements suggest rough contours of a new dialogic order. The investigation is important, for more than anything else, we find in the problematic of autonomy, the congealed presence of several outstanding elements impinging on the present politics of dialogue, namely, the minority question, the issue of self-determination, globalization, and the twin strands of rights and justice. In terms of demonstrating the need for a new dialogic order, it is further necessary to show, as I suggest in the first two chapters, how constitutionally bound dialogues and arrangements become inadequate in facilitating resolutions of conflicts arising out of the negotiation of claims for recognition by a state that builds up its political power on the basis of producing majorities and minorities, and configuring almost continually an ethnic core in order to push others to the margins.

One of the critical political issues facing the states in the region of South Asia today, as we all know, concerns demands of political recognition and the multiple claims for self-determination by communities and groups in the society. Communities marked by language, religion, place, and others associative factors have been vocal about such recognition from the state. Particularly to the minority communities such claims and demands appear as effective defence against political irrelevance, exclusion, and marginality. In the old imperial forms of rule this was not an acute problem (at least, not to the present degree), as the state was not seen as relevant to the identity of communities. Or, ruling groups and the ruled communities both considered the state as the guarantor of an arrangement under which all communities lived. But with the rise of modern nationalism and democracy, rights of the communities, particularly minority communities, are not only an expression of social diversity and differences. Their singular marker is that they are against the state, and indeed, only by being against the state they are seen as capable of effectively protecting diversity and difference. Minority rights are therefore seen increasingly as belonging to the category of the rights of self-determination; they form the core of group rights; the demands that they place upon the polity exceed the terms of the liberal agenda of 'protection' (of minorities), and community rights are seen as redefining citizenship.

In the received theory of democracy, citizen is the individual with whom the state enters into contract. Individuals collectively form the people in whom sovereignty resides. The democratic state emerges from that, but ironically, the democratic state recognizes the sovereign only in the repository form of an individual. In this case, the response has been in the form of the emergence of a type of communitarianism that challenges the liberal version of democracy. The argument for respecting social diversity is extended to one of redefining democracy in terms of communities. To these two we can add a third characteristic of the situation. Since minority rights emerge only in the wake of the spread of the ideas of nationalism and democracy, minority rights never appear as an isolated item in a nationalist agenda. They characterize an entire political region. Thus identity politics is articulated in terms of extra-territorial linkages, ethnicity (which by definition can not be co-terminus with national territoriality), and the breakdown and formation of new states. In short minority rights are considered to be challenging the state and the states-system. Against such a background, for critical theory of dialogue the imperatives are complex.

At a formal level, one may say that while liberals have given primary importance to individual rights, communitarians have maintained that recognition of the individual is not enough; groups and communities have to be granted rights. Only in this way, democracy can be made tolerant and dialogic, and diversity can become the cornerstone of democracy. The strength of the discourse of minority rights has been that it is capable of drawing attention to the weaknesses of modern liberal constitutionalism which acknowledges only individuals, legally guaranteed institutions and sometimes federal units (for example, states in India) as political actors. Communities are not accepted as legitimate actors. The defence of communities in the arena of political rights, as opposed to this, swings from economic arguments to cultural explanations. In the former, community rights are defended in terms of the imperatives to save the weaker communities and indigenous people from marginality in the wake of policies ushered in by the state, big business groups and trans-national economic forces bent upon pushing global standards and establishing global compulsions and linkages. In the second, better known as the multicultural option, diversity and pluralism acquire importance as the key to freedom and equality. Since multiculturalism has been offered by liberal theory as the answer of democracy to the charge levelled against it that it is organic, non-tolerant, and non-dialogic, it is necessary to see at least briefly, how far such multicultural option goes in meeting the requirements of justice and dialogue.

In this connection, it is interesting to note that there has been of late a meeting of the two arguments. Liberalism has embraced, and many liberals can say that this was the case from the beginning, the argument of pluralism, diversity and equal opportunity in public sphere. Cultural minorities and other groups have no reason to fear that the state would deny them fairness and justice. Liberal theorists speak of civil society as the guarantor of justice, accountability, fairness, and most important of all, plurality. The philosophical defence of liberalism takes the form of universality of the principles of equality of opportunities, and the universality of the principle of dialogue among constituents of civil society, also between these constituents of civil society and the state on the issue of equal opportunities, fairness and justice. Communitarianism has also come forward half way to admit that differences cannot always be confined to civil society, they may impinge on individual rights, and group assertion may at times threaten public sphere. We have thus a reworking of the problematic in the field of democratic theory. This is through 'going few steps back' by both in negotiating the issue. The concern today is about securing a better deal for the 'worst off class/group' through positive discrimination, and advocating a theory of justice that admits in its fold a notion of public goods and overlapping consensus in the public sphere on procedures of dialogue and accommodation. In this reworking we are also assured that the groups traditionally excluded from civil society are included, because the public sphere now consists not only of themes of individuality, but theme of identity also. In short, we have now the marriage of classic liberalism and a pluralist version of democracy. Under this new notion, pluralism becomes both a normative theory and a political agenda. For, the sovereign state must now respect the freedom of communities, reorient its foundations to accommodate a diversity of associative groups, and reconcile individual freedom, group claims, and power of the state.

The reworking however remains unsatisfactory for three reasons. First, it does not adequately address the issue of inequalities within groups. Second, it has no reply to the critics who contend that with this modern democracy has become in effect an interest group theory, which means green signal to groups becoming non-dialogic with their entrenched self-interests and with little inclination to converse except to bargain on interest. Third, all this notion of tolerance and multiculturalism depends too much on civil society's capacity to democratize politics by injecting plural values, while the question remains, who will democratize civil society?[1] In short, even the re-worked liberal theory remains a liberal one that builds on harmony, and not contention and contest. It wants to embrace the values of political conversation, but is not prepared to accept that dialogues are contentious practices of society.

The new interest in liberal pluralist democracy as the site of dialogue that borrows from communitarian discourse, albeit selectively, has come with the collapse of old multinational states in Eastern Europe, and now with exacerbated violence in many parts of the globe. The renewal of civil society has been to some the guarantee to group rights, including minority rights. Yet, the gloss over this supposed guarantee is quickly wearing off. The politics of cultural identity surrogate to global forms of politics including the nation form has displaced the basis of minority rights. From being an agenda in democracy, this has become linked to neo-nationalist and racist community based organizations in the newly discovered space of civil society. It appears, groups do not encourage tolerance; they produce hatred. Equality of rights is equality in hate acts. The theory of self in politics fails to settle the political question of self. Repression, assimilation, denial, exclusion, and marginality are too strong evils to be wished away by the theory of tolerating differences. Ironically the 'subject positions' position themselves as un-reconciling to each other; they do not aspire to become the 'decent self' aspiring for cultural recognition. It is not civil society that is at stake here. What is at stake here is political society – its character, goals, constituencies, and its own self.[2]

This, indeed, is the issue at stake. How minorities arise, how they refuse to reconcile permanently to the status of 'minority', and aspire to be recognized as 'people', how they have to survive in modern democracies invariably marked by majoritarianism, is not a cultural question, but a political question; more significantly it involves not an interrogation of *civil society*, but *political society* – our ideas and ideals of reorganizing the political society along dialogic lines which draw stimulus from notions of justice, and a more critical approach to politics. And such dialogic reorganization calls for widening and restructuring social and political space, reorienting the 'habitus' of democracy. As I show gradually in this chapter through a long survey of juridical-political practices of accommodation in the form of arrangements of autonomy – through recounting their limited success and failure stories – that 'civil society' and the 'public sphere', the two of the most celebrated concepts in democratic literature, cannot take us far towards the restructuring of political society along dialogic lines. They may encourage involvement of the 'public' in 'public affairs'. But while such involvement can be good for the rulers who can assure people certain rights, minorities their special rights, and vulnerable people their protection after having identified their vulnerability, such involvement cannot wish away, and can at best reduce, the fundamental asymmetries in power, and the consequent injunctive nature of society. Our examination must therefore be directed to the entire range of political-legal arrangements, which while granting minorities some rights and some measure of autonomy, enables majoritarianism in various forms to rule and thereby succeeds in impoverishing the fund of tolerance and dialogic spirit.

Assured Rights and the Inadequacy of the Rights Language

The conundrum of certain assured rights and the inadequacy of the rights language in addressing in any substantive way the powerlessness of large sections of people in politics is not unique to South Asia or to a particular de-colonized region. The phenomenon is general to democracy in almost all its forms – developed, wealthy, undeveloped, and post-colonial. Human rights in any case began as an issue in individuality. The universal was the man. Therefore the Universal Declaration of Human Rights (UDHR) reflected the first phase of the existing corpus of rights more than it reflected the two succeeding phases. The first phase declared the right of the individual. The second, reflecting the socialist revolutions focused on economic, social and cultural rights of the people. The third phase reflecting the tide of anti-colonial revolutions has focused on national self-determination and non-discrimination. All the three phases are intimately linked with the discourse of democracy. Yet, they show the bind. In each of the phases, the discourse of human rights tried to reduce democracy to human rights. And in each phase the effort produced tensions while significantly enriching our understanding. Rights therefore went on perfecting themselves. 'Making the rights more democratic' became the agenda, though the discourse of rights never really escaped the tension. The concept of minority rights has been no exception. Thus, the International Convention on Elimination of all forms of Discrimination (1965) followed the UDHR, International Covenant on Civil and Political Rights (1976), and then the Declaration on the Elimination of all forms of Intolerance and Discrimination Based on Religion and Belief (1981). Finally came the Declaration on the Rights of Persons belonging to National or Ethnic, Religious and Linguistic Minorities (1992) that expanded the scope of minority rights. It was thought that the tension would be solved now. Yet, it reappears today with the emergence of the issue of multinational intervention at a global scale to protect group rights. Power has proved irreducible to rights,[3] and the rights language has proved inadequate to address the critical need of a dialogic politics in the context of a unilateral order and powerlessness of large numbers of populations everywhere.

Not surprisingly therefore, successive sessions of the UN Working Group on Minorities have tried to break out of the closed circle that the rights language presents before the minority question. It introduced new issues, broadened old ideas, added new dimensions. The fourth session emphasized the role of treaty bodies such as the Human Rights Committee, Committee on the Rights of the Child, and the office of the High Commissioner of for Human Rights itself. In the same vein, the seventh session asked for enlargement of the powers and functions of the working group. The fourth session raised the role of media in respect of minority rights and issue relating to forcible displacement.[4] The fifth session raised the issue of legacies of slave trade throughout the Americas.[5] The sixth session went further. It surveyed the levels of participation by minorities in decision-making procedures, the extents of multi-cultural existence, and the fundamental problem of recognizing minorities ('recognition of a minority is a fundamental step both for protection of the rights of a minority and for restraining ethnic conflict'). Even more fundamentally, it then spoke directly on the need for dialogue on conflict prevention in situation involving minorities.[6] The seventh session raised the issue of the fundamental perspective in which rights would become operative – integration or separation, though in its wisdom, so concerned was it with conflicts, that it did not mention another option, namely co-existence. Nonetheless, the growing realization is evident that one must walk beyond the legal discourse of rights.[7]

The Conference (later on renamed as Organization) of Security and Cooperation in Europe (CSCE, now renamed OSCE) in its Final Act, Helsinki, 1975, upheld the principles of inviolability of borders (Art. 1.2), territorial integrity of states (Art. 1.3), fundamental freedoms of conscience, religion or belief (Art. 1.7) and self-determination of peoples (Art. 1.8). The subsequent political developments soon made clear that these provisions did not happily sit together. The concluding document of the Vienna session of the CSCE, 1989, hence emphasized the humanitarian obligations of the European states in the context of the 'new Europeanism' that involved rights of minorities (Arts. 13.2, 18, 19, 20, 21). Yet the Basle Conference Resolution (1995) felt the need to go further in reinforcing the minority rights. Rights of minorities and of self-determination (Art. 1) were again put together. It spoke of federal structures, diverse forms of autonomy, and a peace order based on federalism (section 3). The final text of the framework convention of the Council of Europe (1995), opened up for signature on 1 February 1995 was wise in the light of the perennial realization that rights were not enough, and that they had to be reinforced with ever widening scope. Therefore while enlarging the protection provisions, it avoided defining minorities altogether. It certainly could not get into that complicated act, because in the context of the new Europeanism, what was to be lot of non-European minorities in Europe as distinct from European minorities? It spoke of national minorities. But in the Europe without borders, and in a milieu of European citizenship, how would the minorities be defined?[8] If Europe, which has given the post-colonial world the ideology of liberalism, the notion of minorities, and the concept of rights, has still not found a way to bypass history by legal mechanisms, this only shows that minority rights remain irreducibly contextual.

My contention is that a singular strategy of rights neglects the contingency of history, because such strategy has to focus on legal institutionalization of rights at the expense of the dialogic imperatives posed before a political society by history. As the Indian experience of many ethnic killings and government-inspired and tolerated riots and mass murders notwithstanding the existence of many institutions and juridical provisions shows, such a singular strategy based on the presupposition that rights can guarantee coexistence, toleration, devolution, decentralization, and autonomy, does not inquire if toleration flows from rights only, or whether we require the assurance of minimal justice also, in the absence of which rights produce only legal essentialism and intolerance.[9] The truth is that, rights are not pre-condition for political action; they can be only the end product. When they come as the companion of mutual obligation, the pre-requisite for dialogue, they are secure as the fruits of mutual recognition. Republicanism, in the absence of a philosophy of tolerance, obligation, decentralization, and autonomy, only allows the political rulers to devise ways to rule the nation mainly through efficient administration of inter-group relations. It transforms the ideal of secularism into a field of power. Community survival depending on political actions of the state, and the principle of toleration transforming into principle of secularism, it becomes an example of what Michel Foucault called governmentality.[10] In stressing the politics of dialogue, I am suggesting therefore two things. My first suggestion is that justice, as minimally defined in a concrete situation, is the benchmark that can help devise ways out of the closed circle that power imposes on the language of rights. Second, dialogue driven by the search for minimal justice becomes contentious. Clearly, these two implications suggest one thing more – the need to dislodge 'civil society' as the principle site of dialogue. Surely the point in such an inquiry at the present moment is – do the minorities want in the main some cultural rights? Do they think that incorporation in a majoritarian political structure can be avoided principally by some cultural autonomy? Alternatively, is the recent history of self-determination with outward-pushing ethnic boundaries, producing partitions and sometimes genocide, an evidence of the need for a direct and fundamental political answer?[11]

This is the sign of the legitimacy crisis. Already theorists started noting the retreat of the public from public sphere, retreat into 'civil and familial-vocational privatism', by which the philosopher Jurgen Habermas meant 'an interest in the steering and maintenance of administrative system but little participation in the legitimizing process'.[12] Such 'privatism', he further noted, offered no support in face of basic needs of existence, did not make possible human relations with a fundamentally objectified nature, permitted no intuitive access to relations of solidarity within groups and between individuals, and allowed no real ethic in political and social life.[13] The crisis shows further the declining belief in legitimacy of orders that a liberal system created and issued. Thus, minorities are dissatisfied with existing provisions; the majorities are dissatisfied with the 'inadequacy' of the old provisions of domination. In this situation of legitimacy crisis, compounded by economic crisis, rationality crisis and motivational crisis in post-colonial polities, the new right is the only assured political animal. He does not need society as a moral entity, he does not need any justification for power; he, however, has still need for democracy, but in a redefined form, that will hold out the promise of prosperity without freedom. An effective territorial state with effective bureaucracy, spirit of rule of law, and an exclusionary economic society may tolerate and even promote a civil society, but that civil society will be severely banal, an exhibit in 'boutique multiculturalism'. The situation is indeed worse, and even by Habermas' standards, defies the capacity of 'communicative rationality' to be set right.

In such situation, civil society, which at the best of time, is only a 'polysemic signifier',[14] can only be a silent witness to the strange co-existences in modern time – namely of, civility and barbarity, ethicality and identity, associative freedom and discipline of laws, procedures and mores, singularity of power and globalization, moral indifference and many truths, and finally geopolitical realities and the victory of restoration after the *annus mirabilis* of 1989. Therefore as some theorists had correctly argued earlier, civil society is a contingent historical phenomenon, and not a natural (and therefore moral) given state of affairs conforming to laws of accommodation, rights, and tolerance. Today's state-guaranteed civil societies reflect conflicts, exclusions, and strategies of incorporation. They show that, democracy has no meta-historical guarantees.[15] The *fundamental transformation of the minority problem into a problem of ethnicity* takes place against a background dominated by economies encouraging greed and moral selfishness and disregard for public good. It is this transformation, which brings out the limits of civil society in harsh tones. Fiscal squeeze, technological ruthlessness, and legitimacy breakdown form the overarching frame in which identity politics plays out its games and conducts 'trials of space'.[16] Classical republicanism hankered for a unified public space. Spurred by liberalization, globalization and identity politics, this classical republican space is now differentiated structurally, so that on one hand we have little chance of a neat solution of the minority problem while keeping the republican space intact, on the other hand an endless series of self-determination through bloodshed, where society can wake up only to the 'grey mornings of tolerance', and the ever receding prospect of autonomy in political life.

We must therefore once again work back to the theme of autonomy as form of dialogue and justice. But contrary to the Anglo-American philosophical ethics, this time we go back to politics – by politics I mean the politics of those need dialogue and justice most, *the politics of the ruled.*

After all the last century in the history of this region as elsewhere has been marked above all by gigantic clashes over forms the political society should take. The politics of forms is nothing but the political issue of how society should shape itself, how it is to be governed, in other words *the politics of the ruled is the politics of autonomy, the politics of finding autonomous spaces in their condition of government.* It is this enveloping spirit of autonomy that coupled with republicanism produces the democratic answer to the conundrum, and takes us beyond the sphere of rights to the world of justice and the freedom to converse and dialogue.

The Real and Moral Assumptions of Autonomy

The ruled, we know, thinks of autonomy in the light of the politics of self-determination. As the previous chapter makes clear, in South Asia as in many other parts of the world today, what engages the attention of international relations experts, policy makers, regional and national elite and the global managers of peace is the issue of national self-determination. Disobeying the established borders of states and the established states-system in the region, the issue of self-determination raises its head in every discussion on war and peace and puts to disarray even the best-laid plans for order in the region. Enormous confusion thus prevails. Is it a case of external linkages to internal conflicts, or a case of kin state as the problematic? Or, is the phenomenon of the assertion of the right to self-determination an instance of cross fire insurgency, where the issue of self-determination arises as a result of diplomacy that takes the form of low level support by the neighbouring states to insurgency wars? Or, is it an unsolved problem relating to the minority groups in the process of nation building – a case of failure of autonomy arrangements? In the absence of a universal treaty on minority rights, which will incorporate a right to autonomy, the right of self-determination often provides the route to autonomy for minority groups, though in a limited legal view this right is available to 'people' and not to 'minority'. The trouble to the states-system is thus compounded when a 'minority' refuses to see itself as a minority, reconcile to that status, and claims the status of a people and thus the right of self-determination – not merely 'internal' but 'external' also. Also we are confronted with an intractable question posed by history – is the right of self-determination over with the completion of the process of de-colonization and formation of the nation-state?[17] What happens when it is argued that the process is not complete, that a right of a nation to form itself cannot block similar right of another nation? All these questions go beyond the issue of *right*, and point to the basic issue – whose self, autonomy for what and from whom, and the right to determine what? They also point to the need for a dialogic agenda – at least the need to consider the question, namely, can the post-colonial polities assume that there is to no need to converse on the claims to self- determine?

By now it should be clear that what I am pushing for here is a thoroughly critical approach not only to international politics but to the practice of dialogue itself. Dialogue needless to say is a relational act. Yet when we say that this is a relational act, we assume that the act indicates only the relation between the dialogists of which the act is the reflection, and in which the act is sited. But to say, that this is a relational act means further. It means that the act puts everything in relation, sites everything within its reach in some relation, and through this step of relating, establishes the 'extra' dimension – the third dimension (of things in relation) – that invites reflection, criticism, and lends the critical edge to the thing relating or being related. By critical approach therefore I mean above all critiquing the underlying assumptions of the realist paradigm that takes the 'reality' in a non-relational way. Thus, to the extent I have been able to examine in the previous two chapters some of the forms of conflict and the geopolitical imagination that lends to these forms an ethnic future, and to the extent I examine in this chapter the existing forms of autonomy, my intention is to put some of the most classical components of political theory in a relational concern – namely notions of self in politics, autonomy, and self-determination – and show how current political theory has failed to put a dialogic politics on the agenda, precisely because it has tended to look at these components in a non-relational way. Two instances should suffice.

First concerns the inadequacy of the realist theory in politics to tackle the issue of self-determination. The realist theory cannot locate in any meaningful way, how in today's world, ethno-politics and geo-politics connect, in other words, locate in a meaningful way the issue of the relation between space and the self, in which the politics of autonomy is sited. Similarly it cannot standardize the right to self-determine (for instance, it can not provide the answer to why X nation can be granted the right of self-determination, and not Y, or who will judge and who has the authority to judge whether X is a nation or not). Similarly again, it is unable to provide direction as to what form this right can take, or what the various forms of this right are under various circumstances. Also, it is still unable to say whether by its own standards the issue of self-determination is a problem within a state or strictly between two or three states, or a matter of international community. The 'morality of states' principles speak of mutual obligation of states to support each other's integrity, the right of states alone to engage in violence for a 'just cause', and their duty to honour agreements.[18] Yet, the realist theory based on a 'morality of states' cannot answer to the fact that there have existed other moralities, as real as the morality of states, for instance the fact that as against the tradition of contract (the basis of inter-state relation), there is an equally long, if not longer, tradition of ethics, justice and moral obligation of the authority. Or, for instance, the morality flowing from group existence has reduced through the ages the legitimacy of a central power such as the state or a concert of states. Indeed, the *relational perspective of morality* subverts or transcends the contractual premise of politics, and demands that the reality of other moralities be acknowledged, only on the basis of which the conversation between selves located in various sites can converse among themselves as to how to determine their respective positions.

My second instance concerns the ethical argument of moral cosmopolitanism, which is equally non-relational in its approach to the political question of accommodation, and is consequently blind to the insurgent reality. Cosmopolitanism is an ancient term and has always identified individual human beings as bearers of human values and makers of communities whose identities are defined by moral relations instead of shared values in practice and who therefore make and belong to one global moral community. The application of cosmopolitan principles is based on normative issues in international politics, such as global distributive justice, international humanitarian institutions in responsibility of international equity, and goal-based obligations of states towards issues of justice, equality and recognition. In such theory of international justice, the demand for self-determination has no need or relevance, as the task is to build a moral global community and the institutional path is to build up global moral agencies. If there is gross intolerable injustice that causes the cry for self-determination, the situation calls for international intervention, because it is seen as one of transgression of fundamental principles of justice. Thus moral cosmopolitanism makes the principle of self-determination antiquated, thereby makes the issue of accommodation, understanding, and dialogue, ethically irrelevant and globally incongruent. The case is compounded by the fact that the moral communitarian argument also fails to provide a satisfactory ground of judgement. Mill's idea of people's right of self-determination and the liberal idealist gloss in Woodrow Wilson's vision had failed in providing a criterion of what is just and unjust in wars of self-determination. If nation-state as the mode of self-determination is the embodiment of the ethical value of self, then is there any ground for intervention for or against another nation's right to self-determine? What is then just and unjust according to the ethical basis of collective right? What is the source of the ethical status of the collective right? And how does one combine the normative character of the right of self-determination with other applicable communitarian norms of international politics?[19] Besides there remains the supreme paradox, if nations win the right of self-determination and no body grants it to them, then this is a supreme realist argument, and there is no need for any moral reinforcement of the realist position. The national right of self-determination then returns to Mill's position namely that, the nation-state resembles human growth from childhood to maturity, and its right to grow is for itself and for nobody to give. It also implies a 'thick' moral stand that communities have the moral right to exist as sovereign power. Such communitarianism only effects a return to the vocabularies of realism, and though beginning with much promise fails to address the problem of negotiating the relations among intrinsic rights of peoples, individual rights, existence of states and the existence a global moral community. To the question therefore, whether morality should guide politics, in this case international politics, the answer is that like the realist assumptions, the ethical theory also in both versions fails to address the issue satisfactorily. For, it does not answer in thoroughly rigorous moral terms, makes compromise with realist assumptions, and therefore cannot cope with what I have termed as *insurgent realities*, of which one is the demand for self-determination.[20] The 'morality of states' and the 'ethics of international relations' thus share the same ground, namely universalistic ground of political conduct that rules out the particular urge of peoples and nations to self-determine their political future, and any need to converse on the form/s that political societies take or should take.

But we must remember that the demand for self-determination, dialogue, and partition have been always closely interconnected, as they were in this sub-continent in the forties of the last century, though we shall require a separate treatise on that interconnection to expound the dynamics of dialogue in a contentious situation. Muhammad Ali Jinnah invoking the principle of self-determination and claiming for the Muslims of the sub-continent the right to self-determine, commented in his presidential address to the Muslim League in Lahore in 1940,

> Babu Rajendra Prasad...only a few days ago said, 'Oh, what more do the Musalmans want?' I will read you his words. Referring to the minority question, he says: 'If British would concede our right of self-determination surely all these differences should disappear.' How will our differences disappear?
> ...The word 'nationalist' has now become the play of conjurers in politics.
> ...The problem of India is not of an inter-communal but manifestly of an international character, and must be treated as such. So long as this basic and fundamental truth is not realized, any constitution that may be built will result in disaster and will prove harmful not only to the Musalmans, but also to the British and Hindus...
> Musalmans are not a minority...Musalmans are a nation according to any definition of a nation, and they must have their homelands, their territory and their State.[21]

Today we have two strange reactions to that event. One, it is a reaction in the form of amnesia, as if formation of Pakistan had nothing to the internationally recognized right of self-determination. That was a nightmare, which the nations in South Asia today are determined not to repeat. But ironically, the second reaction contrary to the first, is that today religious denomination as one of the certificates of popular self-determination has become 'natural', in other words the principle has been successfully naturalized. It is important to note this ambiguity. In the discourse of state-formation, the right of a nation to statehood cannot be denied. But the state discourse being a hegemonic one, it cannot allow the little discourses of nationhood. The ambiguity leads to a naturalization of the principle of self-determination – beyond the 'pale of law', the 'morality of states', the 'ethics of cosmopolitanism', and even beyond the pale of collective right – the principle to be decided as one decides of nature, beyond reason, assuredly in the horizon of given. In such a historical situation we have on one hand a sort of return to the age of natural right, on the other hand, a reinforcement of authority, for in the case of a right not defined, it is always best to leave it to the authority to define, sanctify, authorize, institutionalize and execute the right.. In short, the principle of *natural right* in an ironic way leads to centralization of power, primarily juridical power, because natural rights in the past, and the rights language in general today, cannot but depend on juridical power to become actual. If, as Immanuel Kant had said that self-determination was not an end but a means to other ends, the point surely would be how the right to national self-determination enshrined in the UN declaration of human rights, would relate to other rights – collective and individual. We are back, thus, to the historical irreducibility of the principle of autonomy that appears as a moral one.

In this double appearance of a historically predicated principle of autonomy, the moral and the real, we have the clue to its intriguing relation with a dialoguing politics. It lies in the fact that since rights (the real) always appear accompanied by justice (the moral), they cannot be exhausted even by their most extreme juridical recognition and institutionalization. Justice, un-exhausted by rights, acting as its *supplement*, provides the grounds for re-opening the conversation – the dialogue on the future of political society. The right of self-determination, which is both right and ethics, and whose core is the principle of autonomy with its spatial and spiritual dimensions, thus invites dialogue. The implications of this argument for the decolonized region of South Asia (probably elsewhere also) in terms of a dialogic politics are great and need to be mentioned even if briefly, before we see how polities of the region have tried to subsume the right of self-determination in various institutional forms of autonomy. We can speak of five implications.

First, the experiences of one partition (1947) and one war of self-determination (1971) have made the discourse of state-formation aware of the cracks and fault-lines, and have impelled the official discourse of the state in this region to accept devolution of power as an important part of it. Autonomy, while provoking dialogue, has become in this way a part of governmental technique. Each demand for self-determination is no longer judged in terms of nation building but in the mirror of territorial integrity. 'Nation's geo-body', as the Israeli political scientist Adriana Kemp calls it,[22] produces its own discourse. It becomes the most important part of the discourse of the state, and all other concerns of the state become subservient to the concerns of nation's geo-body

Second, as it has happened in India, .the discourses of rule and government become one in emptying the principle of autonomy of political content. Group rights are thus acknowledged solely as cultural rights, and all economic, political and democratic substance of governance is distanced from the grant of certain group rights.

Third, the strategy of a constitutional devolution of power, through chamber dialogues, as a form of granting autonomy fails disastrously in the face of the claim to self-determination which assumes the form of demanding *homeland*. The idea of *homeland* is markedly different from a claim for devolution of power. The fact, that the demand for homeland has made itself known, shows in most cases that sovereignty has been internalized. A state has emerged or is about to emerge within a state, that the ethnic turn of the self is an accomplished fact. And, the self is ready to achieve its homeland even at the cost of expelling other selves, by little and big mass murders, by driving out neighbours, turning them into the stranded minorities, and committing genocide. Yet, the idea of homeland, as experiences of this region show, does not cause breakdown of the state. It follows and not precedes, the breakdown of the liberal state that had thought that liberal ideas of statehood were enough to accommodate different selves and their desires under an overarching constitutionalism that from its beginning sheltered majority rule, made minorities the victims, and occasionally granted devolution of power.

Fourth, the insistent urge for self-determination, and the equally insistent strategy of the state to subsume this urge in the techniques of devolution make the situation beyond the reach of both a liberal discourse of governance and an ethical discourse of toleration and multiculturalism.

Fifth, the closed situation makes the politics of reconciliation possible, and of making *justice* a part of the politics of recognition an alternative to the discourse of the state. As I make clear in a subsequent chapter of the politics of friendship, it is this opening that impels the dialogue of the political selves, which goes beyond rights, and makes the task of accommodation and understanding a significant theme in the political agenda.

To put all these even more briefly, factors such as the politics of recognition, the cry of the self for the right to determine its agenda, the contractual basis of the state, the widespread internalization of sovereignty, territorialism, the demand of 'homeland' as the route to self-determination, and the consequent *ethnocide* make the situation closed. In such a closed situation, self-determination is redeemed only by the notion of justice that is always suffused with elements of virtue, ethics, truth and reconciliation, in short the values not of state, but of society. Edward Said ended *The Question of Palestine* with these words,

> The Palestinian people today constitutes a nation in exile...For too many years, the official policies of Israel and the United States...assumed that the Palestinians would fade into the Arab world, that Jordan would become Palestine, that Palestinians would accept permanent subservience under a Bantustanlike 'limited

> autonomy' (or as the Likud formula has it, autonomy for the people – not for land), that the people may even be willing to perform an act of politicide on itself and declare itself null and void...Nothing less than Palestinian self-determination will do; and only that will ever defuse the already far too explosive Middle East. Yet some Israelis and non-Israeli Jews have in fact understood that if Israelis and Palestinians can have any decent future it must be a common one, not based on nullification of one by the other. In 1988, we Palestinians as a people took a giant step towards reconciliation and peace. We now await a corresponding gesture from the Israeli people and its government.[23]

Political Forms of Autonomy in South Asia

We now begin the rapid and broad survey of the political forms of autonomy in South Asia, and find out the lessons that it holds for us in the light of contradictory features of the dialogic situation outlined above.

Bangladesh

In Bangladesh the British had enacted for the Chittagong Hill Tracts (CHT) the CHT Regulation Act of 1900 that had been preceded by the Act XXII of 1860 whereby they had introduced indirect rule in the area by declaring it an administrative district. Introduction of such autonomy provisions was not uncommon in colonial rule when the colonial authorities had combined a variety of methods for administering the country. Thus, inner line regulations, declaration of excluded and partially excluded areas, quasi autonomy in land regulations and dispensation of justice were some of the methods that accompanied the introduction of rational administration over the entire sub-continent. In the CHT the powers of the local chieftains were gradually whittled down and by 1937 the safeguard on immigration was withdrawn, in 1937 the role of hill chiefs was reversed, and they now only advised the government on policy matters. And finally in 1964, the special administrative status of the hill tracts was repealed by an amendment to the constitution of Pakistan. The constitution of independent Bangladesh does not recognize minorities as groups distinct from the Bengalis; everyone is a 'Bangladeshi'. And though Article 23 enjoins upon the state to conserve cultural traditions and heritage of the people, again the purpose is to enrich 'national culture'. It was a situation of severe loss of autonomy compounded by the complexities of partition of 1947. The construction of the Kaptai dam affected about one hundred thousand people, inundated 400 square miles including 54,000 acres of arable land, rehabilitation and compensation was minimal, and the demographic balance also changed in a short period dramatically. In the Hill Tracts whereas in 1947 indigenous population was nearly 97 per cent, in 1991 the rise in Bengali population was from 2 per cent to 48.5 per cent. The re-introduction of provisions of autonomy through the Chittagong Hill Tracts Agreement in 1997 following the almost twenty year long insurgency goes halfway in ensuring the right of the people there to self-determination.

District local governments in the three districts of Rangamati, Khagrachari, and Bandarban and the regional council were introduced after the enactment of the Implementation Acts in 1998. But the enactments do not have constitutional guarantee. They have deviated from the agreement, the Acts has incorporated almost all Bengali settlers, and the government has refused to amend the acts to ensure the provisions of the accord. The regional council does not enjoy legislative power, except on traditional 'tribal' laws. The stipulated land commission to undo the wrongs of the past regarding the ownership of land and hills by those who had illegally settled and occupied has not yet started functioning. A great number of settlers have become 'non-tribal permanent residents'. Fiscal and economic autonomy is limited to taxation of local trade and items such as land and buildings, holding tax on industry, part of royalty from forest resources, fishing, supervision of local health, education, transportation. Autonomy in cultural affairs is predicated by the use of Bangla as the only official language. Yet, one must remember that this limited measure of autonomy, legally but not constitutionally ensured, is not available to other minorities spatially concentrated or dispersed. Thus, for indigenous people in other areas of the country, or for religious minorities dispersed in the country, (as in the hill areas) the constitution does not provide any measure of economic, cultural, or political autonomy. On the contrary, the country still retains the act related to vested (earlier enemy) property whereby an unaccounted amount of land (nearly ten per cent of total cultivated land) has been transferred by force or threat of force from the Hindus to many Muslim landowners/peasants/speculators and profiteers. In an overwhelmingly Muslim and Bengali country whose constitution proclaims Islam and Bengali as two main characteristics of the nation, the state only exceptionally recognizes rights and identities of other groups. The land rights of plains and hill indigenous people are often unrecognized. Independent surveys complain that the number belonging to these groups is often underestimated. The 1981 census estimated the number to be 6,23,216; the figure was revised in 1989 to 8,97,828, and the census of 1991 estimated the figure as 12,05,700 – 1.13 per cent of country's total population. The land wealth (often the only wealth) of these groups further is now whittled down by market forces, such as the ADB (Asian Development Bank). Displacement, deforestation, dispossession of land, and the very weak nature of decentralization symbolized by the frail local councils, provide the backdrop in which autonomy remains provisional, exceptional.[24] Certain powers have been given or conferred and devolved. But the power to self-rule, the essence of political autonomy, remains unrealized.

Nepal

In Nepal, about 20 per cent of the population belongs to hill groups such as Magar, Tamang, Rai, Gurung, Limbu, Sherpa, Lepcha, Chepang; there are the indigenous people of the Terai, the *janajatis* of the foothills, such as Tharu, Dhimal, who may constitute another 10 per cent of Nepal's 18 million population. The Muslims also live in the Terai region. Religion-wise, Hindus constitute 86 per cent; the Buddhists are 8 per cent and Muslims about 4 per cent of the people. Nepali is the first language of about 50 per cent of the population, and in 54 districts out of a total of 75, it is the first language.[25] Yet in this polyglot and poly-ethnic society, dominated by the upper castes and the Kathmandu valley, there is little presence of any notion of autonomy. Autonomy from Indian domination for the nation of Nepal, yes, but the idea or any arrangement of autonomy for the regions, groups, and nationalities is absent from the constitution. And this is important to note that the movement for democracy and the democratic exercise in constitution making did not usher any sense of autonomy for various sections and group rights. Many demanded a secular state; there were also demands for making the constitution-making process more broad-based by including members of various nationalities, but the transition to democracy represented a typical liberal process oblivious of the problem of plural selves in society, and largely reminiscent of the Indian constitution making process. One observer has reflected on the process with these words,

> The politics of constitution making was characterized both by an unprecedented euphoria and by a sense of frustration among ethnic communities...though among the nine members of the Constitution Recommendation Commission...six were Bahuns, and three were one each from the three ethnic communities – Newar, Lama and Tharu. Yet, they could not be called as ethnic representatives because of their own social status and recognition as politician (Lama), Attorney General (Tharu) and Chief Secretary and Chairman of Public Service Commission (Newar).[26]

The constitution of course declares Nepal as a 'multi-ethnic, multi-lingual, democratic, independent, indivisible, sovereign, Hindu and a Constitutional Monarchical Kingdom'. But even the inadequate recognition of plural selves is diluted further in several articles such as Article 2 that declares the people to be united by a bond of common aspirations and faith in the integrity of the nation, or Article 20 that says that the state while maintaining diversity will adopt a policy towards strengthening national unity. There are several administrative arrangements for the improvement of the *janajati* communities. But these are welfare measures, at best they succeed in producing a 'creamy layer', at worst they institutionalize corruption in one more way. Exceptionally the can become part of a devolution strategy. But all these have nothing to do with autonomy or federalization. Though by standards of the region the degree of tolerance and consciousness about pluralism in Nepal is more than that in other countries in South Asia, yet with no decentralization and no devolution of power, and with extreme conditions of deprivation, the movement for self-determination has already begun. It is difficult to tell how the political ideas of the popular insurgents will develop on this issue, but there can be no mistaking the fact that apart from the classical promise of rights of various nationalities, *this movement is itself one of self-determination*. In conditions of semi-feudalism in the vast countryside, the immiserization of small peasantry, the exploitation of rural poor, and the absolutely non-productive presence of a comprador business class, the battle for extending democracy to the regions, sub-regions, and villages cannot be but one of self-determination. By drawing democracy out of its parliamentary limits and making agrarian democracy the key item in the agenda of the democratization of the country, the principle of self-determination will assume a new form. It will enrich the experiences of autonomy in South Asia.[27]

Pakistan

In Pakistan, typical of the state of contemporary political vision, aspirations and demands for democracy have been viewed as 'ethnic' demands; in more sophisticated but actually shallow language, it has been termed as 'rise of ethnicity'. Take again the plural composition of the country. Punjabis are 68 per cent, Sindhis 13 per cent, Pathans about 9 per cent, and the Baluchs form about 3 per cent of the country's population. In terms of political units, Punjab has 56 per cent, Sind 27 per cent, North Western Frontier Province (NWFP) 13 per cent, Baluchistan 5 per cent and the Federally Administered Tribal Areas (FATA) have 2 per cent of the population.[28] Yet Pakistan carries the legacy set in 1955 when various provinces of West Pakistan were integrated in one single unit to counter the Bengalis who constituted 55 per cent of the Pakistan's population, and to kill democratic movement in West Pakistan also. It is a legacy strengthened by the suppression of the movement for self-determination in Baluchistan, the repression and massive killings of the Bengalis who demanded autonomy and later on independence, and the repeated take over of power by the military oligarchy.

The constitution of 1973 failed in protecting the autonomy of the minority provinces such as the NWFP and Baluchistan. The Council of Common Interests (CCI – Article 152) apart from helping some inter-state dispute on water sharing failed in protecting the autonomy of provinces. The National Finance Commission (NFC), another constitutional body set up under Article 160, similarly failed in protecting financial autonomy of the states. In any case from 1973 to 1991 the CCI and the NFC did not have any meeting, and the constitution itself was not functional for much of this time. And reminiscent of the experiences of India and Nepal, articles in the constitution ritually spoke of preservation of language, script and culture (Article 28), protection of minorities (Article 36), promotion of social justice (Article 37), at the same time the same constitution ensured loss of provincial autonomy, domination of one language group and one province, and an unprecedented centralization of power – a process capped by repeated assumption of all powers and authority by the armed forces. Therefore whatever limited provincial autonomy was given in the backdrop of the struggle for self-determination of the Bengalis became even more limited by the absence of any measure of autonomy for the minorities. Indeed, the constitution does not even recognize the non-religious minorities at all.

Today the enlightened political class in Pakistan speaks of a 'new social contract'. But the question is, contract with whom, and in what form? Will it be contract with various groups, provinces, and constituencies of the democratic masses, or a contract with sections of the political class to undertake once again the failed task of nation building?[29] And what will be the form of this contract? If federalism were enough, then why could it not ensure the rights of the different categories of minorities? Again, how will the regions get autonomy? And finally how will the country be protected of its democratic autonomy from the repeated invasions of the army-bureaucracy-comprador class? The country's 1973 constitution was more liberal than the preceding attempts – but this poses two things of critical concern. First, the repeated failures to have a stable constitution show that without democracy for the parts and groups as well as for individuals, constitutional stability remains an uncertain prospect. Second, liberal constitutionalism that the 1973 constitution wanted to usher in will not be the path. An overriding truth of South Asia's political history is that democracy comes through the liberal path, and the limits of the liberal constitutional path soon endanger democracy itself. No wonder that many Balochs, Pushtoons and Sindhis have argued that not federalism but con-federalism has to be the form of political organization where the republic would control defence, foreign affairs, communications and currency, and the confederative states would have the rest and would have their own militia with also the constitutional right to withdraw from the con-federal union if the republic or its armed forces violated the principle of the confederative structure.[30] Whether such a structure will be the most suitable form that would embody democracy is difficult to tell. Surely it would be an advance upon the present state of affairs. However, future will tell if that would be sufficient to ensure democracy within the unit and guarantee against a strong local bully. But the Pakistan experience certainly shows the multiple ways in which the principles of democracy, federalism, autonomy, self-determination and minority rights cut each other's path, and that a neat congruence can be a rare thing in actual political life.[31] The experience of 1971 was imbued by the nation, nationalities and classes each in their own way. What prevailed over many realizations and therefore many discourses, was that East Pakistan in any case was an aberration with the two parts of the country thousand miles from each other, and after all the Lahore Resolution had envisaged many Pakistans. And now with the aberration removed, though unhappily for the nation, chances have come to build a strong integrated country. No doubt, there are other visions, but at present the dominating discourse seems to be this.

Sri Lanka

In Sri Lanka the rulers have traditionally thought that if liberal principles of multi-party democracy, parliament, independent judiciary and a written constitution are reinforced by a devolution package, then the issue of self-determination will be successfully negotiated. Indeed, the fundamental question arising from Sri Lanka's political history is, can constitutional reform become a successful instrument for settling demands for self-determination? In cases such as Sri Lanka where the national groups are not given an equal status and are not a party to decision-making process, and a certain unilateralist politics prevails, what worth are the constitutional reforms? Constitutional reforms that do not come as part of a reconciliation and peace process are exactly like the imperial reforms that the colonial authority brought to India from time to time, failed to satisfy the urge for autonomy or self-determination, and in fact fuelled the urge further. The Morley-Minto Reforms, Montague-Chelmsford Reform, recommendations of the Simon

Commission, the India Act of 1935, and the Cripps Mission – all these were constitutional reforms through the imperial process. The post-colonial polities of the region continue with the imperial tradition of a unilateral constitutional process. Thus consociationalism, confederacy, autonomous political regions, and other forms of shared sovereignty cannot come without a vigorous political process which is essentially an extra-constitutional process but which jurisprudence can allow for the former to achieve the status of a basic law. All major constitutional amendments require some special forms of process that can range from referendum, two-thirds majority in parliament, and special ratification by a specific number of states, and judicial advice. Constitutions being mostly conservative documents and asymmetrical documents, they represent interlocking structures. In other words, amendments to the constitutions to secure minority rights, or rights of different groups to justice are extremely difficult to achieve for either the constitution in question does not have symmetry among groups and estates of the society, or an amendment requires an entire set of revisions in the structure. Therefore it will be wise to remember the lessons of the constitutional history of this region in the context of the emerging calls for expanding democracy, that a revision of a political process precedes that of the constitutional process, and only a revision of the moral foundations of the constitution can propel successful constitutional reforms. The histories of engagements with the principle of autonomy will point to this truth that remains neglected owing to the massive power of the governing class, which retains huge interest in maintaining the basic law in the way that has brought it to power. Indeed, the constitutional history soon reaches a crossroad in the wake of the persistent cry for self-determination – either a constitutional counter-revolution that happened in Pakistan and Sri Lanka, and may now take place in India, or a revision of the moral basis of the constitution in the direction of democracy followed by reforms.

In Sri Lanka, constitutional denial of autonomy and political, cultural and physical repression on the non-ruling nationalities, particularly the Tamils went together and made the situation intolerable for the latter. The Sinhalese language became the sole official language in 1956. Before that the Soulbery constitution, that guided the country from 1948 when the country gained independence to 1972 when the republican constitution was introduced, in the name of equality of all communities had made positive discrimination of any community very difficult (Section 21, sub-section 2). Thus protection of minorities was completely dependent on the rulers from the beginning. Even the thirteenth amendment to the present constitution brought in the wake of the Rajiv Gandhi – J.R. Jayewardene pact in 1987 entailing provincial constitutional councils and a limited devolution of powers was cancelled later. Thus the constitution reflects no awareness that this is the basic law of a society where the Sinhalese are 74 per cent (of this 74 per cent, almost 70 per cent are Buddhists), Tamils 18 per cent (of this 15.5 per cent are Hindus) and Muslim minority groups constitute 7 per cent of the population. Of the Tamils constituting 18 per cent of the country's people, 12.5 per cent are the so-called Sri Lankan Tamils and Indian Tamils 5.5 per cent. The Senanayake government had curtailed citizenship and voting rights of the Indian Tamils in 1948 and 1949 through a series of legislations. The new republican constitution of

1972 abolished whatever safeguards had been accorded by the Soulbery constitution. It not only declared the Sinhalese language as the only official language, it granted special status to Buddhism. It abolished the second chamber one of the reasons for the foundation of which was that the body could act as a restraint against discrimination of minorities. The system of nomination of representatives of under-represented groups in the Lower House, somewhat like the Indian system, was also abrogated. The 1978 constitution gave Tamil the status of a national language, accorded statutory recognition to its use in various fields of administration and education, abolished the distinction between a citizen by descent and registration that helped the remaining Indian Tamils to some extent. But all these concessions continued with unabated violence against minorities, riots, the pogrom of 1983, and full-scale military offensive against own citizens with the help of a foreign power.

Even the very limited devolution under the thirteenth amendment was whittled down in conditions of extreme executive centralization in Colombo, the political manipulation of various Tamil groups by the government in the name of running the councils, the revocation of the sixth amendment of the constitution, and then the dissolution of the council in the north-east. With virtually no autonomy and fruitless clamour for democracy for years, the Tamil demand for self-determination arose. The history of the Tamil self is as important as the foregoing story of the constitutional denial of the recognition of the self. The Citizenship Acts of 1948 and 1949 had made the vast majority of the Indian Tamils stateless. Out of eight lakh Indian Tamils, only 16 per cent got citizenship, rest had to melt away in the vast plains of the south Indian countryside under the Sirimavo-Indira pact and through other means. Tamils formed the Federal Party, which objected to a single national flag, and called for a federal union with an 'autonomous state' for the Tamils. The deliberate and non-deliberate settlement/migration of the Sinhalese in the north and the east, the language policy, statelessness of a huge number of people, racial discrimination, and riots completely destroyed faith of the Tamils on the successive constitutions. The policy of the Sri Lankan state to make domestic politics ethnicity-oriented in the name of redressing balance between various groups bore the poison fruit by early eighties of the last decade. The Bandaranaike-Chelvanayakam Pact that had provided for recognizing Tamil as a minority language and the dominant language of the Tamil areas and had envisaged a plan for regional councils was withdrawn on the pressure of the Buddhist clergy. Again, we have a similarity here with a less-discussed chapter of Indian history when the C.R. Das-Fazlul Haq Pact, popularly known as the Bengal Pact was repudiated by the Hindu-dominated Congress soon after Das' death in the second half of the twenties of the last century, and that abrogation marked the irrevocable phase of communalization of Bengal politics. The Federal Party withdrew from the government in 1968, the radical Tamils censured and threatened the Tamil M.P. s who had collaborated in the enactment of the 1972 constitution that denied any recognition to the minorities, and Tamil consciousness reflected in the merger of the Federal Party and the Ceylon Workers Congress in the form of Tamil United Front and later the Tamil United Liberation Front.[32] Even this was not enough. The radicalization of the Sinhalese youth reflected in the struggle of 1971 had impact on Tamil youth also. The collaborationists were denounced, TULF had to give the call for Tamil *Eelam*, and this spurred the Tamil youth all the more for decisive action doing away with rhetoric into which the constitutionalism of the country had pulled the Tamil politicians and had hitherto subsumed. The rest of this history is well known.

What is not acknowledged is the presence of a pattern, the uniformity of practices of self-determination in the region, and the sordid history of constitutionalism in countries of South Asia. Elsewhere, as in Sri Lanka, constitutionalism was a failure. Elsewhere, as in Sri Lanka, democracy was perceived in the ethnic glass. Elsewhere too, the state was most responsible in conducting politics with the ethnic baton. Everywhere it was and is still thought that a constitution, party system, some devolution of power would solve the issue of 'separatism'. Everywhere, under these conditions, the self has become an ethnic self and the right to self-determination has thus become an assertion of ethnicity. Thus we have greater Nagaland, greater Mizoram, or a greater Tamil-land consisting of the north and east of Sri Lanka, or greater Pakhtoonistan. And everywhere this assertion is accompanied by killings of 'moderates' by 'radicals' in the movement – a fury against constitutionalism, then a solidified ethnic divide, a determined campaign to secure 'homeland' by killing some and expelling others from the promised land, and everywhere then it is a closed world.

What is also not acknowledged but is implicit in the political discussions is that, constitutionalism and the political demand of the self to determine to its destiny run like the hare and the hound never able to catch each other until one is able to meet up with the other for a murderous engagement. To take a leaf out of Sri Lanka's recent constitutional history again, Article 9 of the present constitution of Sri Lanka says,

> The republic of Sri Lanka shall give to Buddhism the foremost place and accordingly it shall be the duty of the State to protect and foster the Buddha *Sasana*, while assuring to all religions the rights granted by Articles 10 and 14 (1e).

What does Article 10 say? It declares, 'Every person is entitled to freedom of conscience and religion, including the freedom to have or adopt a religion or belief of his choice'. In other words, Tamils have the right to be Hindus, but the state will be Buddhist. Yet this is not surprising. The amendment bill to repeal and replace the constitution towards correction of historical inequalities in the Sri Lanka society, win back the confidence of the Tamils, and drafted on the basis of concurrence of the Sinhala political class, redrafted the above article thus in the form of Article 7 (paragraph 1),

> The Republic of Sri Lanka shall give to Buddhism the foremost place and accordingly, it shall be the duty of the State to protect and foster the Buddha *Sasana* while giving adequate protection to all religions and guaranteeing to every person the rights and freedoms granted by paragraphs (1) and (3) of Article 15.

And then it said further as if stung by conscience (paragraph 2),

> The State shall, where necessary, consult the Supreme Council, recognized by the Minister of the cabinet of Ministers in charge of the subject of Buddha *Sasana*, on measures taken for protection and fostering of the Buddha *Sasana*.

And what do these two paragraphs of Article 15 say? Paragraph 1 says,

> Every person is entitled to freedom of thought, conscience and religion including the freedom to hold opinions and to have or to adopt a religion or belief of the person's choice.

Paragraph 3 says,

> Every person is entitled to the freedom, either alone or in association with others, and either in public or in private, to manifest the person's religion or belief in worship, observance, practice and teaching.

So, constitutional patronage of the majority religion remains as unambiguous, only the freedom to practice one's religion is announced in a more detailed manner. In such situation the fate of a constitutional politics in its mission to win back loyalty of other selves to the state of Sri Lanka should not cause any surprise. Thus, each round of political activism is followed by a corresponding round of constitutionalism that only serves to impel the next round of activism to press its demand of self-determination. And this double narrative, as the experience of Sri Lanka and more loudly the Indian experience show, is strewn with broken promises, broken accords, umpteen revisions of the constitution, with the general lowering of the level of trust in the country.

Two related issues here merit critical analysis. One, constitutionalism is so much *naturalized*, that is to say, appears as so much natural, that we take autonomy as a lower tier whose higher one is self-determination. Thus if arrangements of autonomy succeed, there will be no need for the latter, and if they fail, then the latter course becomes regrettably inevitable. In this way it becomes the pretext for the continuing text of constitutionalism, also our excuse of not analysing rigorously the entire right itself whose manifold form calls for investigation.[33]

Second, and this is the more significant issue to think over, what would have happened if accords held forth, if promises were kept, if the incipient experiments at accommodation were given chance to grow? This question is actually one of critically looking at the 'narrative closure' of the process that has already happened. It appears, as if this is an 'accomplished history', rather than an open 'historical process', that is to say, open towards the future. Such a narrative closure holds sway over the common perception of that history and prevents us from seeing possible alternatives. This is a sort of 'squeezing' or 'erasure' of the present dimension, that is of the life experiences (and of the political subjectivity and citizenship) of people who have lived through those events. In the constitutionalist narrative that closes the possibilities of the process, the experiences of political subjectivity count for nothing, time-count starts from zero, fifty plus years of life of this region wiped out by such selective memory, which is what allows (among other factors) for the repetition of the cycle of violence and constitutionalism. Once legitimated, violence reproduces itself. Violence also produces the soon to turn into a stereotype the division that the 'community' is just, and the 'nation' that brings in its wake disorder, riots, violence, and irrationality is unjust. The more this categorical distinction takes root, the more the nation-state wants to become the community by making the polity communitarian, communal. Society, a multi-dimensional open space, loses out from the discourse. Community, often a vertical non-democratic construction, wins. Communal violence is justified in the name of the community that is seen as trying to break away from the 'national state', but to produce what? The answer is, state of their own nation that has grown out of their community. There is almost no way to stop this fatal movement, ironically presided over by constitutionalism, towards killings, genocide, and divisions, other than by investing in critical appreciations of other possibilities, which I shall be terming as 'contra-history', which is the other name of historical possibilities residing within history to go against that history.

India

In India the political history of decentralization offers the widest variety of constitutional forms of autonomy in this region, the capacity of these forms to contain the urges of many selves, but also their utter incapacity to go beyond their 'container' role and become instruments of democratization of the country. Indeed, the entire paradox of self-determination, narrated above, is nowhere more evident than in the Indian experience. The Indian experience is most instructive because of its diversity and range, the full extent of colonial innovations, multiple forms of autonomy, the complex path of constitutionalism, a wide the variety of accords, the persistent cry for self-determination, and an unyielding and innovative state determined to keep the destined nation intact while keeping others from becoming nations. In an earlier work, *A Biography of the Indian Nation, 1947-97* I had described the political-constitutional way in which the minorities had been negotiated by the state – the legal forms and the forms of accords.[34] It will be instructive to see in that context at least the rough contours of the double narrative of constitutionalism and the insistent demand for self-determination.

The constitution provides for special status for certain states such as Jammu and Kashmir, Nagaland, Sikkim, Assam, Manipur, Arunachal Pradesh in Articles 370 to 371H. The constitution also embodies the principle of non-discrimination in Articles 14, 15, 16, 19 and 29. It assures freedom of conscience in Article 25 and freedom to manage religious affairs in Article 26. Article 30 ensures right of minorities to establish and administer their own educational institutions. Under the special protection clause in Article 371, tribal customary laws, procedures, and land rights are protected. Part XVI ensures special provisions for scheduled castes, scheduled tribes and other backward classes. There are arrangements for zonal councils. The States Reorganization Commission ensured statehood for major linguistic groups. There is provision for autonomous district councils in scheduled tribe dominated districts. The 73rd and 74th amendments to the constitution ensured devolution of powers at village and town level. Similarly the constitution arranged for financial autonomy of the states through constitutionally prescribed division of resources and the National Finance Commission. Apart from creating new states (some very recently created) and autonomy for some states in particular, a range of accords and unilateral measures on Darjeeling, Bodoland, Leh, North Cachar Hills, Karbi-Anglong district, Khasi district, Jaintia Hills district, Tripura Tribal Areas district, Chakma, Mara and La districts in Mizoram, created autonomous areas and district councils under the fifth and sixth schedules.

The pattern of combining nationhood with exceptional autonomies is significant. Is autonomy part of the basic features of the constitution that the Parliament should not touch? There is no clear answer whether the provisions of autonomy are inviolable or not in the context of the erosion of Article 370 providing for autonomy of the state of Jammu and Kashmir. In any case, autonomy never captured the republican imagination. Thus the Supreme Court never had autonomy in mind when commenting upon the in/violability of the basic features, that primarily meant fundamental rights, which have only one among it dealing with autonomy.[35] The republican mood sways the overwhelming part of rights. Thus provisions such as Articles 14-16, (again combining exceptional discrimination on positive grounds), Articles 22-23, Article 25 (combining exceptional right), Article 29, Articles 38-39 defining *common* welfare, securing *common* good and indeed laying down the constitutional basis of a welfare state, Articles 46-47, Articles under Part IX (the *panchayats*) – create a polity based on republicanism that while allowing autonomies will be the ethos of one nation, one people, and one land. Therefore provisions such as Article 244 (administration of scheduled areas – fifth and sixth schedules) cannot counter the wave of majoritarianism that arises out of the republican spirit.

There is more to this unequal co-existence of nationalism and autonomy. For example, there is no uniform civil law, on the contrary a variety of personal laws, and linguistic autonomy in some measure. Indian constitutional and political system has evolved through at least seventy years history of a range of forms of autonomy – administrative, cultural, religious, fiscal and legal-juridical. Yet, demands for right to self-determination ranging from more autonomy to secession have arisen frequently, and if some have mellowed, others have persisted and have grown insistent notwithstanding massive state-suppression and loss of lives. It began with the Muslim demand for self-determination in the pre-independence time and continues in various forms and at various levels still today. The constituent states have said that their legislative, administrative and financial autonomy is inadequate or has diminished. Kashmir says its autonomy is fictive. Insurgents in the northeast have said that grant of statehood is a ploy to subsume them in Indian polity. Religious minorities say that they are under unprecedented attack of the fascist communal forces belonging to the majority community backed by the state. The scheduled castes and tribes say that their deprivation, poverty and disempowerment have only grown. The legal-administrative measures for protection of autonomy such as the Minorities Commission, Human Rights Commission, Women's Commission, are severely limited in their powers. These national commissions have their state counterparts even more limited in powers and functions. So are weak and inadequate the commissions in the states for protection of minority languages and cultures, interests of scheduled castes and tribes. In short, we have in the Indian instance the most extra-ordinary juxtaposition of measures of autonomy and a relentless centralization – seen from another angle, the most relentless constitutionalism and the most insistent cry for the self to achieve recognition. It is also a narrative of how and when a group refuses to accept at some historical moment the identity of a *minority* and claims the status of a *people*, a *nation*.

This however should not cause us surprise. For along with the discourse of modernity and liberalism we received from the West the category of the minority also, a gift from Europe wrestling with the mode of dismantling the Austro-Hungarian empire, and settling once for all the 'eastern question'. At a time when our people were learning how to build a popular policy of anti-colonialism, building our nationhood, we also learnt how to be majorities and minorities, and thus we produced not one nationalist discourse in the colonial world, but several nationalist discourses. In the midst of this double learning, we arrived at the historical moment of de-colonization. We shall need thus a detailed narrative to show how constitutionalism in the beginning held promises of accommodation, reconciliation and friendship, how the trust broke down quickly because the arrangements collapsed under the pressure of state-discourse, then the long defiance, state suppression, equally resistant and often brutalizing struggle, escalation of homeland demand, and a renewed attempt to contain all these within constitutional confines. The official discourse of the state looks at all this as the republic's engagement with forces of separatism and the attempt of liberal democracy to innovate ways of nation-building. However, to the discourse of those who want their homelands at any price or to part ways completely, it is a battle with constitutionalism which to them is the rock-bed on which renegotiation of the polity has floundered. Of course, most have the feeling that this is a road that meets the 'closed' notice sooner or later. Hence is their anguished appeal for justice, their mistrust on institutions of autonomy, their plea that the urge of the self to gain recognition be judged in the light of democracy, their despair at the double-edged weapon of ethnicity so ruthlessly turned against them now by a superior adversary, and hence is the cry for a better understanding (which most think can come only from the international actors who are non-state and hence 'disinterested') of the genealogy of the bind that has now trapped them, the inferno that they know threatens to consume them, burn them all. The Indian experience of the double narrative is thus also one of return of the theme of justice and democracy to the heart of the argument of determination.

In short, 'the Indian paradox', to borrow a term from late Myron Weiner, has been most evident in the tortuous history of legal negotiation with the notion of autonomy on the basis of certain constitutional principles which have not been adequate for that purpose. Marc Galanter had commented thirty years ago,

The modern legal system has transformed the way in which the interests and concerns of the component groups within Indian society are accommodated and find expression. In traditional India, many groups (castes, guilds, villages, sects) enjoyed a broad sphere of legal autonomy, and where disputes involving them came before public authorities, the latter were obliged to apply the rules of that group. That is, groups generated and carried their own law and enjoyed some assurance that it would be applied to them. In modern India we find a new dispensation – the component groups within society have lost their former autonomy and isolation. Now groups find expression by influence in the political sphere, by putting forth claims in terms of general rules applicable to the whole society. The legal system, then, provides a forum in which the aspirations of India's governing modernized western-educated elite confront the ambitions and concerns of the component groups in Indian society. In this forum the law as a living tradition of normative learning encounters and monitors other traditions of prescriptive learning and normative practice.[36]

We can argue that nationalism and democracy has broadened the public sphere and the disadvantaged within the group has now access to justice previously denied to them. Also it can be said, that in Galanter's argument only one kind of autonomy looms large. Primarily concerned with battles within 'Hindu' society for legal justice, he neglected the issue of autonomy of other kinds of groups. But the predilection is clearly shown. Do we have a differential system of justice, or an equal system of public justice with one or the other group dominating the public and turning its norm into its favour? The predilection further is, if we have a differential system, will it be competitive or cooperative? Since a modern constitution is based on the republican system of an equal public sphere that sees autonomy as an exception, how will such a basic law inhere autonomy as an integral principle of democracy? Also important is the question, how will that equality be pursued in a land of hierarchy?

Indian law permits different family laws on religious lines, even permits different public laws according to different religions on matters like religious trusts, permits compensatory discrimination in favour of disadvantaged groups, and is sometimes extremely solicitous of religious sensibilities. The broad regulative powers that the state has ('subject to public order, morality and health' vide Article 25.1, and Article 25.2a) are rarely comprehensively enforced. The result is a paradox: we have on one hand a publicly equal system with broad state powers to regulate practices of separate identity so that they do not go against equality, we have also differential provisions to help the disadvantaged, and then besides these two features we have a public system accessible to a group determined to impose its values in a large or total measure thereby almost equalizing the *public* and *group* interest. In such situation, as Galanter shows, the stress on judiciary is excessive. The political failure is sought to be compensated by judicial activism to the extent where a judge is compelled to define 'who is a Hindu', or the boundaries of faiths.[37] Galanter goes further. In analysing what he termed as 'symbolic activism', he shows, how compensatory measures sit so unhappily with the broad doctrine of equality as a fundamental right. Should these measures be seen as defining equality, so that the court should now force compensatory measures, or should they be seen as guidelines to the state towards making un-equals equals?[38] The implication is that, if we take the first view, a view that bifurcates Indian society into two broad groups – the scheduled castes and tribes, and other dispossessed minorities on one hand, and the rest on the other, we cannot have a republican constitution promoting public politics. If we take the second view, that is prescriptive policy to the state to provide succour to some disadvantaged groups, we herald an unending series of demands for classification on the basis of which the discrimination will be made. Both possibilities show the dilemmas of a liberal constitution trying to grapple with the issue of inequality and autonomy. This engenders what I term here as 'constitutionalism' that ironically produces even more inequality, dispossession, and clamour for autonomy. The *politics of recognition* has to be seen in this context.

Finally, one can take the case of caste as an example of the impossible politics of a constitution producing an enduring form of autonomy in India. Though the constitution finally did not include a whole series of draft enunciations relating to lower castes and tribes that aimed at defining them as minorities, and had suggested an entire separate part for this in the constitution, clearly the text as we find it today aims to define the place of caste in Indian life and the role of law regulating it. The constitution takes a dim view of the place of caste in Indian life, is ostensibly not concerned with that, the ties of ascription remain beyond its domain; but it sees itself as the fundamental instrument to ensure that these ascriptions do not lead to hierarchy, inequality and invidious treatment in public life. Caste therefore with its own internal order and rule promulgating powers and functions may continue as an *autonomous association*, but this autonomy is supervised so that this does not spill over into public politics. Doctrine, ritual and culture – all remain outside the juridical bounds till these affect the constitutional mission of promoting equality. If they do, court and the law become active. In such a perspective, caste becomes both a religious body in the sense that its own prerogative on such matters such as marital rituals, devotional methods, or representation to bodies like relevant commissions is allowed, and a non-religious body in the sense that the constitution tries to detach it from the wider perspective of the Hindu society and determine its character (advantages and disadvantages) among Muslims and Christians as well. But the institution of caste securely tied to the 'Hindu world', few can opt out of a caste-fold and receive protection. The entire controversy over the right to be converted to another faith shows the limits of a constitution in upholding the autonomy of an individual or a group to choose faith.[39]

In short, the Indian instance in an acute manner shows the South Asian experience, namely that, modern law is not a self-fulfilling prophecy working towards the satisfaction of the political subject. Faced with asymmetries of power it can manage at best the co-existence of various normative orders, and at worst it can become a willing accomplice to the manipulation of the public principles by a particular group. It also shows that beyond the given territorial form of autonomy, South Asia by and large has not been able to discover other forms of autonomy that will combine both spatial and spiritual dimensions. The resultant situation displays the existence of two political idioms. There is a liberal republican idiom of democracy originating in the constitutional exercises of 1946-49 in India (elsewhere in this region at different times) that tolerates autonomy and dissent to a defined extent. There is also a politics of recognition fundamentally expressed in territorial form, pre-existing and now reinforced by the same liberal idiom, which threatens to break out and indeed sometimes breaks out of the imposed confines, resulting in a renewed phase of constitutional frenzy. These two are the quarrelsome duo, one not being quite able to displace the other. Together they have produced the uneasy reality of accommodation, which is not tolerance, but an accommodation of competing realities till one succeeds in pushing the other out of existence. But this situation, as I have indicated earlier, impels the issue of justice to surface again and again, places the question of a new dialogic order at of the core of the issue of self-determination.

Experiences of the Political Forms of Autonomy Elsewhere

In the context of the South Asian experiences, we need to ask a relatively innocuous question, why do some movements for self-determination persist and others vanish, that is to say, why do some forms of autonomy (at least seem to) work, and some prove unworkable and collapse? In other words, can we have a broader understanding of the practices of self-determination in terms of a politics of accommodation? At the same time, we have to remember that, what is a success-story today may not remain so tomorrow. Kashmir was a success-story in sixties and seventies and but not in the nineties of the last century. The union of England with Scotland was till recently considered a success-story, but has to be re-negotiated anew. What is important is at times people may start demanding the redress of historic injustices that they perceive as still continuing in various forms, and when they demand, whether the scope of re-negotiation remains. Nothing is a closed chapter in democratic politics; nothing is beyond dialogue, reappraisal and renegotiation.

Let me briefly consider the success-stories belonging to my first group – the stories of *global compulsion behind local arrangements of autonomy.*

The principality of Andorra, a landlocked small unit of 450 square kilometers between France and Spain, in the eastern Pyrenees in Southwestern Europe, is by all accounts a success as a co-principality of France and Spain. It has own relation with the European Union (EU). Its general council determines all substantive matters relating to its administration and identity, jointly protected and guaranteed by the co-states, while the ceremonial sovereignty resides in both of them. It is a tax haven, a duty-free state. Again, who does not know of the arrangements of the Gibraltar or of the Holy See? Gibraltar with its strategic location in Southwestern Europe on the southern coast of Spain remained a Spanish possession until the war of Spanish succession, when Britain seized the territory under the Treaty of Utrecht in 1713. The treaty had an optional clause to the effect that Spain would have the first choice if Britain relinquished the custodianship of the state. The physical shape of the rock, the landmass, the location with its roughly 5000 strong population loyal to British rule, is the symbol of how colonial possession can be naturalized. The Spaniards and the Genoese left long ago. Now with a governor sent from London, its own assembly, council of ministers, and a court, the autonomy arrangement that fails elsewhere is successful in Gibraltar. The scope of self-governance covers election of political representatives, determination of citizenship, foreign affairs, its participation in international and regional organizations, defence, where nothing can be done that goes again British sovereignty. Legal experts call it an 'outdated anomaly'. But the rules of geopolitics ensure the continuity of the Gibraltar phenomenon. Another such successful arrangement is in fact one more instance of 'outdated anomaly', it relates to the Holy See, the sovereign entity of the Roman Catholic Church. Independent from Italy since 1929, armed with vast extra-national power and legitimacy, it is autonomous in its secular and religious matters, exercises 'sovereignty' over the believers in vast stretches across the globe, has laws on its own affairs in the city and (in some cases) elsewhere, protected by a political treaty and a concordat, and enriched by a financial convention by which Italy has paid the Holy See a definitive settlement of financial claims following the loss of its territories and property. Its power is to bar any election of political representatives, any outsider claiming citizenship, and to conduct foreign policy without being bound by any international law, for it is not a member of the United Nations or the European Union. This 0.44 square kilometre area is again an example of how autonomy arrangements are successful when occasioned by geopolitical or strategic considerations. In each case, the area is tiny and its importance vastly disproportionate to its human population or natural resource. In each of such cases, it is an arrangement to exercise sovereignty elsewhere, the *local site of a global arrangement*.

Probably, for a similar contextual reason the autonomy arrangement for Hong Kong may turn out to be a success. Though a part of the China since ancient times, China may find extremely difficult to gain Hong Kong back completely which went over to Britain in 1840 and is now China's door to global capital, information and technology. The principle of 'one country, two systems' may have more rationale than what catches the legal eye in the larger context of globalization, the rationale of the restructuring of the state evident everywhere, and the flexibility of the political forms of rule not consonant with earlier theories of sovereignty. Thus though the legal framework of Hong Kong was passed in 1997 by the National People's Congress of China dealing with the general principles, relationship with the Central authority, fundamental rights and duties, political structure, education and science, economy, external affairs and interpretation and amendment of the basic law, the story of Hong Kong's special nature may not be over. While formally, in election of political representatives, determination of citizenship, right to change political status, foreign policy, military affairs, customs, border control and immigration, and participation in international and regional organizations – the central authorities have ensured that nothing happens that brings ill fortune for China's sovereignty, Hong Kong's independence in taxation, monetary policies, education, economy and social services will ensure autonomy for the land in the foreseeable future. There are other instances where the global imperatives of geopolitics, political strategy, finance capital, skill and information have made awkward arrangements successful, and apparent anomalies relevant for current ways of rule over land, people and sometimes the globe.

My second group of success-stories of autonomy arrangements belongs to what I call the *affordability factor*, sometimes buttressed by environmental factor, which highlights durable arrangements for autonomy.

We have the case of Greenland, the largest island in the world with a land area of 2,175,600 square kilometres and a population of 55,000 people of whom 87 per cent are Greenlandic-Inuits and Greenland-born whites, and rest Danish and other Scandinavians. Inuits and the Vikings had settled there in the eleventh century and for about seven hundred years the island was under Danish control. The Greenland Home Rule Act of 1979 introduced home rule whereupon the island has become a federacy of the Danish crown. Greenland has substantial autonomy, in return has a minimal role in the governance of Denmark and the relation can be dissolved by mutual agreement. The High Commissioner represents the queen, a legislature elected by direct elections legislates, organizes local government, supervises taxation, planning, conservation, governs trade and competition, regulates work hours, and the protection of the environment. Greenland does not have judicial power. People of the island are Danish citizens and are Greenlanders. Till the Inuit people of other countries (such as Canada, USA) do not press for a political reunion of the Inuits, human rights concerns will remain centred around issues of protection of the culture of indigenous people, fishing and whaling rights, indigenous land rights, and resisting assimilation with the dominant culture of the powerful neighbours.[40] As on this time, Denmark's control over Greenland has not been seriously challenged because lack of a substantial indigenous population, absence of a past history of fierce colonial battles over possession of the island, harsh climate, lack of arable land, high cost of establishing a major industrial site and extraction of raw material. The circumpolar politics is marked by a naturalization of colonization practices till now beyond serious challenge by other powers or the indigenous people, probably because destabilization of the situation will harm all except the indigenous people. Again, then, a global arrangement in a particularly local form maintains itself as an 'exotic variety'.

The reason for tolerating 'exoticism' is not far to seek. Think of the difference in the publicity of the two claims of self-determination – of the Nunavuts (which means 'our home' – the Canadian Inuit nation) and the people of Quebec. The former is devoid of the publicity that the latter issue has. But on the significance of this, we can wait a little. Approximately one quarter of the Inuits are in Canada, and the total land area they inhabit in Canada is 1,968,400 square kilometres. The Nunavuts Land Claims Agreement of 1993 was the result of twenty years of negotiation by the Inuit representatives under which 772,000 square kilometres of land or roughly 20 per cent of Canada becomes the new territory of Nunavut. And approximately 85 per cent of the 25,000 inhabitants are indigenous people. Under the agreement and the subsequent Nunavut Act, the Inuit ceded ownership and control of nearly 80 per cent of ancestral lands and received title and ownership rights to little more than 350,000 square kilometres of the Nunavut territory and 36,260 square kilometres of mineral and sub-surface rights. The Crown retains title to all other surface and sub-surface rights in the Nunavut territory. The Inuits also received one billion Canadian dollars as compensatory cash payment, fishing and hunting rights within Nunavut territory, government subsidy for establishment of local self-government and social programmes, and guaranteed participation in decisions on environment and natural resources. Emphasis was given on decentralized, community-based government. Yet high unemployment, lack of education, economic opportunities remain difficult issues. The British by the Royal Proclamation of 1763 had established their exclusion policy in Canada as they had pursued everywhere. Under the proclamation, land was reserved for the exclusive use of the indigenous people in areas outside of the the-then established Canadian government. In the 1950s the Canadian government undertook a programme of resettling them in government-built villages. The Inuits had no right to vote until 1962. The negotiations began in this background. Transfer of control of different matters covered under the agreement is to complete by 2008. The decades of the seventies and eighties of the preceding century witnessed legal and constitutional steps following Inuit activism. These began with a court decision in 1973 upholding the existence of an indigenous land claim, followed by an upsurge of similar claims culminating in the Constitutional Act of 1982 and subsequent agreements protected by the Constitution. The Land Claims Agreement deals with clarification of ownership and use-rights for the land, participation of the Inuits in government decision-making on water and land management and conservation, wild life harvesting rights on land and water, financial compensation, self-reliance and cultural and social well-being. The Nunavit Act establishes a self-legislative body, a supreme court of Nunavut, an elected commissioner, and limited right on immigration and citizenship. Thus self-determination of the Inuits in this case is limited to the Inuits of Canada, and though they participate in Inuit Circumpolar Conference of the Inuits of Russia, Alaska, Canada and Greenland, and cooperate with other indigenous people of the arctic and semi-arctic lands on issues of global warming and transportation of pollutants, modernity has taken its toll. In the near future there can be very little politics of self-determination by the indigenous people of the arctic and semi-arctic lands *as a whole*, underdeveloped economic state will continue as fur and other such trades decline, and a fishing and hunting life transitions to a wage economy. In Nunavit economy, unemployment is around 20 per cent, private sector hardly thrives, and there is marked dependence on federal subsidy for housing and other social issues.

One can say that the Canadian management of the issue is an instance of the best possible management of the question of autonomy, at the same time the instance of the limits under modern political conditions of rule.[41] The issue concerning the Saami peoples of Northern Scandinavia also illustrates the extent and the limits of the most liberal management under modern political conditions. Spread in Norway, Sweden, Finland and Russia, of a total of about 100,00 strong population, the Saamis have parliaments in Norway (established in 1989), Sweden (1993) and Finland (1972). Certain linguistic and cultural rights have been ensured. The parliaments can review the acts and actions of the central authorities concerning Saami lands and affairs. The Saami Council, a recognized body and founded in 1956 by the Saamis of Norway, Sweden and Finland, works towards common goals of the Saamis. Extensive forests of birch, spruce, pine and the vast mineral deposits of iron ore and nickel have turned land ownership and land rights the most contentious issue faced by the Saamis in Scandinavia. More than a thousand year history of reindeer herding, hunting and fishing, and migratory movements across lands now politically defined as separate entities has made them disqualified as owners of land in the eyes of the states which regard the forests as state property. Ironically thus, the land question supposed to be a mark of the vast underdeveloped landmass of the third world has now left stamp of its intractability under conditions of modern state-sponsored democratic regimes.

This is not to say however that the *improved political and economic conditions of liberal democracy* have failed everywhere in integrating the principle of autonomy in the very structure of rule. The third group of success-stories involves countries, such as Belgium, Catalonia, Liechtenstein, Scotland, Quebec, where in varying degrees autonomy has been durably inscribed, or has been renegotiated, and the conflict is enacted within the theatre of dialogue. Land of little more than ten million people, Belgium is divided in to main groups – Flemish (55 per cent) and those of Walloon (35 per cent) and three linguistic groups – Flemish (56 per cent), French (32 per cent) and the German (1 per cent), while the remaining 11 per cent are legally bilingual. The Belgian constitution recognizes language regions and territorial regions, and three separate consociational or non-territorial communities are as important as the territorial regions as legal-constitutional entities. Flemish nationalism grew in the space created from the rivalry between the French and German nationalism, and grew as a remarkable political force in a country, which was separated in little way from France during the nineteenth century. Constitutional progress towards a multi-community polity began from early 1970s and the whole process culminated in the constitution of 1993 that adopted consociationalism as the guiding philosophy. Thus federal polity, cultural communities, territorial regions and the Brussels regional commissions act as four pillars of the structure in which each community or region self-governs, each having its own legislative assembly. The assembly known as the community council elected by the people of the communities controls the cultural community governments. Territorial regions have power over all issues not reserved to the federal government or the communities. The judicial system is single. Citizenship, right to change political status, foreign affairs, military matters, customs, border controls, passport, postal and telecommunications, currency and monetary policy – in all these the federal government is supreme or the sole authority. However, treaty-making powers are also gradually devolving on the territorial regions. Also Flemish nationalism has not exhausted itself, so that we may see yet more federalization, and a still more complex and unique structure. Who is thus a Belgian is not yet settled; but the current identification of Brussels with European union and unity, the Franco-German axis will ensure the space for the growth of the Flemish as the Belgian albeit in a consociational framework.

Or take the case of Catalonia, the home of the Catalan autonomous community and the historic region of Spain. Constituting 13 per cent of Spain's population, and the richest and the most highly industrialized region of Spain, Catalonia has an area of about 40,000 square kilometers. The kingdoms of Catalonia, Aragon and Valencia had created an advanced constitutional system in middle ages. It was always more European than Spanish and had led earlier a separatist movement to break away from Spain. It had won its autonomy in 1932 to be deprived of that by Franco in 1939 to regain it in form of the autonomy statute in 1980 after the dictator's death. Catalonia had played a prominent part in the Spanish Civil War and its suffering under the dictatorial regime of Franco could not lessen its desire for autonomy. The present constitution proclaims the indissoluble unity of the nation, but recognizes the right to autonomy of the nationalities and regions that comprise the state. Basque, Galicia and Catalonia are the historic nationalities, but the extent of self-government that these regions have achieved remains ambiguous. For while public works, social welfare, municipal boundaries, town planning, forestry, environmental protection, housing and many other matters are under the authority of the autonomous communities, the constitution permits central intervention and makes it obligatory for the autonomous communities to exercise financial autonomy in a way that ensures coordination with policies of the central government. Autonomy however does not satisfy Quebec, though we do not as yet see a civil war in the rich northern hemisphere. Firmly rooted in the French language and tradition, and historically a French colony, Quebec nationalists want sovereignty and do not believe that recognition of Quebec's distinct identity can be achieved within a federal framework. It is the largest and oldest of the ten provinces of Canada and is the home of 26 per cent of country's population. By 1980 Quebec's relation with rest of Canada was in crisis and tense. The Constitution Act of 1982 was never ratified by Quebec. And the last referendum for secession was defeated by as narrow a margin as 50.6 per cent in favour of union and 49.4 against. Another case of (relatively so compared to outside, but not within the union) wealthy nation having not taken the extreme step is Scotland. The Act of Union of 1707 united Scotland with England and Wales. Under the act, it retained a separate legal system, established church, national bank and fixed percentage of representation in the British parliament, home rule in respect of local government, and its own flag. From 1978 the Scottish and Wales members of parliament tried to force the path of constitutional, economic and social reforms in Scotland and Wales. In 1990, a constitutional convention in Scotland discussed the general plan of a Scottish parliament within the union with exclusive control over traditional federal subjects such as foreign affairs or fiscal matters. And in 1998, following under the verdict of the referendum on devolution on the basis of the Referendum Act of 1997, the Scotland Act was passed paving the way for a separate Scottish parliament. The Scotland Act empowers the Scottish parliament and a Scottish executive headed by a first minister to legislate and administer law, home, justice education, health, education, agriculture and environment. The defence of the union, relation with Europe, national security, transport safety and some other related matters rest with the United Kingdom. It is difficult to say, if Scottish nationalism based on a long history of separate identity, would remain satisfied with a parliament only. Reforms began with 1885 when the office of the Secretary for Scotland was created. The reform continued through 1926, 1939, 1977 to the Act of 1998. Labour victory, persistent Scottish demand for recognition, sufficient economic integration, and the need to modernize the political system of the union coupled with a sense of safety in the ruling class of the harmlessness of the devolution programme have combined in a contradictory way in paving the path towards federalization of the polity. But how fissures and fault-lines would disappear for good or widen is unclear now. But certainly the juggernaut of autonomy has rolled on.

These are then some of the success stories. They do not necessarily imply a resolution of the problematic. But they signal the success of constitutionalism, of containment, and of keeping the clash between the majorities and minorities, dominant and subordinate identities, state and nationalities, forward and backward regions, within a framework of dialogue. Melissa Magliana in her study of the highly visible success of the framework of autonomy of the province of South Tyrol in northern Italy cites some reasons behind the success of the South Tyrolean model. She cautions that some of these are culture and history specific, but says that the lessons can be of some universal interest. According to her, the continuous involvement of the kin-state Austria (South Tyrol has German-speaking majority in its population), the moderating influence of Austria over nationalist South Tyroleans following the Gruber-De Gasperi Agreement as part of a chastened Austria's peace treaty with Italy in Paris in 1946, the role of United Nations and the suggestion that the International Court of Justice could be approached for an impartial solution to the issue of South Tyrol, the role of Europe, the structure of negotiations, the structure of autonomy of South Tyrol (extensive autonomy of legislative and administrative capacities), proportional ethnic representation and a firm commitment to bilingualism, and the economic efficiency and capability of the province – are some of the important causes behind the success of South Tyrol. But, most significant, she says, has been the continuous engagement of Italy and people of South Tyrol on the issue of autonomy and not taking it as a one-time affair.[42]

But what about the stories of failures, of the angry voices that refused to fade out? Why is it that constitutionalism failed in these cases, for example in Basque, North Ireland, Kashmir, Palestine, Tibet, Sri Lanka, or the Indian northeast (the case of the CHT remains unclear)? These belong to my group of failure chronicles?

Take the case of Basque. The Basque people are the first identifiable group on the Iberian Peninsula and one of the oldest surviving groups in Europe. Basque passed its autonomy statute in 1979 becoming first of Spain's autonomous regions to achieve autonomy. The statute established a relationship between the region and the Spanish government for which the powers and functions were delineated under the new constitution. The autonomy statute granted the Basque region financial autonomy and the right to its own police force. But militant attacks have continued despite the grant of autonomy in 1979. In 1989 the Basque regional government passed a resolution proclaiming the right to self-determination and to decide freely on its political, social, economic and cultural status. But the movement for self-determination has remained unsatisfied. The Basque National Party, the mainstream nationalist organization was forged through emphasis on four issues – race or ethnicity of Basque, the Basque language, Basque's Catholic heritage, and the union of all Basque people in all of Spain's and France's sub-divisions. Franco's rule destroyed all existing institutions of self-rule. Use of Basque language in schools, any ecclesiastical or any other public setting was banned. The Basque Homeland and Liberty Organization (ETA) was born in such environment in 1959 and is still active today, as it remains dissatisfied with the creation of regional autonomous governments under the new Spanish constitution of 1978. The Statute of Autonomy of the Basque Country (1989) provides for a Basque parliament with power of taxation, all matters pertaining to internal security, amendment of the statute, institutions of historic territories and justice. Yet lack of real powers in matters of security, relations with other regions and countries, attacks on ETA members, and the lack of federalizing the union's relation with a historic region that has proclaimed the right of self-determination (1989) have been responsible for a situation where the struggle for freedom from control by the Spanish state continues. Limited grant of autonomy has failed in Northern Ireland also where a protestant unionist majority (63 per cent of population) has discriminated against the Catholic republican minority (37 per cent). Just as in the case of Scotland, Ireland was united with Great Britain under the Act of Union in 1800. Home rule demand started in late nineteenth century and early twentieth century leading to the Easter Uprising of 1916 and the creation by Britain of a 26 county Irish Free State with the remaining six Irish counties incorporated within United Kingdom in 1921. In 1968 peaceful civil rights movement started, in 1972 violence erupted, the northern Irish parliament resigned and the region was placed under direct British rule. Three successive attempts to re-establish the Northern Ireland national Assembly failed in 1974, 1975, and 1982. The Anglo-Irish agreement of 1985 stipulated that any change in the political status of Northern Ireland would only come about with the consent of the majority of the population. But the agreement was meaningless in view of continuing direct rule, unionist coercion on the republicans, and high handedness of the British army. In 1990s events including international mediation assistance, pressure from the United States, and Labour government in London, resulted in a series of multi-party talks concluding with the Good Friday agreement of 1998. But peace remains elusive because of mutual distrust, unwillingness to allow political moves towards Irish unity, discord over surrender of arms, and lack of concrete policies of partnership, equality and mutual respect. In these cases, autonomy (it cannot be anything except *limited*) has failed.[43]

I think that, these instances show deep similarity of patterns across the board in the politics of autonomy. The fundamental lesson is that there is no universal message of success of constitutionalism; on the contrary the failure stories convey the limits of such a revisionist strategy. Almost the same type of constitutionalism was employed in Basque, the Indian Northeast, Kashmir, Northern Ireland, Palestine, Xinxiang-Uighur region of China, Sri Lanka, and in several other places – and yet constitutional package failed in these cases. Either the constitutional measures were diluted soon, or political agreement did not precede the reforms or were not reflected in legal measures that soon after busted. In Chittagong Hill Tracts the agreement was diluted in the measures adopted to implement the agreement, the principal issue of land was not addressed, the measures for autonomy were not provided with constitutional guarantee, the peace agreement and the subsequent enactments kept substantial sections of the people of the CHT out of the politics of peace; political prospects for peace, accommodation and autonomy have hinged consequently on the thin rope of the good wishes of the rulers. In Kashmir, whatever constitutional guarantees of autonomy were there, were diluted and soon virtually abrogated. In northeast India political commitments of accommodation of various layers of sovereignty were violated, and the constitution was a counter-revolution in terms of federalizing practices there. In Sri Lanka, constitutional politics of autonomy became a part of various manoeuvres of the political class and therefore carries no legitimacy for those for whom the package of devolution measures is intended. Similarly in Palestine constitutionalism has failed, for here again political accommodation on behalf of the rulers has been lacking, and peace diplomacy is seen only as part of a counter-revolution in form of complicated measures of limited autonomy. In the two cases of Tibet and Xinjiang Uighur Autonomous Region, once again a policy of political accommodation has been lacking in the policy of the rulers of China.

Why has constitutionalism failed in these cases? The answer, as I suggested earlier, relates to the limits to the constitutional path to autonomy of the self. We can recall the lessons of the South Asian experiences, borne by experiences elsewhere. These lessons can be re-told in a slightly different way:

First, everywhere rule by law replaces rule by men.[44] Thus, political accommodation becomes subordinate to constitutional process and legal realism. Constitution is invoked to ensure a smooth passage to political accommodation, which is defeated by the fundamentalist nature of constitutionalism itself.

Second, imperatives of globalization make the daily practice of political accommodation difficult and subject it to universalized constitutional practices that push the political-ethical task of accommodation to the margins of political rule.

Third, constitutionalism does not offer any way out of a contentious situation, because it does not offer a way of escape out of the entrenched positions of power achieved through arms.

Fourth, by relegating to the margin the task of guaranteeing minimal justice, of political accommodation and reconciliation, and by clinging on to a framework of 'protection of ethnic minorities', constitutionalism cannot but representing *gradualism* and can only help retaining the old framework of political power that had given rise to the dissenting self. The federalization and con-federalization of politics becomes an ever-receding task.

Fifth, there is a paradox. Economic integration makes the task of special protection of minorities non-threatening to the political class. Thus, it facilitates resolution of old conflicts. But this integration creates new ghettos, therefore new exiles, new selves, and a declining legitimacy of the political rule.

The most fundamental limit is on the issue of violence. If the politics of self-determination provokes violence, and if from violence to constitution is the acknowledged way of reinforcing democracy, modern rule as yet does not know how to successfully engage with the problem of *passage*.

In short, constitutionalism cannot replace a politics of reconciliation based on achieving minimal justice. Subordinate to a just reconciliation that goes beyond the language of rights, constitution can help codifying the notion of justice at least minimally and thereby help politics address the issue of self and the recognition of self. It is rightly said that there can be an ethnic problem, but there can be only democratic solution. But this implies recognition of a dual inefficacy. Just as ethnicity cannot become a language of solution, the received language of democracy cannot be the language of understanding a problem. The limits of the success-stories are indeed the limits of the modern state to achieve compassion, reconciliation and minimal justice.

These relative successes and failures of the politics of autonomy suggest the requirement of guarantees only which can pave the way for reconciliation and justice at a minimal level; also they show the presence of historic obstructions on a large-scale in that path. These obstructions are evidence that autonomy, federalism, pacts, legislative measures and new constitutionalism, cannot be enough to usher in a dialogic order. The politics of dialogue must reckon with the way in which modern power is exercised, and meets its limits. But before we try to locate these guarantees, which lend form to the principle of minimal justice in our time, and towards which a dialogic politics is working itself out, I shall recall for one more time, but briefly, the historical perspective – this time not through case histories, but through a discursive review of the history of the debate over autonomy and self-determination. It will make clear why I speak of guarantees to ensure minimal justice towards which the politics of dialogue and reconciliation is moving on in large parts of the de-colonized world.

The Legacy from Versailles to Dayton

In trying to understand why the politics of self-determination refuses to be finally 'settled' and continuously generates dialogue, we must remember that the principle of autonomy, now anointed as the gist of self-determination, and with distinction being made between 'external' (secession and independence) and 'internal' (autonomy) self-determination, has always lacked a concrete definition due to its historically contingent nature. To Woodrow Wilson, self-determination was the route to lasting peace in Europe, while to Lenin this was key to redistribution of power, and socialism and democracy through grant of independence to oppressed nationalities and freedom of peoples subject to colonialism. Article 1 of the UN Charter speaks of one of the purposes of the United Nations is to 'develop friendly relations among nations based on respect for the principle of equal rights and self-determination of peoples, and to take other appropriate measures to strengthen universal peace'. Article 55 again refers to the principle. In both cases it was not referred as a right, but as a principle worthy of respect. It was in 1966 that the UN recognized self-determination as a right in the two human rights covenants, the International Covenant on Economic, Social and Cultural Rights (ICESCR) and the International Covenant on Civil and Political Rights (ICCPR), in both of which it was declared, 'All peoples have the right to self-determination. By virtue of that right they feely determine their political statures and freely pursue their economic, social and cultural development'. This was made more specific in Declaration on Principles of International law Concerning Friendly Relations and Cooperation among States in Accordance with the Charter of the UN (1970), where it was stated,

> The establishment of a sovereign and independent State, the free association or integration with an independent State or the emergence into any other political

> status freely determined by a people constitute modes of implementing the right of self-determination by that people.[45]

And yet we know that things are far from clear. Autonomy and independence – are these two points on a sliding scale or are they different and mutually exclusive? Are there other forms that override this distinction and create new structures based on shared sovereignty? Is there an unbridgeable gulf between 'minority' and 'people'? In the context of these historically unanswered questions, what strikes is the marginality of the success stories, recounted in the previous pages, in this history of the engagement of modernity with self in international politics, its determination and the limits of a pure constitutional path for resolution of conflicts around that demand. Arising in the wake of anti-colonialism, formation of the national states, the emergence of a fictive ethnic core at the heart of the national state, and therefore the rise of majority-centric politics as a way of dominating the subjects, and the coupling of minority and ethnicity in the politics of recognition, the issue of self-determination now refuses to be confined in the prescribed framework of autonomy, the mark of Helsinki style democratization, at least in the post-colonial parts of the world. The minority-discourse and the discourse of self-determination had been always interconnected. Yet they had been kept apart in the international system through the legal fiction of 'autonomy for minority and self-determination for people'. In any case this fiction derived its existence from Central and East European experiences of the last half of the nineteenth century and the first half of the twentieth century. But today the fiction finds it hard to sustain itself in as much as the fiction of every self determined to get recognition at all costs has proven to be a non-sustainable phenomenon. Later in this book I shall argue how trust is important in reconciliation of selves and how a politics of accommodation flows from a history of altruism, compassion, and understanding. But this does not flow and cannot flow primarily from constitutionalism. I have already shown why and how constitutionalism that claims for itself universal applicability in terms of maintaining the distinction between *minority* and *people* (and therefore the *difference* between the respective principles of autonomy and self-determination) is actually the experience of Europe, and therefore a provincial experience that now seeks to be generalized. It is important therefore to have a brief look into the historical legacy from Versailles to Dayton to appreciate the conundrum, and the imperatives for a politics of dialogue.

Self-determination has deeply influenced international relations, the process of state-building and national politics for at least one century. The geo-political map wide across the globe has been repeatedly changed as a result. Each European empire at one time sought to attract the minorities in a rival empire in order to hasten each other's collapse. The phased break-up of the Ottoman empire along ethno-nationalist lines fuelled by national aspirations and backed with great power sponsorship dismantled piece by piece a once united and far-flung empire. With each success in this long act, the number of rejected people and territorial and non-territorial minorities grew. The anti-colonial uprisings and the socialist policy of self-determination of peoples that hastened the end of both wars (ironically people note the link only in case of the first war, and not the second) determined world response to the issue, which had been sought till then to be framed in primarily in Wilsonian terms. Consequently, the Versailles peace treaties were not so much a confirmation of the politics of self-determination as a forced confirmation of the democratic limits of polities featured by substantial number of minorities, and where the recognition of the ideal of nationalism had been considered to be a solution of the 'question' to which minorities had been reduced in a nation-state centric framework.[46] National animosities increased as mass societies appeared and fascist tendencies appeared among national groups in the large ethnically mixed area east of an imagined line in Europe from Elbe River to Trieste. Unending wars of secession started thence. Rejected peoples were lured with false promises of security by rival regimes that established arbitrary lists of loyal and disloyal subjects. The end of second war brought into light a second approach that had been running parallel for quite some time but unnoticed – the approach of a graded policy of recognition, of federalization, of combining self-determination with autonomy, secession with voluntary union, and people's democracy as counter to liberal democracy. The Soviet Union, Yugoslavia, Czechoslovakia – with variations symbolized this parallel approach to self-determination. But if the former line propounded by Wilson failed in the East European quagmire, the alternative also did not succeed in face of mass bureaucratization, the preponderant role of state in determining social priorities, economic collapse, globalization, and the growth of a federation that barely concealed a majority-centric politics. The *desire* to grow, to become as big as the majority, to become a nation, to form a national state was the other of this strong, authoritarian, patriarch state that claimed to know what was good for peoples, nationalities, and minorities, and therefore substituted their selves with its own knowledge. Thus in the second case also democracy became an ethnic democracy. The demos could not be demos unless it was an ethnos. During the cold war, there was a process of mutual appeasement that allowed the formation of the CSCE (Council of Security and Cooperation in Europe, renamed later as OSCE – The Organization of Security and Cooperation in Europe). The Helsinki agreement included in its final declaration the protection of minorities, self-determination and secession among categories of human rights, and the recognition of the high priority of the 'national question' in relations between socialist countries and liberal democracies.

But history sometimes is so forgetful and proceeds so fast. In less than two decades, statesmen and politicians forgot the background – the logic of rapprochement, compromise, the plurality (at least the duality) of vision, and therefore the need for recognition of a plurality of forms of democratic acknowledgement of the issue of identity and recognition. With the history of the alternatives to received notions of democracy coming to an end, like tin-pot soldiers rushing into gunfire the great powers and global political managers congratulated each other over break-down of non-liberal states, encouraged one nationality, silenced another, recognized haphazardly new states, provoked secession, and came out with accords, declarations, speeches, missions, till it became clear to all except to the emperor without robes, that the deployment of the national question to the service of cold war and great power rivalry had brought back the issue of *alternatives* – precisely the alternative whose demise they rejoiced. The great powers could now only invent another myth, after the first myth had perished in the inferno, the myth of sustaining states with humanitarian interventions, privileging the humanitarian aspect of the crisis to bring about a consensus on solution, which meant another round of *humanizing* an act of undeclared war against non-liberalism, a new cold war with not always cold means. Never in the history of the *human* its twist has been so bizarre.[47]

The classic form of law between nations considered war as a freely disposable means within politics – *liberum jus ad bellum*. Viewed legally, war and peace were both just. However soon it was argued that war had to be just, and if states had to resort to war, laws of war doctrine had the function of checking the excesses of war and laying down specific rules. Modern international law has made a break; war in principle has been proscribed in favour of peace, so that there is no equality between war and peace. Article 2 (paragraph 4) of the charter of the United Nations has declared, 'All members shall refrain in their international relations from the threat or use of force against the territorial integrity or political independence of any state, or in any other manner inconsistent with the Purposes of the United nations'. The United Nations had however visualized collective compulsory measures, individual and collective self-defence permitted and promoted by the collective body of the nations, the UN. This has taken a bizarre turn – a group of nations, or conceivably one nation alone can also take coercive measures against a state if it thinks that the latter has violated international and national norms as envisaged in the charter and other subsequent declarations on human rights. Humanitarian international law has acquired a more differentiated nature. NATO military measures against Yugoslavia on grounds of humanitarian intervention (avert humanitarian disaster, defence of human rights) were justified on the argument that Serbia's conduct was an infringement of international law, that the rights of the ethnic Albanians in Kosovo had to be protected and the killings and outflow of the Kosovars had to be stopped, that protection of minorities could be ensured only stopping 'the genocide', and that therefore international security law permitted such action. Exactly one century ago Ottoman Turkey had been dismembered on the ground that the Christian subjects needed protection, which the 'unspeakable Turks' could not provide. The elastic interpretation of international humanitarian law has not only justified such military intervention, it also envisions further curtailing of state sovereignty on various grounds, such as, whether a state can stop internal displacement (UN Guiding Principles commonly known as the Deng principles), or states have provided autonomy or not. In this sudden twist in international law, what strikes is not the valorization of humanitarianism, but the graded nature accorded to it. Thousands of tons of arms and ammunitions were provided to insurgents in Afghanistan both during the presence of Soviet troops there and after their exit on the ground that the communities and their faith were in danger of extermination. Now after the country has been ravaged by civil war, the interventionists have withdrawn to another theatre leaving the humans dying in the quagmire of hunger, physical disabilities, war, and vicious orthodoxy.[48] Humanitarian aid poured in Kosovo. Afghanistan and other countries in greater need for humanitarian attention remained neglected.

Clearly the absence of a framework for humanitarian assistance in such conditions of war over self-determination allows the state, great powers, and international financial institutions to get away with providing little aid.[49] Real assistance little, hyperbole much, military intervention if necessary, with new technology (Revolution in Military Affairs – RMA) – this seems to be at the core of the humanitarian agenda today. Humanitarian concern over minority issue remains today as much a weapon of the political zealots to reform the world after their own dreams as it was eighty years ago when the map of entire one half of Europe was drawn and redrawn by a club of political managers. The myth of democratizing states through humanitarian interventions on grounds of protection of minorities sustains the zeal. But the gloss on this myth is wearing thin as global concerns over lack of success mount.[50]

As policy makers bid farewell to the twentieth century, they knew that the success stories (of regulated autonomy and protection of minorities by humanitarian interventions) were getting more and more restricted in their relevance in face of the massive tide of what they termed for lack of better words, 'ethno-nationalism'. The reports issued out of the Helsinki process are worried and complex, reflecting almost exactly the worries of a world that began one century back. Ethnic politics whether through more autonomy, or merger with neighbouring ethno-kin polities, or through outright secession, haunts the great powers and global political managers. With a century of understanding of 'people' as a narrow collective definition of a people, these managers do not know how to combine the two principles – of protection of minorities and self-determination (including the right of secession) of the people. Wilson's self-determination was allied to *jus solis*, deriving from Locke's tradition of universal rights. This was mixed with the more romantic tradition of a nation marked by cultural commonalities of language, religion and history, and was further mixed up with the tradition of *jus sanguinis*, and the principle of collective rights of groups of peoples large and small who were now invited to participate in the dismemberment of the multinational empires and unions.[51] Large parts of the globe thereafter started pursuing the still unrealized dream of achieving polities based on homogeneous population. That has been Europe's legacy for 'the rest of the world where minorities clamour for self-determination in face of state-induced ethnic intolerance, persecution and genocidal politics, ranging from mass expulsions to mass killings'.[52] Theoretically, at the heart of the problem lies the failure of Locke's principles of universality to take roots. With each break up of states and formation of new ones, new minorities emerge. One has to only take note of Russian minorities in Central Asia where out of 25 million Russians some 4 million have left for 'home', others do not know where their destiny would take them. Geo-political frontiers have been bolstered by ethnic frontiers and ethnic peripheries, in the process creating landscapes of inequality.[53] And yet, in face of the evident failure of the *universal* principles of rights and equality, the principles of self's right to recognition and creation of states on the basis of self-recognition, we are still thinking of somehow crossing this phase of 'blood and belonging' after which the world of equality, universality and brotherhood would emerge. Stefano Bianchini has remarked,

> Actually, in spite of the fact that ethnic cleansing has been sanctioned by democratic public opinion, particularly during the Yugoslav secession war in the 1990s, on many occasions ethnic cleansing has been considered, and still is, a 'deprecate, but inevitable event' on part of the policy makers and scholars of democratic regimes. Their hope, or illusion, is that ethnic, language, and religious homogeneity might settle the question of self-determination and secession, stabilizing the political map once and for all.[54]

Dayton agreement was thus the ghost of Versailles. If Versailles is remembered for the ghost of Locke that it carried on its shoulders, Dayton carried the ghost of Wilson. During the cold war, the Western world had toyed with the possibilities of the use of secessionism in order to quicken the collapse of its adversary. But now all the discourses are entangled in the mire of a master-discourse of security. The resultant paradox is clear as from the Helsinki Declaration of 1975 on inter-state relations. It declares the inviolability of borders and rejection of violence in changing state borders. At the same breath, it declares the right of self-determination as a 'universal collective'. In this entanglement, there is an admission that local autonomy as a container of tensions falls short of being the mark of a 'constitutive people'. We are witness therefore, on one hand, to the quest of an impossible mission, on the other hand to 'the economically attractive business (of) self-determination'.[55]

And here in lies my central argument in terms of the agenda of a dialogic politics, namely, that it is the closure of a discursive situation on rights that provokes the new conversation. After all, does not this history of self-determination, which is much local and aspires to take on universal relevance, and fails repeatedly in realizing that aspiration, become in this way instrumental in new dialogues? To the neglected peoples and nations, self-determination came through anti-colonialism and not secession or a resolution of the minority question (which was undoubtedly a related issue but not the fundamental). Anti-colonial leaders such as Ho Chi Minh and Sun Yat Sen had pleaded with Wilson in Versailles to help them gain independence, declare a process of de-colonization, and allow them to invent their own democratic forms. Wilson expectedly could not oblige. With a mission to rival Lenin and stabilize liberal democracy, he could be only discriminate. Finally when with the end of the two wars with most of the erstwhile great powers considerably weakened, independence arrived in large areas on earth, it came as a fruit of a poisoned tree. Democracy and nationalism were de-coupled. Independence came in form of partition. Minority issue became ethnic issue. Massive cross-border population movements ravaged homes, hearths, and settlements. Soon national independence turned into majority-centric politics and the long history of co-existence, tolerance, and accommodation was (at least temporarily) submerged in the new history of national hatred, forgeries,[56] wars, and global determination of the relations between various post colonial selves in both cold-war and post-cold war times. This history has relation to, but different to a great extent from the history that Versailles had set in and had reached its apogee in Dayton. The forms of autonomy we surveyed in the preceding pages grew up in that core period of late nineteenth century to late twentieth century, more concretely in the age of Versailles to Dayton, when history moved in areas inhabited by democracies, wars and peace were concluded there, and the colonial and the subsequently de-colonized world was kept outside this history.

In this period, particularly in the thirty-five year period of decolonization, democracy was conceived primarily in terms of protection (reflected best in the development of international human rights and humanitarian law). Democratization of relations was pushed to the background. Thus the Convention on Genocide (1948), provisions on non-discrimination in the UN charter, the Covenant on Civil and Political Rights (1966), the Optional Protocol to Article 27, the Helsinki Final Act (1975), the ILO Convention on Indigenous Populations (1957), and various other conventions and enactments, identified the weak and dispossessed, and sought to give them protection including providing them with varying degrees of autonomy, without attending to the fundamental problematic of a process that was continuously engendering clusters of weakness and dispossession. In South Asia, as elsewhere in the de-colonized world, the strategy worked only partially therefore. New nationhood lapped up the protective strategy, because this released it from the obligation of a thorough democratization. Against this closure brought about by almost one hundred years' history from Versailles to Dayton, democratic politics is re-appearing in regions and lands vastly distant from each other – South Africa to Palestine to Korean peninsula – whose features are not conflictive assertions, but a politics of understanding, justice, reconciliation and restoration of trust. This is a politics, whose roots are in a different history and whose contours are barely visible, but whose power is sufficient to lend imagination to a different project and to inspire the victims of discrimination to move along a different political morality. Typically historians are lagging behind to unearth the history of this alternative spirituality. Possibly when the tragic closure will be over, historians will be able to write on it, for it had by that time passed irretrievably into history.

There is one more reason why we must move away from the received discourse of the history of the self, recounted in the preceding pages. This discourse had all the while proposing a protection-based solution pitted that solution against a constructed discourse of terror, insurgency and repression. Ronaldo Munck citing a delegate of the Spanish state to the Council of Europe who declared that 'terrorism (was) not only a threat to the stability of democratic institutions, but represented a threat to western civilization' shows how this discourse reduces all complexities of the situation to fundamentalist dualities – civilization/barbarity, multiculturalism/hatred, and democracy/satanic cults.[57] This is the classic discourse of a sane civilization banishing the non-compliant to the region of madness. Thus, Kurdish Workers' Party, Irish Republican Army, the Hejbul Mujahideen, the NSCN (National Socialist Council of Nagaland) or the LTTE[58] (Liberation Tigers of Tamil Eelam) – all are hooded entities to be lured into negotiation so that sanity is restored and the sane practice of protection can continue producing ever new and more serious needs of protection. In this discourse of insanity conveniently forgotten is the need to democratize the polities. Defamation, penalization, and subjection to disproportionate retaliation go hand in hand with this discourse of insanity whose other mark is a stunning silence on the large-scale impoverishment, destitution, dispossession, and powerlessness which Edward Said calls in the Palestinian context, 'an inured world consciousness'.[59] To make even the first step towards dialogue, reconciliation and accommodation, the way is to disconnect the problematic of autonomy and self-determination from the discourse of defamation and insanity that arrogates to it the right to decide who deserves the right to protection and who does not. In other words, the dialogic path is away from that of imposing penalty, infamy, and subjection.

Therefore, as the issue is put in chapter ten more extensively, given the vast political experience of the post-colonial world, the inquiry of this book is, on what issues does the dialogic agenda revolve? Or, in what way does the search for a way out of the predicament posed by these histories of selves and solutions lead to the politics of dialogue? And, how does dialogue become a key factor in the shift from rights-based language to justice-based language, in as much as the inquiry is, how does this shift become a key factor in pushing the politics of dialogue to the fore? As I have admitted earlier, clearly I am pushing here for a relational stand – in reviewing rights, and arguing for a politics of justice. There are no a priori principles to be espoused here.

Five Rules of Minimal Justice

More important from the point of view of a dialogic politics is that, the shift mentioned above transforms justice itself. Justice moderated by acknowledgement of rights is achieved, every time it is conversed and contested, minimally. It reflects in the passage from retributive justice to restorative justice. The problem with the protective strategy was that it wanted to confront asymmetries with make-believe symmetries. In a world inhabited by bullies, policemen and conformists, where hard ways and calculations replace the conversation of the dialogists, justice can be only a matter of continuous contention and conversation. Minimal justice, as the adjective informs us, cannot do away with all injustices at one stroke, but by subjecting justice to dialogue makes several transitions as a way out of the closure – from a calculus of self to empathy, history to political criticism, violence to trust, rights to justice, virtue to understanding, judgment to dialogue, accommodation and reconciliation with this (another) soul.

It can be now seen that the success, semi-success, and failure stories of autonomy are actually a narrative of five broad rules of minimal justice – the guarantees I mentioned earlier.

First, these stories reflect the requirement of the *principle of compensation* for past injustices, wrongs, and gestures on behalf of the national state to convince the *other* of its desire to continue with the principle of compensation.

Second, they tell us of the need for the *principle of supervision.* This means deciding on the right and agreeable way to supervise the introduction of autonomy.

Third, they suggest the *principle of custodianship*, which is crucial to settle the balance between the territorial sovereignty of the state and the autonomy of the dissenting community.

Fourth, and this follows from the preceding principle, important is the *principle of guarantee* – guarantee against future erosion of autonomy, guarantee of a mechanism for continuous conversation.

Fifth, there is the *principle of innovation* of federalizing and autonomic practices along non-territorial lines towards federalizing the political society.

Yet we must understand that these principles are not abstract, they are the products of the experiences mentioned in some details above, and will remain historically conditioned. Minimal justice is 'minimal', not only because it does not make claim to achieve the redress of all wrongs, but accepts the fact also that justice is always historically produced, and therefore contingent. These principles require a dialogic order at an increasing scale, because in this age of globalization they require international guarantee and flexibility of forms of accommodation. This becomes possible when the state is irreversibly linked first, to an *order* that propels such an evolution of forms of shared sovereignty.

Minimal justice is 'minimal', also because it does not claim that liberal democratic state and politics will be over once these minimal rules are recognized. All it does is to lend a critical edge to politics that is attempting to come out of the closure placed before the masses, and dramatically evident since the miraculous year of 1989. It may very well be that with economic integration helping the re-colonization of the 'excluded' areas through autonomy packages and devolution, and with the historic growth of a political class that sees the merits and more important the possibility of a revision of rules of governance, these principles will be in practice in many parts of the world, and the chronicle of the success and semi-success stories of autonomy will be seen as one of a phase of revision of liberal political rule worldwide. Another fifty years later political and constitutional historians will say that the principle and practices of self-determination occasioned the revision of rule. The *ancien regime* was succeeded by a more multi-layered state with improved methods of ruling its subjects, and with a transformed notion of sovereignty whose features will be the capillary forms of legal power sometimes transcending territorial boundaries, and constitutional practices of co-existence – all these helped by the transition from an archaic form of rule to a con-federal one, and (contradictorily) economic integration. The issue of justice thereby will not be exhausted; politics is a matter of self, and will continue its vocation of being *just*, that is seeking new standards of justice. For after all, the theory and the reality of the juridical form of sovereignty in form of a territorial democratic state exists as a 'state of nature' whence all things follow and to which all things return. It is this immanent problem of a critique of a thing, which bases itself on that thing, that perhaps calls for greater attention in a critical theory of dialogue.[60]

To understand the nature of that double bind, we shall have to investigate the mechanics of dialogue with which political societies have wanted to escape the trap of exclusivities – exclusivities they can be never sure of escaping permanently. Our investigation begins with the subsequent chapters examining mechanisms of dialogue through a close look at its manifold form – dialogue between state and the rebels, between states, and between popular solidarities.

We begin with the dialogue between the state and the rebels.

Notes

1 The literature on the various views is quite substantial, and the debates have nuances. For the sake of my arguments I have somewhat simplified them. Interested readers may seek the labyrinthine paths in John Rawls, *A Theory of Justice* (New York: Cambridge University Press, 1972) and *Political Liberalism* (New York: Columbia University Press, 1993), Charles Taylor, *Multiculturalism and the Politics of Recognition* (Princeton, NJ: Princeton University Press, 1992), Michael J. Sandel, *Liberalism and the Limits of Justice* (Cambridge: Cambridge University Press, 1982), Seyla Benhabib, *Situating the Self* (Cambridge: Polity, 1992), Robert Dahl, *Democracy and its Critics* (New Haven: Yale University Press, 1989), Will Kymlicka, *Liberalism, Communitarianism, and Culture* (Oxford: Clarendon, 1989), Etienne Balibar and Immanuel Wallerstein, *Race, Nation, Class – Ambiguous Identities* (London: Verso, 1991), Michael Walzer, *Spheres of Justice* (New York: Basic Books, 1983), Amy Gutmann (ed.), *Multiculturalism* (Princeton, N.J.: Princeton University Press, 1994), and Bhikhu Parekh, *Rethinking Multiculturalism – Cultural Diversity and Political Theory* (Hampshire: Palgrave, 2000). On Indian writings, again the literature is substantial. One can see, T.N. Madan, *Modern Myths and Locked Minds* (Delhi: Oxford University Press, 1997), Ashis Nandy, 'The Politics of Secularism and the Recovery of Religious Tolerance', *Alternatives*, 13 (2), April 1988, Akeel Bilgrami, 'Two Concepts of Secularism – Reason, Modernity and the Archimedean Ideal', *Economic and Political Weekly*, 9 July 1994, Partha Chatterjee, 'Secularism and Toleration', *Economic and Political Weekly*, 9 July, 1994, Sumit Sarkar, 'The Anti-Secularist Critique of Hindutva: Problems of a Shared Discursive Space', *Germinal*, 1, 1994, Rajeev Bhargava, (ed.), *Secularism and its Critics* (Delhi: Oxford University Press, 1998), and Rajeev Bhargava, Amiya Kumar Bagchi and R. Sudarshan (eds.), *Multiculturalism, Liberalism and Democracy* (Delhi: Oxford University Press, 1999).

2 On the debate on cultural identity, see the excellent summary by Sarah Joseph, 'Liberals, Communitarians and the Protection of Cultural Diversity', chapter 5, in her *Interrogating Culture – Critical Perspectives on Contemporary Social Theory* (New Delhi: Sage, 1998). My only criticism on that excellent essay is that she does not question the *politics of rights* in her presentation of the rival arguments.

3 James Massey indicates the tension in his discussion on minority rights in India. See his, *Minorities in Democracy – The Indian Experience* (New Delhi: Monohar, 1999). For a theoretical formulation of this tension, see Nalini Rajan, *Secularism, Democracy, Justice – Implications of Rawlsian Principles in India* (New Delhi: Sage, 1998).

4 'Report of the Working Group on Minorities on its Fourth Session, Geneva, 25-29 May 1998', Paragraphs 94 & 96, E/CN.4/Sub.2/1998/18, 6 July 1998.

5 'Report of the Working Group on Minorities on its Fifth Session, Geneva, 25-31 May 1999', Paragraphs 59-60, E/CN.4/Sub.2/1999/21, 24 June 1999.

6 'Report of the Working Group on Minorities on its Sixth Session, Geneva, 22-26 May 2000', Paragraphs 31-37, 85-95, E/CN.4/Sub.2/2000/27, 30 June 2000; also indicative of the new emphasis on participation by minorities in decision-making procedures, is the MRG report based on an analysis of constitutional provisions in several countries such as Canada, Fiji, New Zealand, India, South Africa, Yash Ghai, 'Public Participation and Minorities', Minority Rights Group International, London 2001. The report discusses several types of participation and suggests how to strengthen constitutional processes with political measures. See also the report of the Council of Europe on the 'Participation of Minorities in Decision Making' to the Commission on Human Rights, E/CN.4/Sub.2/AC.5/2001/CRP.6 – 30 March 2001.

7 Several issues taken up for discussion in the seventh session of the Working Group make the trend clear. For example, 'The Peace Process in Northern Ireland', E/CN.4/Sub.2/AC.5/2001/CRP.3, 5 March 2001; 'Autonomy in the 21st Century – Through the Theoretical Binoculars', E/CN.4/Sub.2/AC.5/2001/CRP.1, 5 March 2001; 'Autonomy and Minority groups – A Legal Right in International Law?', E/CN.4/Sub.2/AC.5/2001/CRP.5, 5 March 2001; 'Final Text of the Commentary to the Declaration on the Rights of Persons Belonging to National or Ethnic, Religious and Linguistic Minorities', E/CN.4/Sub.2/AC.5/2001/2, 2 April 2001. The Commentary authored by Asbjorn Eide, the Chairperson of the working Group, makes the effort to circumvent the limits of rights by enriching them clear.

8 On this problematic see some recent thought provoking essays by Etienne Balibar, 'Algeria, France: One nation or Two?' in Joan Copjec and Michael Sorkin (eds.), *Giving Ground – The Politics of Propinquity* (London, Verso, 1999), 'Is European Citizenship Possible?' in James Holston (ed.), *Cities and Citizenship* (Durham, 1999); 'The Borders of Europe' in Pheng Chea and Bruce Robbins (eds.), *Cosmopolitics – Thinking and Feeling beyond the Nation* (Minneapolis, 1998).

9 The stark description of the massacre in Nellie in Assam in 1983 in Sanjoy Hazarika, *Rites of Passage – Border Crossings, Imagined Homelands, India's East and Bangladesh* (Delhi: Penguin, 2000), pp. 25-74, is a pointer to the barbarity that this transformation of the problematic of minority to one of ethnicity causes.

10 Partha Chatterjee makes the point in 'Secularism and Toleration', though he misses the obvious conclusion that a community framework can only give in to a statist strategy of rule through ruling the communities, and that it calls for a political response. Neera Chandoke tries to combine the ideal of secularism and the moral obligation towards minorities in *Beyond Secularism – The Rights of Minorities* (Delhi: Oxford University Press, 1999). Yet she too undervalues the principle of political action in making democracy work.

11 The complexities of ethnic self-determination are borne out not only in parts of Africa or Eastern Europe, but closer, in our region. See in this respect, Sumantra Bose, *The Challenge in Kashmir – Democracy, Self-Determination and a Just Peace* (New Delhi: Sage Publications, 1997), particularly the sub-section on 'The Complexity of Self-Determination – The View from Jammu', chapter 4, pp. 88-95, chapter 7, pp. 186-194.

12 Jurgen Habermas, *The Legitimation Crisis*, trans. Thomas McCarthy (Cambridge: Polity Press, 1976), p. 75.

13 *Legitimation Crisis*, p. 78.

14 John Keane, *Civil Society – Old Images, New Visions* (Cambridge: Polity Press, 1998), P. 36.

15 I again draw from the insights in *Civil Society – Old Images, New Visions*, p. 46; also pp. 65-78.

16 Henry Lefebvre in *The Production of Space* (1974), cited in *Civil Society*, p. 188.

17 For details of the international legal arguments both in favour of and against extending the right to self-determination, see Geoff Gilbert, 'Autonomy and Minority Groups – A Legal Right in International Law?', paper for the Seventh Session of the UN Working Group on Minorities, E/CN.4/Sub.2/AC.5/2001/CRP.5, 5 March 2001.

18 For an exposition of the morality of states arguments, see N. Dower, *The World Ethics – The New Agenda* (Edinburgh, Edinburgh University Press, 1998).

19 On this see, Michael Walzer, *Just and Unjust Wars – A Moral Argument with Historical Illustrations* (New York: Basic Books, 1977). For a critique of Walzer's exoneration of Israel's aggressive wars, Richard Falk, 'The Moral Argument as Apologia', *The Nation*, 25 March 1978, pp. 341-343; Noam Chomsky, 'An Exception to the Rules', *Inquiry*, 17

April 1978, 23-27; and Edward W. Said, *The Question of Palestine*, new edition (New York: Vintage Books, 1992), pp. 205-207.

20 Kimberly Hutchings asks in Hutchings, *International Political Theory* (London: Sage, 1999), Chapter 5, in the context of our present discussion, whose self and determination of what?

21 Reprinted in Mushirul Hasan (ed.), *India's Partition – Process, Strategy and Mobilization* (Delhi: Oxford University Press, 1993), pp. 44-45.

22 Adriana Kemp, 'From Politics of Location to Politics of Signification – The Construction of Political Territory in Israel's First Years', *Journal of Area Studies*, 12, 1998, pp. 74-101; also her, 'The Frontier Idiom on Borders and Territorial Politics in Post-1967 Israel', *Geography Research Forum*, 19, 1999, pp. 78-97.

23 Edward W. Said, *The Question of Palestine*, p. 244.

24 On the incomplete nature of the accord and the implementation acts, see the deliberations of the peace audit on the CHT in Jehan Perera (ed.), *Report of a Peace Audit on Nagaland and the Chittagong Hill Tracts*, SAFHR Paper 4, Kathmandu 1999; on land rights of non-religious minorities, see the detailed discussion in Philip Gain and Shishir Moral, 'Land Rights, Land Use and Ethnic Minorities of Bangladesh', SEHD, Dhaka, 1998.

25 Cited from a study conducted by the Centre of Nepal and Asian Studies, Tribubhan University, on the basis of the 1991 census, in Lok Raj Baral, 'Ethnicity and Constitutional Reforms in Nepal' in Iftekharuzzaman (ed.), *Ethnicity and Constitutional Reform in Asia* (Dhaka: University Press Limited, 1998), pp. 84-107.

26 Lok Raj Baral, 'Ethnicity and Constitutional Reforms in Nepal', pp. 95-96.

27 The Nepali sociologist Krishna B. Bhattachan discusses this in the context of rising violence in Nepali society and the agrarian armed struggle launched by the Communist Party of Nepal (Maoist), in a long essay, 'Possible Ethnic Revolution or Insurgency in a Predatory Unitary State Hindu State, Nepal' in Dhruba Kumar (ed.), *Domestic Conflict and Crisis of Governability in Nepal* (Kathmandu: Centre for Nepal and Asian Studies, 2000), pp. 135-162. He quotes the leader of insurgency, Prachanda, as saying,

Along with the development of the People's War is a new consciousness for fighting for their own rights and liberation is spreading amongst many oppressed nationalities of the country such as Magars, Gurungs, Tamanga, Newars, Tharus, Rais, Limbus and Madhesis. People's War has spread up the process of formation of various national liberation fronts and expansion of nationality organizations. Similarly, today along with the development of People's War, a wave of organization and struggle has been created among Dalit castes at a greater speed and a wider scale. The Dalits are today rebelling against inhuman tyranny perpetrated upon by the feudal state of high caste Hindus. (p. 144).

See also Mahendra Lawoti, 'Exclusionary Democratization – The Politics of Minority Domination in Nepal', (unpublished) paper presented to the 'Dialogue on Democracy and Pluralism in South Asia', Jawaharlal Nehru University, New Delhi, 14-17 March, 2001. Lawoti shows how the political issue of autonomy has become more insistent in Nepal, after Nepali polity became a parliamentary democracy and state more majoritarian.

28 *Pakistan Year Book, 1992,* Karachi, 1993, p. 6.

29 On this, Yunus Samad, *A Nation in Turmoil* (Delhi: Sage Publications, 1995); also Charles Kennedy, 'Managing Ethnic Conflict – The Case of Pakistan', *Regional Politics and Policy*, 1 (3), Spring 1993, pp. 135-139.

30 *A Confederal Constitution for Pakistan – An Outline by Sindhi, Baluch and Pushtoon Front* (n.d.), pamphlet cited in Moonis Ahmar, 'Ethnicity and Constitutional Reforms –

A Pakistani Perspective' in Iftekharuzzaman (ed.), *Ethnicity and Constitutional Reforms in South Asia*, p. 114.

31 Syed Jaffar Ahmed, *Federalism and Pakistan – A Constitutional Study* (Karachi: Oxford University Press, 1990).

32 There is an ever-increasing volume of literature on the political history of the Tamil movement. For my purpose that seeks to see the constitutional history of the country and the evolution of many selves, see, W. Howard Wriggins, *Ceylon – Dilemma of a New Nation* (Princeton, N.J.: Princeton University Press, 1960); C. Manogaram, *Ethnic Conflict and Reconciliation in Sri Lanka* (Honolulu: University of Hawaii Press, 1987); S.U. Kodikara, 'The Separatist Eelam Movement in Sri Lanka – An Overview', *India Quarterly*, 37, 1987; and R.N. Kearney, 'Ethnic Conflict and Tamil Separatist Movement in Sri Lanka', *Asian Survey*, 25 (9), 1985.

33 My debt to Etienne Balibar continues. Balibar in an extremely concise essay, 'What makes a People a People? Rousseau and Kant' in Mike Hill and Warren Montag (eds.), *Masses, Classes, and the Public Sphere* (London: Verso, 2000), pp. 105-131, shows how law constitutes people into a people, and yet masses, whose elemental existence creates fear, produce a supplement in this production of a people. Constitutionalism is thus at once a question of freedom but constraint also.

34 *A Biography of the Indian Nation, 1947-97* (Delhi: Sage, 2001); also, see, chapters 5 & 7 of this book.

35 The 'basic features' of the constitution cannot be amended through the exercise of the power of amendment under Article 368. The Constitution 42nd Amendment Act, 1976 had inserted in Article 368 (5) a provision that there was no limitation on the constituent power of the Parliament to amend the constitution. Though the Supreme Court invalidated this, ambiguity still remains. See Keshavananda Bharati v. State of India (AIR 1973 SC 1461), Minerva Mills v. Union of India (AIR 1980 SC 1789), and Srinivasa v. State of Karnataka (AIR 1987 SC 1518).

36 Marc Galanter, 'Hinduism, Secularism, and the Indian Judiciary' *Philosophy East and West*, 21 (4), October 1971, reprinted in Marc Galanter, *Law and Society in Modern India* (Delhi: Oxford University Press, 1997), pp. 237-238.

37 Galanter examines in this context the history of a case (1966) that involved the Swaminarayana sect and went to the highest court of the land where eminent jurists such as Gajendragadkar, Wanchoo, Hidayatullah had to decide who was a Hindu and the various sects traditionally within the Hindu fold could opt out and claim rights of minorities. Ibid. pp. 237-256.

38 In the context of judicial activism, one can read, R. Dhavan, R. Sudarshan, and S. Khurshid (eds.), *Judges and the Judicial Power – Essays in Honour of Justice Krishna Iyer* (Bombay: Oxford University Press, 1985).

39 On the legal complexities arising out of the position that caste occupies in public life, see, D. E. Smith (ed.), *South Asian Politics and Religion* (Princeton, N.J.: Princeton University Press, 1966); Marc Galanter, 'Law and Caste in Modern India', *Asian Survey*, 3, 1963, pp. 544-559; F.G. Bailey, *Caste and the Economic Frontier* (Manchester: Manchester University Press, 1957).

40 Indeed, as many observers have pointed out that in overwhelming number of cases for autonomy, the crucial question is of land. Ward Churchill in a detailed study has shown how human rights boil down to the issue of land right. See, Ward Churchill, *Struggle for the Land – Native North American Resistance to Genocide, Ecocide and Colonization*, revised edition (Winnipeg: Arbeiter Ring Publishing, 1999), specially Part II. One of the significant aspects of the book is a map (facing preface page) described as the 'indigenous North American view of bioregional secession' (p. 11), which shows the

bioregional configurations of settlements of the indigenous people that could be 'legally defined areas of native nations' (p. 11), distinct from current political-territorial configurations such as states, provinces, etc, that are on 'shaky grounds – both legally and environmentally'.

41 One must read in this context the detailed work by the Jan Penrose, the Canadian geographer, on the Nunavut lands and people, 'The Limitations of Nationalist Democracy – The Treatment of Marginal groups as a Measure of State Legitimacy', *Hagar – International Social Science Review*, 1 (2), 2000, pp. 33-62. See also the report 'Examples of Autonomy in Finland – The Territorial Autonomy of the Aland Islands and the Cultural Autonomy of the Indigenous Saami People' submitted to the UN Commission on Human Rights, E/CN.4/Sub.2/AC.5/2001/WP.5, 8 May 2001.

42 Melissa Magliana, *The Autonomous Province of South Tyrol – A Model of Self-Governance?* (Bolzano: European Academy of Bolzano, 2000), Chapter 6, 'Self-Governance in South Tyrol – Why It Worked', pp. 125-143.

43 Most of these instances of measures and practices of autonomy are based on primarily two sources – first, the 600-page report of the International Committee of Lawyers for Tibet, *Forms of Autonomy – The Practice of Autonomous and Self-Government Arrangements* (mimeo, 1999). I am grateful to Hubertus Von Welck of the Friedrich Naumann Stiftung, South Asia office, for making the volume of survey available to me. Second is the UNESCO report on 'The Implementation of the Right to Self-Determination as a Contribution to Conflict Prevention', edited by Michael C. Walt van Praag, UNESCO Centre of Catalonia, Barcelona 1998. Also, references are drawn from different Amnesty International reports on countries mentioned. See also in this context, forms of autonomy discussed in the new legislations in the countries of East Europe that recognize the limits of earlier constitutional exercises. For example, 'Annual Report of the Parliamentary Commissioner of Hungary for National and Ethnic Minority Rights, 1 January – 31 December 1998', Parliamentary Commissioner's Office, Budapest 1999; same report for 1999, Budapest 2000; 'National Minorities in the Slovak Republic', Ministry of Foreign Affairs of the Slovak Republic, Bratislava 2001, and 'Study on the Use of the Autonomy Approaches in the Russian Federation', report to the Commission on Human Rights, E/CN.4/Sub.2/AC.5/20001/WP.3 – 23 March 2001. The Russian Federation speaks of the National Cultural Autonomy Act of 1996 and analyses as to why earlier constitutional provisions of autonomy during the Soviet period had failed. The report cites lack of a substantive federal structure, political, territorial and social displacement of large numbers of people and no proper rehabilitation, a hierarchical nationality system, and absence of national cultural autonomy as the core of national cultural self-determination. (pp. 15-17).

I have also benefited from the insights of the detailed legal tract by Geoff Gilbert, 'Autonomy and Minority Groups – A Legal Right in International Law?', report to the Commission on Human Rights, E/CN.4/Sub.2/AC.5/2001/CRP.5, 5 March 2001. Jurist Gilbert wrote in his report, 'Autonomy in appropriate measure needs to be seen as a right of all groups within the State, nations, peoples or minorities' (p. 38). It is worth to listen to him, 'The genocide and forcible assimilation of minority groups through the centuries is ample testimony that there never have been mono-ethnic States. This is not to condemn President Wilson...rather it is to condemn international law which has yet to cope with groups of individuals. It is the lack of a voice on the international stage for entities other than states which has made secession seem so appealing and autonomy, viewed as a privilege within the gift of the State, as second best.' (p. 38). He also said that even if it is argued that, grant of 'autonomy within the state is ultimately a political decision', yet this 'does not detract from its status as a right in international law'. (p. 39).

44 For historical insight into the problematic, see Sumit Guha, 'Power and law in Past and Present' in Satish Saberwal and Heiko Sievers (eds.), *Rules, Laws, Constitutions* (Delhi: Sage, 1998), pp. 83-96.

45 Annexure to General Assembly Resolution 2625 (XXV), Approved 24 October 1970.

46 Patrick Thornberry brings out the complexities of that period in *International Law and Rights of Minorities* (Oxford: Clarendon Paperbacks, 1992), Chapter 3, pp. 40-50.

47 For a historical overview, see the collection of essay in Henry Huttenbach and Francesco Privitera (eds.), *Self-Determination from Versailles to Dayton – Its Historical Legacy* (Ravenna: Longo Editore, 1999).

48 On the cold war legacy of Afghanistan, see Barnet Rubin.

49 The 'Report of the Two Day Workshop of the CHA (Consortium of Humanitarian Agencies) on Districts Meet in Colombo, 3-4 April' (mimeo) maps the priority areas and issues district wise in conditions of civil war and brings out the perceptions of the victims on humanitarian aid and needs. I am grateful to Jeevan Thyagaraja for providing me with a copy of the report.

50 As an evidence of such concern, see the report on the roundtable discussion on Sri Lanka, 'War and Peace in Sri Lanka – Policy Options for Canada', South Asia Partnership, Canada, Ottawa, 22 march 2000 (mimeo).

51 On old and new theories of nationalism, Roger Brubaker, *Nationalism Reframed* (Cambridge, MA: Cambridge University Press, 1996). Also, Y. Tamir, 'The Right of national Self-Determination', *Social Research*, 58, 1991, pp. 565-590.

52 Henry Huttenbach, 'The Decline of European Politics – Then and Now' in *Self-Determination from Versailles to Dayton – Its Historical Legacy*, p. 137-138.

53 I borrow the term from the excellent book on Israeli politics of borders and boundaries, Oren Yiftachel and Avinoam Meir, *Ethnic Frontiers and Peripheries – landscapes of Development and Inequality in Israel* (Boulder, Colorado: Westview Press, 1998).

54 Stefano Bianchini, 'The Changing Notion of Self-Determination – Global Repercussions for the International Community' in *Self-Determination from Versailles to Dayton – Its Historical Legacy*, p. 150.

55 Ibid. p. 157.

56 Ivan Ivekovic uses the word forgery to explain the Yugoslav situation in his 'Identity – Usual Bias, Political Manipulations and Historical Forgeries – The Yugoslav Drama' in Stefano Bianchini and George Schopflin (eds.), *State Building in the Balkans – Dilemmas on the Eve of the 21st Century* (Ravenna: Longo Editore, 1998), pp. 251-274.

57 Ronaldo Munck, 'Deconstructing Terror: Insurgency, Repression and Peace' in Ronaldo Munck and Purnaka L. de Silva (eds.) *Postmodern Insurgencies – Political Violence, Identity Formation and Peace Making in Comparative Perspective* (London and New York: McMillan and St. Martin's Press, 2000), p. 1-13.

58 Liz Philipson chronicles in details the politics of making demons in her report, which by no means is a sympathetic account of the LTTE's peace politics, *Breaking Recurring Themes in the Cycles of War and Peace in Sri Lanka*, www.c-r.org. I am grateful to Shiral Rajtilake for pointing out the report to me.

59 Edward W. Said, *The Question of Palestine*, 1979 (New York: Vintage Books, 1992), preface to the 1992 edition, p. xxi.

60 Etienne Balibar refers to such circularity of such reasoning in his classic book on Spinoza, *Spinoza and Politics*, trans. Peter Snowdown (London: Verso, 1998), pp. 31-36.

Chapter 6

Governing through Peace Accords

Death after the Announced Peace

I have no better way to begin a discussion on how accords, dialogues, and administration of peace have become a part of governmental business, and how peace is ensured now through accords accompanied by 'collateral deaths', than by narrating the following incident recorded by Ram Narayan Kumar, the historian of Punjab insurgency and accords.[1]

Fifty-five-year old Sardar Ajaib Singh from village Othiyan in Ajnala subdivision of Amritsar district was a man of worldly wisdom, who handled the problems of life in a calm and calculated manner. These qualities of his character had stood him in good stead, enabling him to preserve his family, property and considerable social standing in an area of Punjab that for a decade had remained locked in the spiral of Sikh insurgency and the State repression. Ajaib Singh was the elected head of his village council. He had three grown up sons: Thirty-five-year old Kulwinder Singh, married with three young children, was employed as the Panchayat Secretary at Naushera Pannua block of Tarn Taran subdivision in Amritsar district and was very popular for the fairness and energy with which he performed his duties. Second son Jagbir Singh, thirty-two, managed the family's twenty-five acres of irrigated agricultural land, which yielded good crops and income. Third son Maminder Singh, twenty-eight, became a registered medical practitioner. Ajaib Singh and his wife Manjit Kaur kept good health. As devout Sikhs, they often went on pilgrimages, and organized and attended with fervour the festivals of the religious calendar. Things seemed to be going as well as they could under the circumstances.

Ajaib Singh had for long been associated with the Congress Party, and had not given up the association even after the army assaulted the Golden Temple in June 1984. He was all praise for Rajiv Gandhi when, after winning the parliamentary elections with an unprecedented popular mandate early in the year, he signed an Accord with the Akali Dal's moderate President Sant Longowal in July 1985. The Accord paved the way to the restoration of a popular government in the State. Although Longowal was assassinated soon after, his equally moderate successor, Surjit Singh Barnala led the party to a thumping victory in the State Assembly elections, and went on to form a government. For a while, the situation seemed to be improving. But the central government was unable to keep the promises it had made in the Longowal Accord. The radical Sikh groups that had been lying low resurfaced. Militancy revived. The slogan of Khalistan was again in the air. In October 1987, the Union government brought Punjab under President's Rule after dismissing the Akali government. Paramilitary forces were deployed to crack down on extremists. Meanwhile Kulwinder shifted his residence to Amritsar mainly for the reason that his young children needed education. Kulwinder daily commuted to his work on his motorcycle.

20 December 1991 was a crisp winter day. Kulwinder left for his work a little late that morning. On the way, one man asked him for a lift on his motorcycle, a TVS Suzuki No. PB02-C-4455. The man later identified as Palwinder Singh Sona was a known militant. It is possible that Sona forced Kulwinder to take him along on the pillion of his motorcycle, or the two may have been actually acquainted. That would hardly be inconceivable in the situation then existing in Punjab. Thousands of young Sikhs had embraced the path of gun to confront the Indian State. Many empathized with their sentiments and helped them indirectly to find shelter and food even when disagreeing with the wisdom of their chosen path. The motorcycle was stopped for a routine check at a barrier set up by Sadar police station of Amritsar on the road across the railway station. Inspector Ajaib Singh, Station House Officer (SHO) of Sadar Police Station, was personally leading the check-over. One police constable at the barrier recognized Sona as a wanted militant and both of them were taken into custody.

By coincidence, the arrest was witnessed by Manjit Singh, head of the village council of Raja Sansi, an influential man with many contacts in the police. He was looking for a taxi near the barrier when the police nabbed the two. Manjit Singh was a friend of Kulwinder's father. Later, Manjit Singh went to Kulwinder's house and informed his wife Rajbir Kaur who immediately sent a message to her father-in-law in his village Othiyan. Ajaib Singh accompanied by his second son Jagbir Singh rushed to Amritsar and met Inspector Ajaib Singh at Sadar police station, who said that Kulwinder would not be released before a thorough interrogation.

The same night, Sona was killed in a supposed armed combat between the police and a group of militants. Punjab newspapers reported the killing on 21 December. The Tribune said Palwinder Singh Sona was a top militant who carried the designation of a Lieutenant General. The report also said that his three accomplices had escaped and that the police had also killed three other unidentified militants in armed encounters in the outskirts of the city.

The report made Ajaib Singh and his family very nervous. The police could easily kill Kulwinder in their custody and report it as a death of an unidentified militant in a combat. Ajaib Singh decided to negotiate Kulwinder's release for money, without wasting any time. He involved some middlemen, including Manjit Singh of Raja Sansi who knew the Inspector well. Inspector Ajaib Singh demanded one hundred and fifty thousand rupees. Borrowing the amount from his relatives, Ajaib Singh sent it across to the Inspector through the broker who had negotiated the deal. By then, the Inspector had changed his mind. The case was no longer in his hands, he explained. Senior Superintendent of Police (SSP) had taken over the investigation. Ajaib Singh now went to Raghunandan Lal Bhatia, a senior Congress leader and former Minister in the Union government, for help. Bhatia talked to the SSP two three times on telephone. The SSP said Kulwinder had to be interrogated. Ajaib Singh then requested Surinder Singh Kairon, son of former Chief Minister Pratap Singh Kairon and another influential Congress leader in the State, to intervene. Kairon talked to the SSP who again was less than responsive.

Ajaib Singh asked the Sadar police station to formally register his complaint that Kulwinder had been illegally arrested. His friend, Deputy Superintendent of Police Davinder Singh called the Station House Officer to recommend the registration. But Sadar police station refused to do so. Punjab was under the Governor's rule. So, Ajaib Singh sent urgent telegrams to the Governor, the Director General of Police (DGP), the Chief Secretary and the Chief Justice of Punjab and Haryana High Court, informing them about the illegal arrest and beseeching them to intervene. Later, he also sent detailed written petitions about the arrest and the disappearance. But there was no response.

Kulwinder had been very friendly with Birendra Singh Kalon, then Additional District Commissioner of Tarn Taran. Approached by Ajaib Singh for help, Kalon found out that Kulwinder was under interrogation and was being forced to identify wanted Sikh radicals in the area. This was confirmed when a month after his arrest, the police took him to the village of Jagrup Singh Dhotiyan, also a known militant. Later, he was again seen by his former colleagues at Naushera Pannua. Kulwinder was unable to walk and his body showed signs of terrible torture. Ajaib Singh pursued the case of his son relentlessly, although to no avail.

In early 1992, Punjab came under the Congress government with Beant Singh as the Chief Minister. Ajaib Singh again requested Raghunandan Lal Bhatia and Surinder Kairon, who had become a Member of Parliament, to help. But no one could ascertain Kulwinder's whereabouts. Ajaib Singh wanted to know whether he was still being held for anti-insurgency operations, intensified under Beant Singh's regime, or had already been killed. Ajaib Singh met another Congress Member of Parliament Jagmit Singh Brar, who sometimes talked about the issues of justice to Punjab. Brar wrote to Union Home Minister Chavan and, later in March 1993, personally met him to pursue the case. Ajaib Singh was also in Delhi to goad Brar into action. The Union Home Minister talked to Punjab's Director General of Police K. P. S. Gill who confirmed that Kulwinder Singh had been killed.

But there was no formal acknowledgement. The family never received the dead body, nor the mortal remains from the cremation, if it had taken place. The Sadar police station in Amritsar did not even bother to hide or destroy his Suzuki motorcycle No. PB02-C-4455, confiscated at the time of his illegal arrest. The police officers openly used the motorcycle. In 1996, Ajaib Singh engaged lawyer Ranjan Lakhanpal in Chandigarh to file a petition for a writ of habeas corpus – No. 324/1996 – before the Punjab and Haryana High Court. The petition was backed with the supportive affidavit of Manjit Singh, who had witnessed the arrest. The court issued notice. The SHO Ajaib Singh, who had taken Kulwinder into custody on 20 December 1991, became nervous about the possibility of his incrimination if the High Court ordered an inquiry.

Meanwhile, Ajaib Singh's second son Jagbir Singh had been employed by the Shiromani Gurdwara Prabandhak Committee (SGPC), as a dispatch clerk, on the recommendation of then President Gurcharan Singh Tohra. Jagbir worked in the head quarters housed within the Golden Temple Complex. Raghubir Singh was the temple's manager, an influential person within the SGPC who was also acquainted with Inspector Ajaib Singh. In March or April 1997, while the inquiry ordered by the High Court was still pending, Raghubir Singh called Jagbir into his office to make a proposal of payment on the Inspector's behalf so that his father would agree to withdraw the petition. Jagbir reported the proposal to his father, Ajaib Singh, who remained silent for a while after Jagbir completed the narration of his meeting with Raghuvir Singh. Ajaib Singh then asked if he wished to accept the proposal. Jagbir said no. Ajaib Singh repeated the question again and again, with Jagbir affirming no compromise. Ajaib Singh was suddenly seething with anger. He was confident that the High Court would do him justice.

In the beginning, the matter seemed to be moving in the right direction. At the time of crucial hearing, lawyer Ranjan Lakhanpal went away to America and Canada on invitation from the Sikh expatriate community to lecture on the human rights situation in Punjab. His junior was unable to attend the court proceedings. The judge had also changed. The new judge dismissed the petition in October 1996 on the ground of insufficient evidence of the claim that the police had abducted his son. Returning from his foreign tour, Lakhanpal promised to take the matter to the Supreme Court, which, in the meantime, had taken cognizance of illegal mass cremations of supposedly unidentified bodies by the Punjab police force. The Supreme Court referred the matter to the National Human Rights Commission for determination, after the Central Bureau of Investigation corroborated the allegations in its report submitted in December 1996. Ajaib Singh expected his case to come up before the Supreme Court, as Lakhanpal had promised. But nothing happened. For the next month or two, Ajaib Singh remained very distressed. Gurcharan Singh Tohra advised him to engage a Supreme Court lawyer in Delhi to file a fresh petition. Lawyer R. S. Sodhi demanded twenty-five thousand rupees, which he immediately paid up. For some time thereafter, Ajaib Singh remained eager with the impression that the hearing before the Supreme Court would soon commence. When it turned out that the court had not admitted the petition, he was crestfallen.

Soon, he started making fresh rounds of Chandigarh where he met the newly elected Akali Chief Minister Prakash Singh Badal. Before the elections, the Akali Dal had promised justice to all victims of human rights violations that had taken place over the last decade. Badal proposed to mark an inquiry about his case to the Police Inspector General of the Border Range. Ajaib Singh said, that the policemen who had murdered his son would not, therefore, allow the truth to come out. Badal then marked the inquiry to the Deputy Commissioner of Amritsar. This was in May 1997. The Deputy Commissioner recorded the statements of several witnesses, and closed the inquiry on 26 June.

It is not clear what happened thereafter, but on 4 July 1997 Ajaib Singh returned home in a dejected state of mind. Apparently, he had found out that the inquiry report was not going to say anything conclusive. For the next couple of days, he did not talk to anyone and remained unusually calm. His wife Manjit Kaur tried to engage him in conversation, and to draw him out of his depression. She suggested that they sell a piece of their agricultural land to raise the resources to pursue their son's case in other ways. But Ajaib Singh remained silent.

On 7 July 1997 morning Ajaib Singh left the house after announcing that he was going to the Golden Temple, Darbar Sahib. Although the family members had been very troubled about his unusual silence, they did not see anything aberrant in his visiting the shrine. It was routine. After doing the round of the temple's circumference and offering obeisance at the main sanctuary, Ajaib Singh sat down on the platform under the northern gate. One relative, who was also visiting the temple, saw him there scribbling something in his diary. The relative assumed that he was calculating or writing something concerning his business as a property dealer. Mota Singh, correspondent of *Az Di Awaz*, a daily newspaper published from Jalandhar, also saw him likewise engaged in writing something.

Ajaib Singh was composing his suicide note. Probably, he had already consumed poison, which he had somehow procured and taken along with him to the Golden Temple. After finishing the letter, Ajaib Singh walked into the premise of the Bank of Punjab, within the temple complex, whose manager Avtar Singh was his neighbour on Sahibzada Zujjar Singh Avenue. Ajaib Singh, who must already have been feeling the poison's effect, told him that he had swallowed powerful toxins and would not live long. Avtar Singh probably could not grasp the seriousness of his situation and sent him home in his car. Ajaib Singh was vomiting and told his son Jagbir that he had taken poison with the intention to die because he could not bear the injustice any longer. Immediately, the family members rushed him to a hospital where the doctors, after examining him, said that he had to be rushed to the main Civil Hospital. Ajaib Singh was already dead when doctors at the civil hospital looked him up.[2]

Some officials approached Jagbir Singh to suggest that he should mention heart attack as the cause of his father's death. Otherwise, there would be consequences, they warned. But Jagbir Singh stated the truth and released his suicide note to the press. The national press blacked out the story. The following is a short translation of the suicide note. It is dated Monday, 7 July 1997 and it says:

> In this house of Guru Ram Das, I seek forgiveness from everyone whom I may have unwittingly hurt or wronged in anyway. Self-annihilation is the only way out of a tyranny that leaves no chance for justice. Tyrants like former SSP Ajit Singh Sandhu, who eliminated thousands of innocent Sikhs and also extorted millions of rupees, also sometimes commit suicide under the unbearable weight of their sins. It is known that Jaswant Singh Khalra had become Sandhu's victim for his human rights work.
>
> My son Kulwinder Singh was picked up by SHO Ajaib Singh of Sadar police station in Amritsar. Neither had he committed a crime nor was he absconding from the police custody. Why did the SHO kill him then? I understand that the time and the place of one's death are predetermined. SHO Ajaib Singh got promoted to the rank of Deputy Superintendent of Police for eliminating hundreds of young Sikhs. As an officer, he extorted millions of rupees from the people whom he held at his mercy.
>
> My grudge is that no one even confirmed my son's death. I did not even receive his ashes. Otherwise, I would not have gone to the High Court for justice, which I never received. I approached Chief Minister Prakash Singh Badal. Some people rightly say that he is not a fit person to rule Punjab. I had approached him for justice. He marked the case to the Deputy Commissioner of Amritsar for inquiry. Badal should find out if he had conducted a fair inquiry. I believe DSP Ajaib Singh gave him a handsome amount of money to muddle it up.

> I pray to Guru Ram Das to send me where my son is. I hope my prayers would be answered. Once again I apologize to the residents of my colony, my village Othian and Gumtala for any inconvenience I may have caused them. I am not in anyone's financial debt. Nevertheless, I authorize DSP Davinder Singh to sell my land to settle any claim of liability that may arise. I am grateful for the support I have received from Manjit Singh, Sarpanch of Raja Sansi, DSP Davinder Singh and Bibi Paramjit Kaur Khalra. I wish to be cremated near the Martyr's Shrine, Gurudwara Shahindan. I do not wish any rituals, except the recitation of the Guru Granth, to follow my death. If my family wants to offer any service, it should be made to the charitable organization of Pingalwara.
> Now the ink in my pen, also my time in this world is running to their end.
> Wahe Guruji ka Khalsa, Waheguru ji ki Fateh![3]

Ram Narayan Kumar, the author of the report on disappearances in Punjab from which the above narrative has been taken, terms the case as a 'paradigmatic case', and the situation as one of 'complex denial'.[4] Kulwinder Singh's enforced disappearance, the life-exhausting and fruitless pursuit of accountability and justice by his father Sardar Ajaib Singh, and then suicide by the psychologically broken man, the trauma and the ruin of the surviving members – none of these is unique to Punjab. Punjab's countryside is dotted with myriad other examples of people who became victims of India's war against the Sikh separatist threat. But, unlike Kulwinder, most victims are to be found among the poor, the uneducated and the powerless, who cannot afford the mechanical and grinding legal process. Yet, everybody denies what is happening. Organs of law deny justice. The political class denies that justice is being denied. The security forces deny that arbitrary killings are taking place. The victims out of fear deny testifying that gross violations have taken place. State denies that it enjoys impunity in its actions and claims that it follows rules of accountability. Militants deny they practice policies of revenge and that victims have turned killers. And everyone thinks that if peace has returned through accords, ruthless policies of the state, and helplessness of the victims, at least peace be given a chance now. Peace accord, the institutional core of peace process, assumes a larger than life, even if it has failed to evolve or decided not to evolve a workable system of state accountability, to take into account public opinion to counter the bid for immunity, to recognize the realities of torture, enforced disappearances, accountability, denial of justice including compensation to victims of abuse of power and other related matters.

The denial as in Punjab took toll not only lives like that of the helpless father Ajaib Singh, but of human rights defenders such as Jaswant Singh Khalra. And even a police officer by the name of Ajit Singh Sandhu, former Superintendent of Tarn Taran police district, imprisoned for few months on charges, established by judicial inquiries, of committing illegal abductions, torture and elimination in custody of people like Jaswant Singh Khalra and Kuljit Singh Dhat, a relative of Bhagat Singh the famous revolutionary from the pre-independence era, had to commit suicide later by throwing himself before a running train reportedly after consuming alcohol, and driving to the railway track in his own car, leaving a short suicide note behind that said, 'it is better to die than to live in this shame'.[5]

As already mentioned, India's Prime Minister Rajiv Gandhi and the President of the Akali Dal Harcharan Singh Longowal had signed on 24 July 1985 the Punjab Peace Accord. Akali Dal, representing the spirit of the Accord won 72 out of 117 seats in the elections to the State Legislative Assembly in September 1985. The Sikh militants had already assassinated Longowal on the ground that he had compromised with a central government that was guilty of destroying the Golden Temple and of conniving in the Delhi massacre of the Sikhs in November 1984. But the Akali Dal under Surjit Singh Barnala stood by the terms of the Accord, which promised to 'usher in an era of amity, goodwill and cooperation' between the Sikhs and the Centre. The Union government however dismissed the elected Akali government in Punjab in October 1987 on the ground that it had failed to safeguard the Hindu interests in the State against the Sikh militant attacks. The Accord had delineated several points of common consent. It had promised to (a) transfer Chandigarh to Punjab by 26 January 1986, (b) set up tribunals presided by Supreme Court judges to adjudicate the river water and territorial disputes, (c) refer the Akali Dal resolution for provincial autonomy to a Commission appointed to recommend changes in the 'Centre-State relationship to bring out the true federal characteristics of our unitary constitution', (d) to inquire into the Delhi killings of November 1984, (e) withdraw the Armed Forces Special Powers Act from Punjab, (f) and restore the rule of law and human rights in Punjab.[6]

In the event, the Central government failed in implementing the Accord. Chandigarh was not transferred to Punjab as promised. The Commissions on the river waters and territorial disputes were scuttled. Guilty for the November 1984 massacre in Delhi remained unpunished. On 26 January 1986, a large congregation of Sikhs had gathered under the leadership of Bhindranwale's followers at the site of the demolished Akal Takht to review the political developments in the State. Four days earlier, the court that had been trying the case of Indira Gandhi's assassination handed out its judgement sentencing all the accused to death by hanging.

The political resolution adopted by the congregation said that if the assassins of Indira Gandhi are hanged, they would become the first martyrs of Khalistan. The congregation also appointed a five-member panel called the Panthic Committee to guide the Sikh struggle to its goals, which the incumbent Akali government had betrayed. A repeat of the congregation, called Sarbat Khalsa, held on 13 April 1986, adopted a political resolution that asked the Sikhs to break the shackles of slavery to India. A fortnight later, the Panthic Committee declared the 'formation of Khalistan', also creating its own army called Khalistan Commando Force that would fight for the objective.[7] The next day, the government of Punjab sent the army to the Golden Temple to flush out the separatists. But the members of the Panthic Committee had already disappeared. The Accord had within less than half a decade of its birth had run a full circle.

Then Governor of Punjab Sidhartha Shankar Ray and his Police Chief Julio Ribeiro had deliberated with Gurdayal Singh a veteran police officer on their plans to create armed groups that would take on the militant menace without involving the State apparatus directly. Gurdayal Singh, who had retired as the Inspector General of Police for Punjab in April 1966, had been Lahore's Superintendent of Police before India's partition in 1947. As the Deputy Inspector

General (Intelligence) from 1952 to 1956 he had closely followed the movement for the creation of a Punjabi speaking State. One of his main tasks was to neutralize the Communist insurgency in parts of Punjab, which had been led by Teja Singh Swatantra in the years between 1950 and 1960. Because of his reputation as an officer who had successfully handled many explosive situations, the rulers of Punjab still consulted him on how to deal with the separatist violence. Gurdayal Singh advised them not to pursue these plans, as, in his opinion, unscrupulous elements would thrive under official patronage. Apparently, Ray and Ribeiro went ahead with their undercover operations, using informers and infiltrators from the underworld. Ribeiro concedes this in his book 'Bullet for Bullet', which he authored after his retirement.[8] Ribeiro writes with extraordinary candour, 'In Punjab there were some persons with criminal propensities, who were known to police officers at various levels. They were approached and a few of them agreed to form groups, which would move in the guise of terrorists and confront the real militants in their dens...The police did give them financial and logistical support, but their demands grew to an extent where it was impossible to satisfy them within our resources. Besides, they were very greedy people, with a criminal tendency, who began to prey on law-abiding, rich citizens on the assumption that the police were indebted to them and so would do nothing to stop them.'[9]

Meanwhile, the situation in Punjab was witnessing a runaway deterioration, with a steep increase in the daily reports of Sikh extremist outrages and summary executions by the security forces. In March 1988, the Indian parliament passed the 59th Amendment of the Constitution which enabled the central government to extend the President's rule in the State beyond one year; to impose emergency on the ground of 'internal disturbance' and to suspend Article 21 of the Constitution which guaranteed that no person shall be deprived of life and liberty except according to the procedure established by law. The Union government dragooned this unheard-of constitutional amendment through parliament, despite all the special legislation already at its disposal (which not only conflicted with the elemental principles of due process, but also eliminated the existing legal safeguards of free and fair trial). The legislation already in force included the Terrorist and Disruptive Activities (Prevention) Act (TADA), which provided death sentence for terrorist actions resulting in death and the minimum term of five years in imprisonment extendable to life for other offences. Section 21 of TADA commanded the presumption of guilt against the accused. Its definition of 'abetment' in section 2 (1.a) eliminated the proof of criminal intention. The section 15 of TADA allowed a police officer of the Superintendent's rank to record confessions of the accused in custody and to use them as evidence against them. Preclusion of anticipatory bail by 20 (7) of TADA destroyed the protection, which the section 438 of the Code of Criminal Procedure offers to the innocent against manifest abuse of police power. Section 20 (4.b) of TADA allowed sixty days of police custody of an accused under interrogation, and one year of judicial remand without bail. Section 20 (8) prohibited bail even when the prosecution failed to furnish a charge sheet after ninety days of arrest. The section said that no person accused of an offence under TADA would be released on bail unless the designated court was satisfied on 'reasonable grounds' that 'he is not guilty of such offence and that he is not likely to commit any offence while on bail.' The Act did not explain how the accused should adduce the evidence of his innocence in the absence of a charge sheet, or how the judge should authenticate his guiltlessness ahead of the actual trial, and go on to certify that he would not 'commit any offence while on bail.' TADA cases were heard in special courts by executive magistrates who were appointed centrally. The hearings were held in camera, and could be held in locations far removed from the disturbed area itself. In March 1994, the Supreme Court of India upheld the validity of TADA. Apart from TADA and the Terrorist Affected Areas (Special Courts) Act, 1984, there were also other black laws like the National Security Act, 1980, as amended by the Act 24 of 1984 specifically with the reference to '*the extremist and terrorist elements in the disturbed areas of Punjab and Chandigarh*'. The Act provided for detention without charge or trial for one year in all parts of India, and two years in Punjab. Also in force was the Armed Forces (Punjab and Chandigarh) Special Powers Act, which empowered the security forces to enter and search any premises, and to arrest any person without warrant. It also allowed the security forces to destroy any place on the suspicion of being a 'terrorist hideout' and to shoot to kill a suspected terrorist with immunity from prosecution.

Thus, with 'political, legislative, and judicial consensus on state terrorism' in the guise of counter-insurgency, peace process in Punjab was declared successful, and the Accord was anointed with killings, impunity, and what Ram Narayan Kumar has called repeatedly a situation of 'complex denial'. Desire for peace, killing for peace, silencing all voices for peace – all were mixed up in this consensus. If in the preceding chapter I have suggested how dialogic politics can lend a critical edge to politics by suggesting new forms of accommodation on the basis of minimal justice, we must remember also, and as repeatedly mentioned in previous pages, that dialogue is a contentious act – in other words, dialogue not only contends and contests the entrenched positions of power, dialogue can be used as a strategy also as part of the techniques of governing society, particularly an unruly society. Nothing brings this out more than the governmental business of establishing peace accords, in which deaths are accepted as 'collateral damage' for 'waging peace' (another phrase in currency), and the deaths of the past are not supposed to upset the peace achieved through the today's dialogues.

Locus of Accord as an Event Caught in Two Desires

Peace dialogues are thus haunted by war crimes of the past, and today's process of dialogue is inhabited by ghosts of yesterday. When speaking of dialogue, it is necessary therefore to take a longer view about the matter, about problems forgotten, and about old dilemmas that today's process of dialogue and peace may confront. I began this chapter with detailed references to events in Punjab including the Punjab Accord, events that followed the Accord, and how deaths put the seal of approval on the Accord, in order to point out the general direction that peace accords have taken in this region. The history of peace accords is worth visiting not only because these accords were aimed at bringing peace, or some of these accords died in the course of the pursuit of peace, or some of these try to make a decisive break with a past marked by conflicts. They are worth visiting also because they show how today's process of dialogue is linked with various ghosts hovering round us. These are the ghosts of past patterns of power present in today's patterns of democracy; ghosts of yesterday when states were being formed through the attrition, and have come back now to haunt the states of today which are trying to reach a plateau of stability through accords with adversaries outside and inside, who may be equally desperate in their desire for a settlement, but acutely aware that peace could come only with dishonour. A study of accords is thus a study of the process of dialogue, which indeed by being so becomes a study of state-formation, patterns of power, and a study of the desire of democracy to run away from a geo-political past from which it eternally tries to escape only to find itself never free. As my accounts in the following pages will show, peace accords are located in the intersection of two contradictory elements relating to the state – its geo-political configuration that has made the state possible, and the popular politics (appearing in form of democracy) in demand for dialogue and peace that confronts the geo-political foundations of the state.

Understanding the location of accords is thus an important part of understanding peace process and the mechanisms of dialogue. It is crucial to understanding the tools of political conversation in a world of asymmetries that rely on the instrumentality of contract to strengthen themselves. The history of accords is like a bunch of documents that lies before us. It contains the texts of treaties, accords, memoranda of understanding, resolutions, and also the enabling acts for their implementation. These are the texts of resolution of conflicts, plus these are protocols of power, scripts of status, arrangements of prestige, and the invitational metaphors for further conflicts and peace. Some have failed as tools of peacemaking. Some have succeeded. Some have passed the verdict of history. Some remain as papers in dusty archives. But they belong to a genre, a specific bunch, with their own rules of formation, their procedural laws. They come as contrived, deliberate efforts of agencies, and then they dispose things their own way. They are the result of practical efforts. The agencies contracting them impose a practicality on them, then these treaties as events find their own principle of practicality and regularity.

Treaties and accords are not only understandings belonging to a particular discursive shape. They are also interpretative documents of the current reality of conflicts. But if we watch them closely, we shall find that their rules of operation make them *acts* on the succeeding pattern of conflicts. They sometimes reverse the order of power, at times they discontinue certain practices; they are specific as events of conflicts. By invoking a particular genre in order to identify themselves, they take recourse to a generality that hides their specificity as external events (for example, they are not treaties, but accords take the form of treaties giving rise to illusions). Thus they are considered as internal to the organism of the conflicts. They appear as the outsider. There is no inherent guarantee of their 'success'; also it is not clear what we mean by *success* in this case, no internal logic, and no nucleus towards which the treaty will move to fulfil avowed aims. The actual occurrence of a treaty/accord is compromised by the external conditions of its possibility, by its locus as an *event of consent and compromise*. They are driven by a desire, a desire for peace. Yet its location as an event of conflict (where compromise is being sought for) makes that desire competitive inside and outside. The serpent in the garden is always there. To this morphology of desire, then, we move in an effort to understand accords as an institution – an institution perched on a conflict between desire for peace and desire for power. The rules of formation of this body of documents are the key to comprehend the process of dialogue, the troubled soul of peacemaking in the region.[10]

In short, accords are documents that contain the secret politics of the two desires I have just referred to – the innocent desire for peace and the culpable desire for power that masquerades as the desire for peace. Together, they make accords the perennial but impossible invitation to peace. Hostage to the vagaries of statecraft, and therefore hospitable to the idea that 'peace with justice' is impossible the accord remains the 'other' to both war party and peace party. The war party thinks that accord is surrender; the peace party thinks that the accord is a tool in the hands of the war party for the renewal of hostilities. Ridiculed, maligned and reviled, the institution of accord in the politics of the region is never fully legitimate, ironically not even to the contracting parties. It remains the eternal outsider to the politics that had projected the institution of accord on the ground of the swelling desire for dialogue for peace.

Yet the nation cannot do away with them. As the supreme community claiming allegiance of the highest level from its members, it cannot de-legitimize other communities, or at the least it cannot banish the idea of other collectives. So, the idea of a nation entering into a compact with solidarities is not surprising. In some cases, the nation is born out of such compact. In some cases, the founding constitution of a nation is in the nature of an accord, full or in part. Thus the Indian constitution may be seen as an accord, or an ensemble of accords. A close reading of the debates in the Indian Constituent Assembly shows how even seemingly very fundamental parts of the Indian constitution were compromises between communities, opinions, and regions. Article 370 was one such accord. There were crucial amendments to the Indian constitution, such as the State Reorganization Act of 1956 that amended the Fourth Schedule and reinforced the contractual nature of the constitution. In these cases, however, there was very little formal arrangement by way of accords. The Indian Independence Act of 18 July 1947, also, in a way was an accord between India and the erstwhile colonial power, Great Britain. The Cabinet Mission, the subsequent decision to partition the country, the Boundary Commissions and the Radcliffe Awards, the agreements with the native states, and prior to that, the round-table conferences, the Ramsay McDonald Award, the Gandhi-Ambedkar agreement were compacts that laid the contractual basis of the post-colonial polity. The accords signed by the imperial power as the ruler of India with other countries on subjects like borders, and with frontier communities on issues like partial exclusion, total exclusion, inner line regulation and partial autonomy remained valid for the successor state. Also, paramountcy may have remained vague and undefined, but the paramount power had become so through treaties, accords and understandings. The politics of accord was thus embedded in the process of de-colonization itself. Moreover, in this region the newly independent nations were obliged to enter into mutual accords soon, as problems of a common past appeared intractable without compacts.

The Institutional History of Peacemaking – Hegemonic and the Modular

But before dealing with the various types of accords that emerged over time, we must briefly look into the institutional history of peace in the sub-continent that has subsumed in it the phenomenon of the *desire* for peace, which I mentioned above. Antonio Gramsci had written that the modern regime of power depended for its existence on a strategy of hegemony that implied a mechanism of securing consent of the ruled.[11] In the political history of post-independent India we must look for the evolution of the appropriate hegemonic forms – forms that made possible the overlap of the two desires – the desire for peace and desire for power, turned peace dialogues and accords into a tool of rule and governing, and adapted the geo-political form of rule to its ethno-political form by making accord the combined configuration of both varieties of modern politics. And for that let us once again go through a story of peacemaking – a fragment from the political history of South Asia.

Fear

When the Naga Club of Kohima had presented a 'memorandum of the Naga Hills' to the Simon Commission on 10 January 1929, all they had wanted was their exclusion from the Reformed Scheme, because the Nagas had lived 'in a state of intermittent warfare with the Assamese of the Assam valley to the *North* and *West* of *our* country and Manipuris in the *South*' (italics mine). They said they had 'no unity among us' (meaning among the Nagas) and it was really the British Government 'holding us together'. They could perhaps boast of two three university graduates, but no common language, no way to 'represent us (them) in any council or province'. The prayer was therefore to the Government to continue to safeguard their rights by excluding the Naga Hills so that they were free from 'new and heavy taxes'. They did not know how they could be represented in a council, or saved from new policies and systems. Here were the promises of an accord with the paramount power to the effect that their inability and difficulties be understood and alternative arrangements made.

Hope

By the time the British India Act of 1935 had placed the Naga Hills outside the Reformed Scheme, making it an 'excluded area', the war was on the doors. Therefore the exclusion led to only further opening. It was an exposure to the forces of war and nationalism represented by the British forces as well as the Indian National Army. The Naga Hills District Tribal Council graduated into Naga National Council (NNC) whose aim, as the NNC constitution was to declare, was 'to provide a way of working for developing what is good of their own culture and customs, and to promote the growth of democratic self-government and the material welfare of the Naga people'. Sir Reginald Coupland suggested a British crown colony comprising the areas on the eastern frontier of Assam and the northwestern frontier of Burma whose inhabitants are of 'Mongol stock'. Colonial officials of that area like Hutton, Parrey and Mills supported the proposal. Hope was on the guardian power, on the successor-state of India and an interim arrangement of ten years. Therefore on 20 February 1947, and again on 19 May 1947, the NNC sent memoranda hoping that the grey area of 'shared sovereignty' would find appropriate institutional forms. In June Akbar Hydari agreement materialized. Not to be left behind in this voyage of hope, meanwhile The Prime Minister of the Indian Interim Government Pandit Nehru had also joined with these words, 'I see no reason whatever why an extraneous judicial system should be enforced upon the Naga Hills'. And in little more clear terms, he had written to the NNC secretary, T. Sakhrie, 'As for separate electorates for the Nagas, I am not clear in my mind as to how this will work. Generally speaking, we are against separate electorates as these limit and injure the small group by keeping it separated from the rest of the nation. But if the Naga territory is given a measure of autonomy, some arrangement will have to be made for their proper representation.' The NNC said, 'our country is connected with India...in many ways...We shall enjoy home-rule in our country, but on broader issues be connected with India'. Thus when on 26 June 1947, barely two months before the denouement, the then Governor of Assam, Sir Akbar Hydari, was sent to Kohima to impress upon the Nagas the reality of the political situation, it was difficult to vouch who was to impress upon whom. Hope overwhelmed the reality of a successor sovereign state, the brutal fact of the unwillingness of the retreating colonial power to do anything substantive for the Nagas, the total lack of a political preparation in finding out appropriate forms of settlement. The Nagas by the 9-point agreement with Sir Hydari were to have judicial powers, executive powers, legislative powers. They were to enjoy autonomy in land related matters, capacity to tax; they were given promises of unification of the Naga territories, and of continuity of regulation and exclusion arrangement. Finally, there was a ten-year guarantee of these provisions at the end of which 'the Naga National Council will be asked whether they require the above agreement to be extended for a further period or a new agreement regarding the future of the Naga people (to) be arrived at.' And welcoming the governor, Sakherie's anguish was also clear – the territories inhabited by the Nagas be united at once, 'old boundary be restored', '...part of Sibsagar or Nowgong district' be returned. The Naga leader added, 'the Naga people have the assurance that Your Excellency is anxious to meet their desires, and that their case will receive Your Excellency's most sympathetic consideration'.

Hope had succeeded fear, never mind that it was short-lived.

Disillusion, Despair

The Nagas had wanted a guardian power only. But nothing concrete seemed to have emerged. The impasse continued. The Constituent Assembly refused to ratify the Hydari accord. Governor Hydari himself showed bad faith and threatened the use of force if after ten years the Nagas still showed intransigence. The premier of Assam Gopinath Bardoloi declared it 'implementable'. And the over all stand of the government was that the Government of India had never agreed to talk of Naga independence or grant it after the ten-year interregnum. On 19 July the same year just a month before independence of India and the 'independence' that the Nagas declared, an eleven-member Naga delegation met Gandhi. The NNC secretary Sakhrie had expressed fears in his letter to Gandhi that the Indian government might send troops. Gandhi had assured the Naga delegation led by Phizo, Viselie, Khrehie and Seto that he would go to Kohima and he would 'be the first to be shot before any Naga is killed'. Gandhi died before he could keep his promise, the Constituent Assembly refused to consider anything, the Nagas declared independence, and then they went ahead with plebiscite on 16 May 1952. Meanwhile the government had offered them only district autonomy, and the NNC was banned in 1952. Nehru and U Nu's joint visit to Kohima on 30 March 1953 was fruitless, Nehru was not prepared to listen to the Nagas, and the Nagas said in that case they had no reason to listen to his and his guest's speeches. Phizo reorganized the set up. 'Naga Central Government' replaced the 'Sovereign Republic of Nagaland' and in 1957 the 'Federal Republic of Nagaland' replaced the 'Naga Central Government'. State violence had put finish to whatever chance there was of a peaceful settlement, perhaps on the lines of the Hydari accord. Disillusion and despair had given birth to anger. The new Naga leadership summarily served notice on the previous generation of leaders to withdraw. The grand man of Naga awakening T. Sakhrie was killed. Jasokie, Silie had to ask for refuge from the Government of India. The 'overground politics' was thus born. From then on, the Naga political society went on producing a middle ground whose *middle-ness* was made of the ambiguous space between the two combatants – the space that was affected by conflict of loyalties and existences, but remained sympathetic to the underground. The political forms also showed the marks of despair, and its origins – fear and hope. The Naga People's Convention entered into a sixteen-point agreement that centred on statehood within the union.

We have of course the easy way to explain all this. We can say that the Naga society threw up a middle class, that the middle stratum occupied the *middle* ground, that these developments led a section of the political society in Nagaland astray from their declared path. All this may be true in terms of a long history of the Naga society. But here we are cutting short the long trails that each history invariably has, and want to enter into the specifics of the dynamics that establish accord as an institution of dialogue between state and its adversaries.

The Denounced Accord

On 1 December 1963, Nagaland was declared the sixteenth state within the Indian union on the lines of the 16-point agreement. In 1964 with the initiative of the Baptist Church that had set up the Nagaland Peace Council, a cease-fire agreement was reached. In 1972 the last trace of the promise of the Hydari agreement was removed with the quiet transfer of affairs of Nagaland from the Ministry of External Affairs to that of Home. The Peace Mission had an unceremonious end.

Rev. Scott had been deported a few years back. The cease-fire had been over and there was massive military suppression. The middle ground had caved in completely. The denouement was in form of Shillong Accord of 1975, in the words of an observer, 'the accord that never was'.[12]

Yet, the point is why did peace process not end with Shillong accord?[13] Indeed, why does peace process never end with one accord, particularly an accord that ostensibly puts the seal of finality on a modality to end discord, an accord that finalizes a compromise, or an uneasy postponement, or say a defeat? 'On the ashes of the Shillong accord', in the words of an insurgent leader, grew the National Socialist Council of Nagaland. But one can say with equal justification that from that 'ruin' again emerges the hope for a *peace with justice* – peace that is not dishonourable, peace that is brought with dignity. Why again today the echoes of peace-parleys, the prayer bells for peace from the churches of Nagaland? Why again the talks, the strains of dialogue? We have in this case the different elements again active in the peace process, the same quarrel *around* the middle ground, the same dialectic, the same search *for* the middle ground, and the same overlap of the innocent desire for peace and the culpable desire for power. The story of desire is thus at once a bunch of many stories – of power, forms, regularity, and breaks. We shall point out just three principles of the dynamics.

One, the middle ground of the conflict situation reproduces itself with increasing eloquence. At one point, the middle ground no longer remains just the middle space to be squeezed out from the two sides. The middle becomes by virtue of its middle-ness the centre around which the contest revolves. The terms of contest thus undergo change – for example, the contest ensues as to who will occupy the central space?[14]

Two, social institutions exert influence on peace process and reactions in civil society constitute an important element of a political society engaged in the process of peace. People are not just receivers of peace. An inquiry into the specificity of the peace process is thus an inquiry into several 'autonomous' domains, into the subaltern nature of the desire for peace.

Third, the form of a settlement, earlier arrived, exercises tremendous influence on later forms of peace.

Thus the question of agenda-setting capacity of one or both combatants is important. Evidently what is called the 'modular form' cannot accommodate all the rival claims, viewpoints, and interests. But being modular it does not allow other form/s to emerge. The present history of peace is perched on that conundrum. But at least awareness is necessary – how does the *modular* form arise? When and how does a form become a *model*?

The linguistic reorganization of states in India, carried out through the Seventh Amendment, established the model of autonomy and federalism. The refrain was, grant statehood if the disgruntled persevere. It also standardized the manifold form of political entities – princely states, excluded areas, partially excluded areas, and directly ruled states. Gaining statehood in the union was the agenda that captured political imagination. It symbolized federalism and access to resources and power. It made possible a compromise between the nation-state and ethnicity. It extended the manageable limit of democracy. In short, it became a perfect instance of 'governmentality'. But more important to note, unless the form was repeated, it could not become a model. Therefore, the interest of the state has been in making the form viable, acceptable to parties, populations, and combatants. Of course the form itself had to be made flexible. Therefore, Article 371A (or Article 370) had to be added, though with that came the Armed Forces Special Powers Act in 1958 and in amended form in 1972, the Assam Maintenance of Public Order (Autonomous Districts) Act in 1953, or the Assam Disturbed Areas Act in 1955, that made a mockery of the grant of this very statehood. Also, the old colonial practice of inner line regulations became a model to be repeated. The accords were thus founded on a combination of two principles: statehood, and a measure of more autonomy on the basis of some regulation. Once the original form was arrived at through the long constitutional practice of Reforms Acts, Act of 1935, and then the constitution of independent India, accords gained a modular form. Behind the accords thus stands the institutional history of governing the country, the received mode of accord, and beyond this, against this, the desire for justice and peace which is reflected no doubt in a grotesque way in the accord. That is why an accord is never free from the supplement, often dangerous.

Yet the institution of accord remains the *outsider* to mainstream politics. Partly because since the early accords were not formalized, they were *natural*; none looked at them as contrived or deliberate. With the naturalization of accords, sovereignty as the spirit of the political society became the core of the nation-form. But this begs a question. Was it because the earlier understandings were not brought to bear by pressure of arms, that they did not seem to have impinged on the sovereignty of the nation-form? Or, was it because, these were the building pieces of the 'sovereign' nation-form, that they had to be naturalized, so that the nation does not remember them as accords? In any case, the post-1975 accords in India and the accords in the pre-independence era and the early years of the republic are two different bunches. One set related to the time of transition, was part of the particular technique of de-colonization, and was considered as the natural building bloc of the new nation-state. The second relates to the time of consolidation of the state, is part of the encounter between the nation, insurgents and rebels, and considered as the disrupting element of the post-colonial nation. The nature of this *difference* is crucial to our understanding, first, because as we can see, with the former bunch of accords *naturally* melting into the second, people are persuaded to think that accords start only with the appearance of insurgency in the post-colonial history of the nation. Second, the latter bunch is seen as the marker of the surrender of the Indian state and its principles of territorial integrity and national integration to the militants and secessionist communities. Third, such a fragile difference makes the genealogical task of critiquing the politics of accords, difficult. Fourth, the accords belonging to the latter bunch are formal documents, and therefore subject to certain rules of application, examination, acceptance/rejection by sections of population, the political classes, and the state. In short, the first group produced *consensus*, the second produces discord. And, now, we can see, why accord as a politics and as an institution remains outside the mainstream politics of the nation.

Connected to this, is the even more relevant inquiry – does accord as a particular institutional politics create its own public? Again, as an answer to this we may look back to the way the first set of accords was backed up by a public, or created a public of its own. The States Reorganization Commission and the seventh amendment to the Indian constitution were preceded by huge political mobilizations, a consensus was established, and the seal was put on it in form of the Seventh Amendment Act of 1956. Ironically, the Act itself has produced in course of time a public in defence of regional autonomy. Today, we can have more states in India, but not less. Even in the matter of centre-state relations there is public consensus that states are not the subsidiary creations of the centre, rather states add up and make the latter. Accord, consensus, public – the trio seems to have led a charmed life so far. This has in fact solved also some of the tension implicit in the Preamble and the closely running first three articles of the Indian Constitution: (a) the principle of popular sovereignty as the phrase runs in the preamble, 'We the people of India having solemnly resolved to constitute India into a sovereign...republic'; (b) 'India is a union of states' (Art. 1); (c) the Parliament is empowered to admit into the Union, or establish, new states on terms and conditions it thinks fit (Art. 2); and (d) the Parliament can increase or decrease the area of any state, or alter the boundaries, or change the name of any state (Art. 3). In achieving a liberal solution to the nationality problem we can say that the unspoken accords, the consensus, and the re-organization of the polity were desirable to the Indian rulers and played a critical role.

But why does this consensus elude the accords in the second bunch? Long after 1956, when the re-organization of the framework of relations between the state and the communities (having territorial implications) in India seemed to have been achieved, the desire for identity/ autonomy/independence re-surfaced. In Punjab, Jammu and Kashmir, in Northeast, sub-Himalayan West Bengal, Sikkim, and in south Bihar, demands persisted for a re-negotiation of the relationship. These demands, as we all know, did not get the sympathy of the nation. In the absence of a national consensus, no significant re-negotiation took place. What made this consensus elusive, and therefore in spite of the accords made peace difficult? If the earlier phase was seen as the vindication of *democracy*, the latter phase is seen as vindication of the phenomenon of *ethnicity*. Conceding *regional* aspirations was democratic, and by that judgement, negotiating sub-regionalism was a political-administrative task. There was no panic. But ethnicity, ethno-centrism and ethno-nationalism – these were different. Ethnicity vandalizes the nation, whereas democracy makes it sacred. Ethnic aspiration is not democratic aspiration. Therefore accords of the second bunch not only did not lead to consensus, they brought more fragmentation in their wake. Clearly, in the cosmology of the nation, these were two different orders.

The question of course is: why are desires for democracy seen today as 'ethnic aspirations'? Self-perception in the nationalist mirror is critical in this case. Thus nation for me, ethnicity for you; likewise we are going global, you are getting fragmented; ours is democracy, yours is violence; we are sovereign and therefore we can contract treaties, your locus is subordinate, so you can have accords. In this unending series of transgression we also find regression when we look at this series of inversion from the point of democracy. The politics of the nation-state makes the desire for similar national aspirations appear as contending. But more important, nation whose secret origin lay among others in the desire for democracy is consumed by the 'spirit of the state' to an extent where nation almost kills the very spring well from which it emerged. It is this transmutation that creates 'ethnicity' – appearing as a dark primeval force and a terrifying phenomenon to every outsider, including the state, the nation and the global players. State is terrified because it finds its profile in it, the nation is terrified because it appears as contending for loyalty, and the global players are scared because globalization is spawning fragmentation. Yet, a nation-state cannot do without it. For, having legitimized the community, creating a certain institutional framework for accommodating various aspirations, and accepting democracy as the standard of the political nation, the nation has to dialogue with the 'ethnic groups'. Through the dialogue between nation and ethnicity, they live off each other. They are the quarrelsome duo, eternally fighting, eternally making up. The politics of accord is marked by this fluidity and if anything stamps it with its presence it is the uncertainty, the absolutely contemporaneous nature of the agreement. What therefore marks the accord is absence of the element, thought to be the most critical part of it, namely, *understanding* – understanding different positions, differences.

The institution of accord thus occupies a significant place in the way in which the politics of dialogue plays itself out. It shows in a congealed form the contradictions in the location of dialogue. It shows how dialogue on peace conducted towards concluding an accord bases itself on an earlier discourse of peace, and how the hegemonic and the modular form of peace indicates the ways in which dialogue becomes part of a strategy to rule. Finally, it shows how the given geo-political constitution of the state becomes instrumental in naturalizing peace politics, making it banal, divesting it of the critical edge in form of the element of justice. Yet, notwithstanding all these, the institution of accord shows ways in which the desire for a 'just peace' surfaces repeatedly, also the ways in which this desire makes the dialogic process an eternal outsider to the politics of rule, with its own specific rules of formation.

Accords, Governmentality and the Acrimonious Language of Peace

We can pursue the point further. For instance, we are faced with the problem, namely that, understanding in politics indicates consensus, but readers can recall that we opened our discussion by noting that consensus eludes the actors of accords, and therefore accords are not marks of consensus but of discord. To put the problem succinctly, if accords signify achievement of understanding, what explains the chief marker of accords, namely, lack of understanding? Are we to therefore put our fingers elsewhere – on the language of dialogue – as the key to the intrigue? The problem in other words is not one of deciding whether dialogue achieves understanding or fails in achieving so, but that as an institution it builds on contentious language, and revolves around contesting conversation, so much so that understanding and bitterness are but two parts of the same process, known as contentious dialogue.

It is significant to note that V.P. Menon, Secretary of the Ministry of States of the Government of India in 1947-51, writing of the integration of the Indian states had commented that behind the agreements with the princely states lay the vision of Sardar Vallabhbhai Patel who conceived them as 'co-architects' and 'partners in the work of national consolidation and reconstruction'.[15] These co-architects and partners, Menon assures us, were not mobilized under duress, so much so that he declared elsewhere in the same chronicle that India had no territorial ambition in Kashmir. And therefore 'I can say in the face of any contradiction that the Government of India would have left Kashmir alone'.[16] Though, he prefaced this line with these words,

> Personally, when I recommended to the Government of India the acceptance of the accession of the Maharajah of Kashmir, I had in mind one consideration and one consideration alone, viz., that the invasion of Kashmir by the raiders was a grave threat to the integrity of India. Ever since the time of Mahmud Gazni, that is to say for nearly eight centuries, with but a brief interval during the Moghul epoch, India had been subjected to periodical invasions from the north-west. Mahmud Gazni had led no less than seventeen of these incursions in person. And within less than ten weeks of the establishment of the new state of Pakistan, its very first act was to let loose a tribal invasion through the north-west. Srinagar today, Delhi tomorrow. A nation that forgets its history or its geography does so at its peril.[17]

We can for the time being leave aside the historical vision of a bureaucrat, and note that consensus is said to build on 'partnership'. Partnership, in other words, is securing of 'consent' within the basic minimum, for beyond that if the nation is going to accommodate others, it would signal 'peril'. In consensus therefore we have the question of power. Accords with those who are not partners, whose consent cannot be ensured, and who symbolize future peril, are thus not documents of understanding, but papers that keep in abeyance immediate hostilities as a strategic part of continuing discord. They are holding off operations.[18]

If accords by nature then do not sit happily with consensus, why does the nation enter into such agreements, whose mark is *fragile durability*? One answer, as I have just forwarded, is that these are in the nature of holding operations. The second answer is also suggested before, though partly. All nations have a fictive ethnic core. If biological determinants marked that core earlier (there we have the origin of racism), today cultural determinants constitute it. Thus we have a perennial figuring of the *outsider* to the nation, without whom the nation cannot constitute its ethnic core, and therefore whom the nation always needs as the *outside* to authenticate the *inside*. But this constitution is never complete. If some wants to opt out, others join or rejoin. Thus, the need is for negotiation, the eternal politics of contract. In the dynamics of the political community whose durability keeps aloft the nation-form, we have the third answer. In the political community what is required is a specific language of civil intercourse that helps politics going.

Given the configuration of the modern nation, we do not have as the basis of that language of civil/civic intercourse contrary to the commonly held idea of 'a substantive idea of common good', but a *public concern*. In this case the concern is about the breakdown of consensus, order, lack of peace, the concern regarding the absence of common security. If we take up the example of the Indian constitution, we do not have there so much an idea of a common good, though that common good is somewhat defined in Parts III and IV of the constitution. We have in the Indian constitution rather a common concern about the forces capable of tearing apart a polity and a concern about how that can be prevented. It is a concern about the rules of the game. Therefore the polity invents a language of contract, of accord – contract between the centre and the periphery, between the three organs, between the parties, between ideology and pragmatism, between business and democracy, between state and the citizen, and finally between the nation and the alien.

But this means, such a language expressing power relations can never be harmonious. Accords are to a great extent things about the rules of discord, asymmetry, tokens of domination and counter-challenges. They show what antagonisms are, how political community is fractured/constructed. Since antagonistic forces will never disappear, at least in the foreseeable future, and conflicts and divisions will characterize politics, forms of agreement reached will be partial and provisional. We should not be surprised therefore, that this partial and provisional character of accords produce discord even within the parties entering into accords. Thus, within the opposition in the northeast, the Shillong Agreement produced discord. Similarly within the Union Government and the ruling party, the Punjab accord or the idea of accord with the opposition in the Punjab produced discord. These discords in turn produce circumstances that make accords difficult to consummate. The 'language of civility' the cherished ideal in the world of liberal-constitutional politics is distant from this world of conflict, discord, and temporary contracts.[19] Accords are thus the rulebook of political association. They show rulers are to be managers, and rulers can be rulers only when they are managers. They manage obligations, associations, entrances and exits. The implication is that accords upheld as the 'new deals' are more or less a permanent feature of the modern state. They are a technique of governing. In a milieu where every relation becomes governable and is a relation only by being governable, the rulers can rule principally by managing a set of norms. Accords are therefore a part of what Michael Foucault calls *governmentality*.[20]

Governmentality is a complex concept, but we need not stop at that. To put it simply, it is the way of governance, a relation of the governors and governed, it is a field that makes society governable, and makes each and every issue in social life a part of the business of governing. Therefore the question is, how exactly do accords contribute to governability, that is, the politics of the governors, and important in this context, the politics of the governed?

In as much the accords represent the process of dialogue, they belong to the domain of power, and show the process of power. By analysing the actors of an accord, we get an idea of the agencies. But this is only a primary statement. We need to look into the fact of who sets the agenda of an accord. Thus in an important accord, the most important subject may be left out. The brevity of the text of the

Shillong Accord of 1975 is killing. There were references to modalities of arms surrender, in fact they needed a supplementary agreement for that, but there was nothing on devolution of power, democracy, and ensuring human rights. Accords thus tell the asymmetries of power. But even these may not be enough. An accord may be between two roughly equal parties, but may leave the society completely out of the concerns it is addressing. The point is who gains by this lack of transparency? What is the inexorable mode by which the accord becomes not a democratic act or a popular contract, but just an act of governance? In such a multilevel analysis we find issues of genealogy, politics, legality and accountability present in an overlapping manner. In fact, this is why accords are mostly provisional, they are seldom given the seal of permanence, they are rarely incorporated in the basic law of the land, they are often what one observer of the Chittagong Hill Tracts accord in Bangladesh said, 'too little, too late, too loud'.[21]

The reason, of course, is not far to seek. The hidden process of reaching an accord excludes a consensus – ironically, it was however a consensual feeling for the need for peace that had led in the first place to the process of initiating an accord. In regulating and managing a conflict, the interaction may be low or high, but level of interaction does not always determine whether the issue is solved. Shimla agreement between India and Pakistan (1972) is one such example. Pakistan decided to have a maximal objective, namely recognition of the issues of dispute, and ensuring the return of the soldiers in captivity. India pursued a minimal agenda, namely, a stress on the bilateral mode of resolution and recognition of the cease-fire line as the line of control in Kashmir. The consequent interaction was sometimes high and sometimes low, but the agreement was ineffective in settling the dispute. If preventing war is held to have been the principal aim of the Shimla agreement, it was partially successful. But when the aim was defined in terms of having a 'relational control' over the adversary, it could not have been anything else. 'Terminating conflict' was out of question. On the other hand we have the example of the peace accord in South Africa. There, in varying degree, the principle of popular accountability, transparency was followed, so that the accord could lead to a process of 'peace with justice' and 'peace with democracy', a likely guarantee that the accord would not become simply a 'technique of governing'.[22] But we have to admit that almost everywhere, the accord seems to be a theological tool of states in negotiating identities – sometimes perceived as such by its adversaries also – thus making an accord a step not in achieving democratic peace, but simply a step in conflict-politics. Through its success of achieving governability, the politics of accord redefines the discursive space of peace. Peace is not then a space to enjoy rights; it is a space to *monitor* quarrels, to limit, to *manage* conflict, to integrate conflicts with the nationalized society. In being a tool of governance, the politics of peace accords is stripped of any radical normative agenda. And we have to remember, that this recoils on the adversary of the state also.

But there is more to the issue of governing through accords. The technology of rule demands that the three elements of a coherent governing strategy be satisfactorily combined in it. These three elements are – the legitimacy given by relevant social solidarities to the 'politics of peace', the communicability of the rationale, and targeted mobilization for the ends of the peace process. The state certainly sees to it that the three elements are present in a strategy of pushing for and ensuring an accord with an 'internal' or 'external' adversary. In some cases the elements may be only symbolically present, in others they are present geo-politically. But the basic imperative to achieve a strategy of rule through accords remains. Ironically, the state alone cannot achieve such a technology. The media for example is a great helper. There are civil advisers of all sorts who turn from being counter-insurgency consultants to peace consultants. The help rendered by the civil society to the state in this job is immeasurable. It shows that an accord is basically a *civil society operation* by the state.

Of course this opens the gate to civic participation in peacemaking. But with that begins another round of contest. The classic example of such a contested process in India is the Anandpur Sahib resolution of 1 August 1977. The resolution was discussed and endorsed by the main organ of the community/solidarity. Subsequently it spread across layers of the society, followed by wholesale repression of the solidarity by the state and the total negation of any consensus with the adversaries of the state. This phase ended with the memorandum of settlement known as the Rajiv-Longowal Accord of 24 July 1985. But predictably, the accord had been laboured to death, even before it began through a policy of high selection of issues. Thus the Rajiv-Longowal accord spoke of compensation of innocent persons killed, disposal of pending cases, the territorial claims, the All India Gurudwara Act, and similar issues. But on the crucial issue of centre-state relations, the text allowed a mere restatement of the Anandpur Sahib resolution and the assertion that it was within the framework of the constitution of the country. The problem of sharing of river water remained. Above all, the accord continued to be impervious to the most important issues in the Anandpur Sahib resolution on economic, cultural and social matters. We can, in fact, deduce a rule from this example. An accord is rarely reached over the main policy demands of some solidarity; it is frequently over the dreadful consequences of the bitter and often bloody contest that follows. In this way the process of peace is subsumed in the process of power. Thus, to go back to the instance of Rajiv-Longowal accord, the peace process did not end there. Sant Longowal was killed, so were other civic participants, also some state functionaries. The reality of discord overwhelmed the politics of accord, till the state had succeeded in completing the civil society operation.

Here we see the political rules of accord. In the incongruent relations among the state, leaders/experts, and the civil society, we find that perceptions become important. There is no way to synchronize the aims of these three, except through a mutuality of perceptions. But what happens when perceptions face the reality of fundamentally divergent paths and values? In the famous Lucknow pact of 1916 (the Gandhi-Jinnah pact), a milestone in Indian nationalism, the perceptions colluded. But soon the policies of the colonial state, the communities and the leaders of the parties collided with each other. Therefore, accords are from the beginning endowed with certain brittleness. The fragility is frustrating sometimes to the state, sometimes to the adversary, sometimes the peace constituencies in the society. Each of these occasionally tries to map out the perceptions ruling the universe of an accord. But it remains futile. Built on the inaccessibility of the accord to the people, the falsity strikes back. While we may argue that by constantly democratizing the process of peace and by reconfiguring the role of the people, the cosmology of accord can be enriched and its inherent brittleness minimized, we have to note the implication, that an accord is never a one-time affair of concluding peace. Peace is a process. Conditions of peace are historically irreducible. With the same actors a country can have peace, can have war. Peace-war is thus a continuum. Structurally peace conditions and war conditions are not essentially different. The same conditions not producing a war can elide into war. It is no use saying only that this is a war-peace condition. It is the factor of historical contingency that brings peace or war and enhances it. Therefore, the question is of re-democratizing peace, of arguing for peace as a process that is invariably marked by both power and popular intervention. Accord is an event in this process that remains characterized by procedural wrangles. The politics of dialogue in this way assumes a contentious character.

In short, since an accord by itself symbolizes neither war nor peace, the best way is to look at it as an event in the continuum of war-peace. Parties enter into a compact to engage the other. Rules of accords are then rules of *strategic* engagement to reduce the *tactical* space of the adversary. The desire for peace is thus punctuated by the realities of such an engagement. This produces varieties of peace. In as much as the hierarchy of causation produces varieties of conflicts, varieties of accords generate a new hierarchy of causality. Varieties of conflict and peace mark the politics of accords. There is, thus, at the bottom of the particularity of each accord the technique adopted to achieve it, the historical specificity of peace.

The theory of *governmentality* thus shows how various elements constitute the process of power/the process of peace. Fluidity of issues and perceptions, the entry of solidarities and the civil society in the reconfiguration of the political society that was till recently the monopoly of the nationalized state, the hierarchy of solidarities, communicability and the limits to the communicability of the conflicts and the solutions thereof – these are the elements of the process of power. They ordain the technology of governance.[23] This should not of course cause surprise because the modern state since the Treaty of Westphalia (1648) has made the right to wage war a crucial attribute of itself. The theory of sovereignty legitimates the right to wage war, and consequentially the right and the obligation to conclude treaties. All subsequent accords and conventions that define the rules of conduct of war such as the Geneva Convention (1948) are built upon that right, which today has come into contradiction with the desire for peace, more fundamentally with the alternative notion of right to peace. In the emerging politics of accord, therefore, we have a contest. It is between the right to war and the right to peace, between a governmental perspective and a democratic perspective, between the politics of nationalization and the politics of plurality, between a strategy that confines accord to a procedural level and a strategy that attempts to redefine accord as a moment in democratization, and finally accord as a monologue that passes as dialogue and accord as a part of a complex of plural dialogues ('multilogue'), that is to say, a federal dialogue. An audit of peace accords in this region will show how a peace exercise is embedded in contest, the desire for peace entangled in the dark process of power. It will show why peace as a process grows out of the process of power. Finally it will show in what way the process of peace reflects on the state of the dialogic process in this region.[24]

Protocols of the Politics of Accord

Yet there is more to the politics of accord perched precariously on the process of power and process of peace. If an accord is a document that lends form to a contest, what lends form to an accord? What are the rules that shape this form?

Rulebook means order of things, a protocol. How are we to go about in finding the order, the protocol by which accords are guided? Protocol, as the *World Book Dictionary* tells us, means 'rules of etiquette, of procedure, a first draft or record from which a document, especially a treaty, is prepared, a formal or official statement of a proceeding or transaction, and also a set of procedures that permit communication between two or more (computer) systems'. Protocol is thus about the rules of enactment, performance, meeting, and of engagement between systems.

The question is that since there is no single pattern of accords in the country and the region, can we find an order, a protocol? It may be convenient, therefore, to begin with a typology.

First, we have the notion of contract, as hinted earlier, underlying the basic law itself. Accords are in this case subterranean, and therefore the idea of consensus subsumes in it the reality of an accord. As the strategy proved successful and the post-colonial state at least in India and partly in Pakistan was formed on that basis, the state would like to see this form repeated everywhere.

Second, accords with peoples and regions that have to be made to accede to the union of India form a distinct category. Its feature is that of a treaty with the formalities and trappings of sovereign negotiations. This includes the Naga-Hyderi accord (1947), the sixteen-point agreement between the Nagas and the Government of India (1960), the Beg-Parthasarathy agreement (1975), and the Mizo accord (1986). Here the accord is loud and establishes a pronounced form, which subsequent accords follow. The features of such a form are the attesting presence of the top bosses and the signatures of top bureaucrats, the detailed format, the subsequent legislative enactment at the union and state level including sometimes constitutional amendments, and subsuming the substantive part with the operative part. In short, the strategy is one of bureaucratization of relations between the state and the adversaries.

Third (and this has similarity with the preceding group), we have accords with 'ethnic groups' that result in their 'empowerment' in the form of the creation of new states, autonomous councils, etc. The form of these accords such as the Mizo accord (which 'conferred statehood on the Union Territory of Mizoram with a view to satisfying the desires and aspirations of all sections of the people of Mizoram' and undid the Twenty-seventh Amendment Act of 1971 that had made the Lushai Hills a union territory), the Darjeeling Accord, and the Bodo Accord, derives from the master-form mentioned earlier. In this case we observe a certain trajectory. From 'nothing' to union territory/autonomous councils, and then possibly to statehood. Clearly, the federal structure, the sixth schedule, and the 73rd Amendment provide the state with the flexibility necessary to silence the discordant voices to death through the invocation of the compliant forms.

Fourth, there are accords that question the given federality of the Indian polity – precisely the federality to whose charm others are just acceding. We can cite the Assam accord or the Rajiv-Longowal accord. The net result here seems a zero from the point of the adversary of the state. Apart from homilies, they get nothing, precisely because the forms of settlement have not been imagined yet. Thus in these cases we have references to details of 'post-trauma care', such as how prisoners will be released, the mode of amnesty, or some restorative grants. One example is, 'Oil refinery will be established in Assam', 'Central Government will render full assistance to the State Government in their efforts to reopen Ashok Paper Mills and jute mills', and 'IIT will be set up in Assam' (Assam accord, 15 August 1985). Another may be, 'The Central Government may take some steps for the promotion of the Punjabi language' (Rajiv-Longowal Memorandum of Understanding, para 11.1). With no increase in executive and financial power, the (non-state) contracting party becomes almost the glorified panchayat. Almost sardonically, the memorandum of settlement of 1988 between the government of India, government of Tripura and the Tripura National Volunteer (TNV) adds,

> Maximum emphasis will be placed on extensive and intensive skill-formation of the tribal youths of Tripura so as to improve their prospects of employment including self employment in various trades such as motor workshops, pharmacies, electronic goods, carpentry, tailoring, stationery weaving, rice and oil mills, general stores, fishery, poultry piggery, horticulture, handloom and handicrafts. (para 3.8)

Fifth, we have the international accords in this region where the state is a party. That also follows a pattern in the region. Trapped in the metaphors of nationalism, these accords and particularly the negotiations that precede the accords, show the failure of diplomacy in this region. To put it briefly, there is as yet very little non-official input into the accord-making process. There is too much of history in the conflicts in this region. There is a predictable closed circle of arguments. Thus is the attitude, *I give in and do not demand much, I do not want you also to go the whole way, so that I retain certain control over relation* – an attitude termed by one observer as the strategy of 'relational control'.[25]

Finally, and as consequence of all these, there is a fixed bilateralism in the approach to accords as a solution to problems, with the result that we have little plurality in the making of a consensus in this region. (See Table 6.1 on Indian accords). This bilateralism, so characteristic of India-Pakistan relations, is a feature present in the accords concluded within the country also.[26]

In fact, contrary to its impaired ability to create consensus with outside,[27] within its territorial boundaries the state 'manages accords' with greater negotiating capacity, and thus the bilateralism is deeper within the territorial boundaries of the state. From a civil society point of view, the growth of public forums generates a constant critique of the accords, and thus forces a democratic viewpoint into the politics of accords. In the regional milieu, one may argue that such a growth is discernible on the horizon. In any case, the typology is not meant to convey a hard classification. Indeed, these types overlap. But they do show a trend. In one type, the accord belonging to another may find its future.

Table 6.1 Internal structures of the various accords – a classificatory scheme

Paradigm	State	Adversary	Society	Institutional process of accord
Foundational module for state-formation	Colonized state undergoing the process of de-colonization	Pre-state and beyond-state solidarities such as caste, religion, princely states	Plural	Constitutional bargaining
Natural	Hegemonic new state	Frontier region/ frontier community	Political-moral	Persuasive
Consolidation	Legal-bureaucratic	Excluded groups	Object of development and nationalization	New legal - territorial forms of accommodation
Integrated polity, plebiscitary-populist politics	Strong state	Constituent states of the union	Source of demands to renegotiate the union	Division among ranks of adversaries, memorandum of agreement without constitutional validity, violent attrition without any accretion of strength of the adversary, accord based on fatigue
Low politics-high functions, unilateralism, cultural-nationalism as the strategy	State aimed at retaining relational control, but locked in self-created positions	Other states and other solidarities having linkages that seem to threaten the relational control of the state	Marked by the growth of regional civil society institutions	Bilateralism and a regulated regionalism and multilateralism

Note: I have left the inter-state accords in the region out of this classification purely out of expediency, though in a larger discussion they ought to come into scrutiny.

The protocol of accords leads us to one more important aspect of the politics of democracy. I have indicated in the preceding paragraph that the failure of the fourth type of accords is in the irony of not finding a form of settlement. Possibly, therefore, the accord in such cases is elaborated in rituals. But this shows that an accord instead of being the mark of the agenda-setting capacity of the contracting sides, particularly the state, reveals the absence or the loss of such capacity.

Johan Galtung in a fascinating essay on the geopolitics after the cold war speaks of the agenda-setting capacity of the various estates in the world today.[28] He refers to the agenda life cycles and the entry and exit points. He reminds us that a hidden agenda may not always serve the aim that an agenda holds. And also, there may be an agenda vacuum at times. In our context the point is, why is there this inability to find a form of settlement that an accord is supposed to encode? Is some kind of agenda vacuum or a weak agenda-setting capacity responsible for this?

After all, in India, the years running through the fifties were one of finding the appropriate nation-form. The country went through the reorganization of states and the early accords that could pass off as natural, not contrived. Then the state form becomes invariable. Those who had not graduated to statehood were allowed to graduate, and those who had no territorial identity as a mark of democratic claim were allowed union territory or autonomous status. Those who had wanted an exit were presented with different versions of the inner line option, the *cordon sanitaire*. Thus, more than a century after the colonial rulers had found this form,[29] the 1993 Agartala memorandum of settlement between the Tripura government and the ATTF spoke in Art. 2 (e) that 'the case for introduction of an inner-line permit system would be taken up with the Government of India (and) the State Government will insist upon the Government of India to approve this demand'. But beyond these inflexible forms, the leaders of the state, and unfortunately often these antagonists, could not think. The result was that while accords and the consequential grant of territorial identity were characteristic of a democracy at work, they showed the limits of a liberal democracy, of constitutional federalism in negotiating the nationality problematic in India. The nation now finds itself in the enigmatic closed circle – *from ethnicity to accord, from accord to further ethnicity*. The Bodo Accord signed in Guwahati on February 20, 1993, said in Art. 3 (a) that the Bodoland Autonomous Council will exercise power over 'villages having 50 percent or more of tribal population' on the basis of scrutiny of a list supplied by the All Bodo Students Union and the Bodo People's Action Committee. Villages having less than 50 percent tribal population will be also brought under the Council's exercise for the purpose of providing contiguous areas. There was no provision on protection of minorities in the Bodo areas, and no provision on harmonizing a plural 'ethnoscape' in that region within the overall framework of the accord. This was an invitation to partition a land on ethnic lines as the talisman of solving a minority problem. The Tripura accord of 1988, referred earlier, similarly said that the

> Tribal-majority villages, which now fall outside the Autonomous District Council area and are contiguous to such area will be included in the Autonomous District

> Council and similarly placed non-tribal majority villages presently in the autonomous District and on the periphery may be excluded. (Art. 3.7)

In fact, the peace process in Tripura is significant in terms of the successive phases it went through along with continuing violence. The 49th Amendment Act of 1984 conferred statehood on Tripura, and also strengthened the Tripura Tribal Areas Autonomous District Council Act of 1979. Then came the accord of 1988 to settle relations with the Tripura National Volunteer to be followed by yet another accord in 1993, this time with the All Tripura Tribal Force. It was as if a master logic of nationalized ethnicity was reproducing itself remorselessly. It exemplifies the limits to the nationalist imagination, which remains confined to territorial solutions only as well as to the agenda-setting capacity of the elite of this country.

A new agenda cannot be forged. Old ones cannot be sustained. In this situation of agenda-vacuum, accords become routine, their lifecycle predictable. Hence accords fail to arouse popular enthusiasm, and evoke respect in its being as a political form. This is the road to further ethnicity. The convergence of several moments defines this road – the moment of agenda loss of the state leaders, of despair of the rebels, of hopes of the power-seekers, of the further growth of formal legal-administrative structures, and finally, the civil victory of an uncivil path marked in overwhelming manner by the armed forces and violence. Protocol is by nature routine. Therefore a banal form – an absence of imagination – characterizes accords. Harbouring the singular aim of developing a rational mode of governance to rule an unruly society, the institution of accord is singularly incapable of placing on the agenda the issue of dialogic restructuring of the polity. This is also true of the agreements that the Indian State has entered into with other States in this region. In their implications accords thus remain confined to the specific issues of agreements. Accords are thus *episodic*. They remain only as a moment of contest – a moment of consent, often yielding nothing, barren.[30]

In short my submission is that accords are a paradox, and protocols of accords reveal that. They exemplify bureaucratic rationality at its best and worst. Legislative approvals of the texts, constitutional amendments, new elections to legitimize the result of the agreements, clemency – to begin with; and then, rank, prestige, secrecy, routine parleys going by the name diplomacy, and (mis-) information rule this rational world. There is the testimony of Mr. B.K. Nehru, the Governor of Assam and the rest of the Northeast for five years (1968-1973), and later on Kashmir (1981-1984). In his memoirs he describes the importance of the question of protocol in gubernatorial functioning in the region, his emphasis is on making the rebels, servants and officials understand the high rank of regency. He unhesitatingly depicts the contempt in which, he as the representative of the Indian State held the 'peace-makers' of the Northeast, such as B.P. Chaliha, Jai Prakash Narain, Rev. Michael Scott and Dr. Aram.[31] He also expresses the intense 'humiliation' that he as the representative of the high powers of Delhi would feel for having to deal with 'Federal Government of Nagaland', the 'Prime Minister' of that government, and the 'commander-in-chief'. And how could it have been anything else? In order to make peace with the Nagas, the state of India could think of sending as its representative someone with no idea of the region. This was typical colonial governance. It was a rerun of the farce enacted twenty years ago when a man by the name of Cyril Radcliffe, who on his admission earlier had never been to the east of Suez, was sent to the subcontinent to partition it within three months. In Governor B.K. Nehru's words,

> As I have related earlier, I was sent to my gubernatorial post without receiving any briefing even on this most important subject of Nagaland. Having no background of Naga affairs I was completely mystified by reference to Aos and Angamis and Semas and Chakasangs and Tangkuls and the other tribes – sixteen in number – who form the Naga community. It took me for a long time to acquaint myself with the history of the Naga people, their internecine differences, their discontents with 'India' and the post-independence history of Nagaland and its insurgency which had resulted in this tiny group of people being given the rank and status of a full state of the Union without, however, ending the insurgency.[32]

By his own admission the Government humiliated the Nagas, demoted them or snatched away their ranks in the midst of the period of ceasefire and negotiation, and all this was done to show the Nagas their place. Not surprisingly, both the Indian state and the Naga rebels considered the Shillong accord of 1975 to be surrender. The protocols reveal the popular impulses that ultimately impel bureaucratic solutions. They show the rules of neglect of early warnings. They indicate that a peace accord is often achieved as a 'managing exercise' by the state after the conflict has already passed through the stage of denial, the stage of ignoring the early warning signals, the stage of surprise, and 'stage of brisk activities'. The democratic political element is always there, *out there*, as the constant, as the pursuing shadow. Attempts are then on to pull the accords out of the stagnant pool. The papers from Dayton or Shillong are to be saved. Through the neglect of the political, a crisis becomes 'humanitarian'. Then you need an accord. An accord thus while being statist, always has in it a *supplement*, a *dangerous supplement*. Out there, but inside – the paradox.

Protocol thus leads us to a strategic point of analysis. It shows the institutional mode of post-colonial governance, participation, partnership and state building. Rights, entitlements, rule of law, and security combine with each other in the historical evolution of the institutions of governance of the post-colonial nation. This, of course, is as Charles Tilly calls a *contentious* history, but also a colluding one. We know that when the received mode of inclusion in the nation (or the region) through the institutions of rights, entitlements, law, and security was sent out for wider legitimacy, many did not take it lying down. In that continent with a more illustrious history of accords, and contracts, the gypsy, the Catholic, peasant, immigrant, the Jew, and the communist rebelled, and had to be pushed around, coerced, tortured, dispersed, hanged, shot, and in the early days of nation-making broken on wheels. Similarly in this subcontinent, the dalits, indigenous people, the 'far frontier' communities, other nationalities, and popular solidarities have been subjected to police, law, army, and to the only available political discourse – that of inclusion or co-option in the given polity and grant of certain rights.

Accords have arrived against this background. And like all other historically available, that is to say regularized forms, this too, becomes the site of renewed contest. With accords now routine at least in India as their texts show, they have come to belong to the domain of public policy.[33] The 'implementation of such a policy becomes the site of exchange between state and the society'[34] that now constitutes itself into a public realm. In the format of provisions therefore we witness the struggle within the adversarial group over the material and symbolic resources of power. Thus we see the rules of inclusion and exclusion. Remember that the Naga signatories to the Shillong accord had no specific affiliation; they claimed they represented all (in the text it was written as 'on behalf of the representatives of underground organizations'). We also witness the tussle between the state and the contesting populations over exclusive rights over territory, an incremental contest no doubt, but a contest. The public policy process of accords that signifies distribution of resources among those whose consent has been secured becomes an issue of confrontation. Notwithstanding the bureaucratization, the policy is then the occasion for countering the predatory power of the state. Sometimes a rational mode of creating new clients, the policy of accord always encounters democracy – that *supplement* which the accord was to still and kill. How else do we look at the whole history of the peace process in the subcontinent – the prehistory of Shimla, Shillong, Colombo, Thimpu, the lake of Zurich,[35] and the history that these created?

Is there a moral question involved in this transformation of the peace question into one of public policy? One can easily cite here the very recent instance (which we recount in details in next chapter) of the extension of ceasefire between the Indian state and the Naga rebels that became a matter of public policy within no time. Possibly the peace question had always been public. But when does it become a matter of a public moral choice? And how does then dialogue become a matter of deployment of conflicting moral strategies? One may say, it happens when dialogue refuses to be imprisoned in the limits of an accord, and becomes thereby 'public', dragging out the politics of accord from its bilateral confinement to the arena of the public. To be sure, the moment peace becomes a matter of public policy it arouses public judgments. Public judgments create own laws for dialogue – it is wrong to say therefore *Inter Arma Silent Leges*: that in time of war the law is silent. We judge war, we judge peace, we censure, we condemn, we comment, we curse, we cry, we pray. We start doing what we had not done earlier when the guns had first started. We become aware of a choice, of possible choices. Our choices become moral options, hence very real. Should we sacrifice our lives so that empires become even bigger? Or, should we try to account for our losses only in terms of our gains in war? Or, in terms of possible gains if those losses were not to take place? Should we condone killings of the innocent? Should we think that everyone on the other side is culpable for what has happened to us? Is there any room for accommodation? Dialogues on public policies on peace become stormy sessions of moral strategies, just as they are of war strategies.[36] To exercise such moral choices does not mean that everyone becomes an angel. It simply means some make bad moral choices. More important, it means war, peace, geopolitics, and a Calvinistic pursuit of faith – all these take place in a hard moral world, which is as much historically constituted as other social universes are. I am therefore not suggesting a morality play in the context of dialogue and accords as the other side of my argument that accords are a part of governmentality. But I am drawing attention to the moral strategies that are deployed in the dialogue on peace accords. These ethical-political challenges to power succeed in posing several important questions to hitherto unquestioned assumptions about theories of just cause (the state has fought a just war), competent authority, comparative justice, right intention, last resort, probability of success, and proportionality. Therefore, as our experiences of South Asian wars show, with each abortive war and abortive peace discussion the issue becomes not one of discovering a new successful strategy, but how unjust the cause has become, how the nature of competent authority to wage war and peace has been impaired, how right intentions have gone astray, how last resort becomes the only resort, how probability of success goes down, and how the principle of proportionality becomes un-executable.

Recasting the Problematic of Peace

Therefore it should not surprise us that these moral choices compel the state to move on sometime from low intensity war to low intensity peace. Accords, as products of historically formed moral-political strategies, create the war-peace continuum through the mechanism of dialogue. They show the paradoxical nature of a process that we know as peace process. They show why there cannot be any presumption that a peace process will succeed in addressing the root causes of conflict, or even in making these root causes the core of the negotiating agenda. Though larger than the agenda of ceasefire and demobilization, the dynamics of power make the peace process subservient to the 'primary steps', and as I show in the next chapter, these primary steps become focal point of all dimensions of conflict, frustrating the process itself. If the aim of the peace process has been is to restore conflict to the political, instead of the military, arena, this aim is more often than not defeated. For, peace process cannot enough revise the rules of the game as to make the restoration viable. Removing the issue of justice, casting away the anchorage, peace building lands in deep paradox. It wants the conflict to be negotiated politically, while it employs the very strategies that in the first place had dragged the conflict into the military arena, and then had made peace a tool of counterinsurgency, and war.[37] The paradox of the current politics of peace process has few lessons. I shall speak here of ten such.

First, these instances show that viability of peace, democratization of structures and a vigorous dialogue reinforce each other. Conversely they retard each other.

Second, the narrower is the definition of democratic transition, which is required for peace, the less viable is the process. Conversely broader the definition, more sustainable is the process.

Third, successful negotiations are carried forward only when the leaders of government and the adversary are considered as legitimate political authorities to effect changes. Thus political legitimacy is an invaluable wealth in peace process. The quicker is its exhaustion the more brittle is the process.

Fourth, the scope of the reform agenda is not solely a product of military correlation of forces, but also a product of political will. Thus, narrower is the base of political will more difficult is the process of achieving peace. Conversely, wider is the base of political will, therefore wider the structure of dialogue, more irresistible becomes the desire and will for peace.

Fifth, the theory of the specialists that stalemate leads to peace and therefore is the ripe moment for peace, and mutually hurting stalemate is the ripest moment is as false as the record of the specialists is in resolving conflicts. The sheer length of duration of certain conflicts shows that adversaries may move to newer ways of confrontation to break the stalemate, rather than exploring possibilities for peace.

Sixth, the role of international community can be only secondary. In fact, the more international intervention (admittedly for peace) at the expense of internal consensus, more fragile is the peace process.

Seventh, greater is the delay in confronting the 'intractable' issues of peace, more limited is the appeal for peace. Thus, though issues of justice, reparations, shame, reform of security apparatus, establishment of accountability and abolition of impunity are held as remote from the immediate agenda, in fact these become the urgent agenda.

Eighth, and this comes from the previous seven lessons, politics of peace redefines democracy, as politics of democracy redefines peace.

Ninth, the linkage between geo-politics, democracy and peace, as never before, lays bare the banality of the entire debate over ideal versus procedure in characterizing democracy. While it shows that democracy, constituted by its geo-politics, is perfectly capable of waging wars, the ethos of justice places new demands on democracy, as neither democracy nor peace by itself guarantees justice.

Tenth, as these lessons show, peace accords remain at the heart of state-formation. Like war and violence, peace contributes its bit to forming the modern state. By 'democratizing the state', effecting ceasefire and making limited co-existence with adversaries possible for the state, by instituting limited liberties without seriously punishing perpetrators of crimes, in other words by creating new versions of de-colonization, the state becomes modern, eligible for admission in the liberal club – in other words, the *ancien regime* leaves the scene.[38]

The greatest irony is that, peace politics notwithstanding these paradoxes remain viable as part of state strategy, and that is because the agenda of reforming the state and governing techniques can go forward on the basis of the discourse of human rights.[39] The strategy is simple: de-link rights from justice, and finish the dialectic of rights and justice once and for all. If we allow ourselves, just little beyond what we have discussed till now, into the theme of the modular form of settlement that marks the contentious history of the peace process, we shall find the problematic of the peace question facing up to the dialectic of rights and justice. Through the peace accords, the state invites the recalcitrant groups to be citizens, as citizens to *equally* enjoy the rights of citizenship. The relevance of this invitation is in form of a paradox.

In order to enjoy the rights of a citizen, they must accept statehood within the union. But while they are invited to claim such rights, they are issued summons of law in form of the special powers of the state. An avowal and a disavowal therefore mark the core of the texts of the accords. An avowal that 'they' will be treated *equally*, and a statement to the contrary that beyond statehood (within the union) there is no other form of enjoying the rights of a citizen, no other form of claiming *justice*. The modular form of settlement is *modular*, because it has an aspect of denial, of punishment, of demonstrating that beyond the model there can be nothing, you can expect nothing. To the slogan of the *right of a nation* (indigenous peoples are nations, we are told) to self-determine, we are presented with the invitation to claim the *right of a citizen* – a right that we can attain only through statehood within the union – a right if we do not claim invites compulsion. What is striking here is that neither the right of the nation nor the right of the citizen even when strategically deployed is adequate to address the issue of peace, for at the core of the issue of peace lies not *right*, but *justice*.

Privileging peace with right closes the journey of peace for two reasons. First, rights do not provide us today with an overarching politics of peace. Second, the discourse of rights itself is haunted by the supplement, *is it just*? Right of course implies both – an authentication (R1) and claim (R2). Perhaps it is an authentication that the claim is just. But the dual meaning shows that there is always *excess* in what is conveyed by right. In order to prove that the claim, that is to say, the right (R2) is just, which means that this is right (R1), we have to say it is natural, like *what is just is natural*. It means acknowledging the limits to the juridical or legal forms of politics. It further means that equality of rights itself constitutes a right or power, which originates from contractual relations between individuals and agencies who then become permanent victims of the power of contract, and which therefore resolves into a principle of equivalence of rights and power eventually to set up a new opposition – that between the right of passion and the right of power. The ghost of Locke must be laughing at us.

Yet, can we say that *justice* is free from the paradox of a mutually contradictory double significance that arises out of the practice of *rights* and then afflicts it mortally, wounding it grievously? For a moment we shall go back to the death of Ajaib Singh. He had run to every pillar, sought all possible assistance, and found justice in ending his life with a denunciation of the unjust, the unjust machine of justice. Justice had denied Ajaib Singh justice. How is then justice linked to restoration of peace, restitution linked to restoration? How is contract, the juridical form of justice, related to peace, which must mean recognition of dissolution of the past contract? Bertolt Brecht writing on the 'The Threepenny Lawsuit' had commented,

> The whole thing (the process of justice) is called 'a reasonable mutuality of interests' – the 'source' of a healthy economy...and constitutes a document of the most decisive materialism. If the opposition between material and incorporeal interests were not resolved – the rape of the former is not a completely satisfactorily solution...Law mentioned at the beginning have been abandoned by the courts in favour of the newly emerging ideas...Dominant ideologies are just as

> conflated as dominant classes. Every idea is beholden not only to one but to several others; it also holds more than one destiny in its hand. It should not be retained or rejected according to its immediate utility, which is, as already stated, multiple, because even the fact of rejection can be harmful. Ideas must be tested again and again in practice, and frequently in isolation from one another. Practice remains young, while the ideologies age, and they are no stronger than their weakest link, the single idea. Chained together as they are for better or worse, they drop an absolutely compromised idea only with the greatest hesitation, after it has changed its name more than once...But at the cradle and coffin of every ideology sits practice.[40]

In situating the paradox inherent in dialogues on the practices of rights and justice, I have not intended to deny the modular forms of peace settlement. On the other hand, my intention has been to point out the need to break out of the modular form/s, and the mines on that road. At this time when we are simultaneously dealing with state, nation, classes, ethnicity, and globalization, the politics of peace waits to be recast. This is a turning point. All the more hence we must understand the *problematic* of recasting – how do we move from the modular form of peace-settlement marked by the discourse of rights and citizenship to more variegated singular forms? How do we escape the two fundamental phenomena that deeply structure the process of dialogue – the phenomenon of a compelling territorial reasoning in finding a solution and that of the monolithic nature of peace process?[41] How do we build up a theory of plural dialogues, how to understand the problematic of dialogue posed by its paradoxical nature – the plurality of the dialogue and the singularity of the act? There is no easy answer. But without the investigation into the practices of dialogue and peace, the problem could not be posed even.

Notes

1 Ram Narayan Kumar, *A Complex Denial*, SAFHR Paper 9, Kathamndu, 2001.

2 The CCDP Report informs us that only *The Tribune* published from Chandigarh on 9 July 1997 carried a lengthy report with the caption, 'Frustrated Father Ends Life'. The story referred to his five and a half years of battle for justice and his disappointment with the entire system, which compelled him to take his own life. The report said that Ajaib Singh had met almost all the Chief Ministers and Police Chiefs of Punjab, but no one had been able to even tell him what happened to his son who had been lifted by the Police Officer Ajaib Singh from a barrier outside Amritsar railway station on 20 December 1991 – Cited in *A Complex Denial.*

3 CCDP Incident-Report Pro-forma No. 00100.

4 *CCDP Report*, p. 55.

5 Ibid. p. 52.

6 On this, Gurharpal Singh, *Ethnic Conflict in India – A Case Study of Punjab* (Basingstoke: Macmillan Press, 2000), pp. 100-124.

7 On this, J Pettygrew, *The Sikhs of Punjab – Unheard Voices of State and Guerrilla Violence* (London: Zed Books, 1995).

8 Julio Ribeiro, *Bullet for Bullet – My Life as a Police Officer*, (Delhi: Penguin Books, 1998), cited in CCDP Report, pp. 117-128.

9 Ribeiro, p. 349.

10 Clearly my point of departure from the established literature on peace accords is where I look at the peace accords as a narrative of contests, more specifically of struggles of democracy to come to terms with itself. For an instrumentalist view, see the writings in *Accord – An International Review of Peace Initiatives*; particularly Christine Bell, 'A Comparative Perspective on Peace Initiatives', *Accord*, 7, 1999, pp. 6-9.

11 The views of Gramsci on hegemony are to be found throughout his *Prison Notebooks*. It is difficult to summarize them in one single formulation. However, one of his most insightful comments occurs in his discussion on 'hegemony of Western culture over the whole world culture', where he writes, ' Even if we admit that other cultures have had an importance and a significance in the process of 'hierarchical' unification of world civilization (and this should certainly be admitted without question), they have had a universal value only in so far as they have become constituent elements of European culture, which is the only historically and concretely universal culture – in so far, that is, as they have contributed to the process of European thought and been assimilated by it.' – Quintin Hoare and Geoffrey Nowell Smith (eds.), *Selections from the Prison Notebooks of Antonio Gramsci*, Indian edition (Madras: Orient Longman, 1996), p. 416.

12 Subir Bhaumik, 'The Accord That Never Was: A Critique of the 1975 Shillong Accord', in the published proceedings of the First South Asian Peace Audit Exercise, Dhulikhel, 8-11 September 1999, Jehan Perera (ed.), *Peace Process in Nagaland and Chittagong Hill Tracts – An Audit Report*, SAFHR Paper Series 5, Kathamndu, 1999. The facts cited in this section are well-known to researchers of Naga political history. I have taken help from Asoso Yonou, *The Rising Nagas* (Delhi: Vikas, 1974); Udayon Misra, *North East India – Quest for Indentity* (Guwahati: Omsons, 1988); M. Horam, *Naga Insurgency: The Last Thirty Years* (New Delhi: Cosmo, 1988), Subir Bhaumik, *Insurgent Crossfire: North East India* (Delhi: Lancers, 1996) and Charles Chasie, *The Naga Imbroglio* (Kohima: Standard Publishers, 1999). All citations are taken from *The Naga Imbroglio.*

13 As one participant in the peace audit exercise at Dhulikhel remarked, 'We Nagas do not want to speak of Shillong accord. It needs an unceremonious burial', *Peace Process in Nagaland and Chittagong Hill Tracts*, SAFHR Paper Series 5.

14 In the Sri Lankan situation, specifically in the current peace politics, we witness the same trend more clearly.
15 V.P. Menon, *The Story of the Integration of the Indian States* (Bombay: Orient Longmans, 1956), p. 467.
16 Ibid. p. 395.
17 Ibid. p. 394.
18 The example of the Sikkim accords is significant. The first Sikkim accord of May, 1973 between the Indian government, the Chogyal, i.e., the ruler of Sikkim, and the political parties of Sikkim, spoke of democracy and establishment of rule of law (para 1). The Government of Sikkim Act, 1974, elaborated the rules of a constitutional monarchy on the basis of strengthening the 'close relationship with India'. The thirty-fifth Amendment Act of 1974 inserted a new provision in the Indian constitution, Art. 2 (A), to sanctify the association of Sikkim with India, and a new schedule, the Tenth Schedule, was added. Finally, peace process was brought to consummation by the Sikkim Assembly Resolution of 1974 that abolished the institution of Chogyal. This illustrates the growth of a civil society and the transition from feudalism to lumpen capitalism via the path of democracy and accord.
19 On the idea of a civil language and its critique, see two essays – Michael Oakshott, 'On Civil Association' and Chantal Mouffe, 'The Political Community: Universitas or Societas?' reprinted in Gurpreet Mahajan (ed.), *Democracy, Difference and Social Justice* (Delhi: Oxford University Press, 1998), pp. 138-158.
20 Michel Foucault, 'Governmentality' in James Fabion (ed.), *Michel Foucault – Essential Works 3, Power* (London: Allen Lane, Penguin, 2001), pp. 201-222.
21 Dr. Imtiaz Ahmed of the University of Dhaka commented on the CHT accord in an unpublished speech in 1998.
22 In this context, the report on the state of non-official dialogues in South Asia is significant. The people are present in the report by being represented in some of the non-official platforms for dialogues. We find references in the report to various public associations; but the role of popular accountability in the dialogue process lies under the heavy weight of the 'non-official' actors. The issue of democratizing dialogue therefore unfortunately does not occur. See Navnita Chadha Behera, Paul M. Evans and Gowher Rizvi, *Beyond Boundaries: A Report on the State of Non-official Dialogues on Peace, Security and Cooperation in South Asia* (Ontario: University of Toronto – York University Joint Centre for Asia Pacific Studies, 1997).
23 I am indebted to Dr. Sanjay Chaturvedi of the Punjab University, Chandigarh, who has pointed out to me that as against such a governmental theory of accords we have a peace perspective also. Such a perspective builds on the desire for peace and human rights, and turns the 'geopolitics of conflicts' to a 'geopolitical imagination for peace'. From imperial legacies the perspective turns to an approach that may be termed as 'peace from below'. The approach of 'peace from below' interrogates accords with a normative criterion of popular rights and contests the pursuit of primacies marking the accords.
24 Some argue that with modern wars causing more civilian casualty than military casualty, the civil society is becoming more interested in peace accords – 'the politics of accords is not just theirs, the princes, the states, the leaders, etc., but our concern too'. See for example Kumar Rupesinghe (with Sanam Anderlini), *Civil Wars, Civil Peace* (London: Pluto Press, 1998).
25 Maya Chaddha, *Ethnicity, Security and Separatism in India* (New York: Columbia University Press, 1997).
26 Disputes over water resulting in intrastate conflicts are really not much different from interstate conflicts in terms of their dynamics. We see the same ineffectiveness of

institutions, the same hold of rigid language, the presence of same constituencies, the same failure in apprehending early warning signals, finally the same event of process of desire caught in the process of power. The histories of inter-state (within India) disputes, negotiations, and accords over waters of the Yamuna, the Cauvery, the Narmada, and the latest, over the Krishna rivers are worth perusing. See Gargi Parsi, 'The Politics of Water', *The Hindu*, 18 August 1996; M. Madan Mohan, 'Another Water Row', *The Hindu*, 19 August 1996.

27 I have written on it in an essay, 'Flowing Waters and the Nationalist Metaphors', *Studies in Conflict and Terrorism*, 20 (2), April-June, 1999, pp. 195-206; also by Deepak Gyawali and Ajaya Dixit, 'Mahakali Impasse and Indo-Nepal Water Conflict', *Economic and Political Weekly*, 27 February, 1999, pp. 553-564.

28 Johan Galtung, 'Geopolitics after the Cold War – An Essay in Agenda Theory' in Vivian De Lima and Carmencita Karagoag (eds.), *Peace, Disarmament and Symbiosis in the Asia-Pacific*, proceedings of the Conference on Peace, Disarmament and Symbiosis in the Asia-Pacific (Quezon City, Philippines: Solidaridad Foundation, 1995), pp. 41-57.

29 It is important to remember that even some of the tenancy acts in the colonial times such as the Chotanagpur Tenancy Act, and regulation arrangements like the CHT Regulation Act were also a type of cordon system.

30 Gyawali and Dixit in the essay on 'Mahakali Impasse and Indo-Nepal Water Conflict', show the agenda-vacuum of the rulers of Nepal with regards to water resources and describe in details how such a vacuum led Nepal into accepting an unequal treaty with India. This essay is revealing in terms of the nationalist dynamics of treaty making in the region, and is a must for every student of diplomacy in South Asia. See particularly, pp. 561-563.

31 B.K. Nehru, *Nice Guys Finish Second* (New Delhi: Viking, 1997), pp. 473-540.

32 Ibid., p. 501.

33 All references to the examples of accords cited here are from the useful compendium prepared by P.S. Datta, *Ethnic Peace Accords in India* (Delhi: Vikas Publications, 1995).

34 J. Echevarri-Gent, *The State and the Poor – Public Policy and Political Development in India and the United States* (Berkeley: University of California Press, 1993), p. 4.

35 These are the places of famous dialogues between representatives of state and the adversaries in South Asia.

36 One need not agree with Michael Walzer's definition of just and unjust wars to take note of a point he makes in that connection, namely that morality is a matter of public debate. See his *Just and Unjust Wars* (New York: Basic Books, 1977), pp. 3-20. Walzer of course remains singularly insensitive to the fact that moral choices are historically constituted, therefore notoriously unstable.

37 This in fact constitutes the war-peace continuum in South Asia. The most illustrative lesson of this is from Sri Lanka. On specifically this aspect of the war in Sri Lanka, Liz Phillipson, 'Breaking Recurring Themes in the Cycles of War and Peace in Sri Lanka', www.c-r.org.

38 On the entire issue of the relation between democratic transformation/transition and peace, see Fen Osler Hampson, *Nurturing Peace – Why Peace Settlement Succeed or Fail* (Washington, D.C.: United States Institute of Peace, 1996); Elizabeth Zelin and Eric Hershberg, *Constructing Democracy – Human Rights, Citizenship and Society in Latin America* (Boulder, Colorado: Westview Press, 1996); on de-linking the two and relating peace to stalemate in contest, see I. William Zartman, 'Ripening Conflict, Ripe Moment, Formula, and Mediation' in Diane Bendahamane and John McDonald Jr.,

Prospective on Negotiation – Four Case Studies and Interpretations (Washington, D.C.: Foreign Service Institute, 1986).

39 To get an idea of how revision of state strategy is linked to considerations of peace politics, see the massive documentation of Latin American studies, Cynthia J. Armson (ed.), *Comparative Peace Processes in Latin America* (Washington, D.C. and Stanford, California: Woodrow Wilson Center Press, and Stanford University Press, 1999).

40 Bertolt Brecht, 'The Threepenny Lawsuit' in Marc Silberman (ed. and trans.), *Bertolt Brecht on Film and Radio* (London: Methuen, 2001), pp. 188-189.

41 I have tried to address the phenomenon of territoriality imprisoning dialogic imagination in Ranabir Samaddar, *A Biography of the Indian Nation, 1947-97* (New Delhi: Sage Publications, 2001), chapter 8, 'Nation at War, Nation at Peace'.

Chapter 7

Two Ceasefires, One Story

Plural context, Singular event

By narrating in this chapter the stories of two ceasefires in South Asia, I want to address therefore the issue of the plural context of dialogues, and the singularity of the dialogic act that often succeeds in effacing the multiple traces. My intention is not to predict the final destiny of these two ceasefires, which still are holding at the time I write of them, but to bring to light an inherent paradox in the dialogic politics (briefly mentioned earlier), that a theory of dialogue has to take into account. Such a theory of dialogue will be then able to enunciate the reality of plural dialogue existing side by side of monologues that pass off as dialogues in this political world.

Possibly we shall require anthropological studies of new kind to retrace the long history of plural dialogues. Anthropologists have studied exchanges, contracts, state-formation, and above all war-making. Even in studying corporate life, the attention has been on the solidarity of the life, and not so much on the dialogic mode and dialogic imagination that makes collective life possible. It has required men and women of aesthetics to speak of dialogic imagination as a phenomenon that acts as the counterfoil to the solidarity, makes plural dialogue possible, and thereby makes collective existence dialogic, and not monolithic or fascist. Probably we shall need in the first place, as has always been the case, criticism of the episteme (of collective life) to find out that history, and anticipate it to write it out.

In terms of historiography, an effective strategy for this task will be to investigate how the plurality of a context is subsumed in the singularity of an event,[1] and find out in the process how dialogues acquire monolithic character suppressing its multiple origins and traces. The event is the congealed context. For peace politics then the question would be how to re-inscribe into an event of peace making, which is singular and monolithic because it is congealed, the plurality of the context. We can recall here events of conflict or conflict taming – events of war and ceasefire, a nuclear test as in Pokhran in 1998, a much publicized arms-control talk, a peace talk in Camp David, in Sharam-el-Sheikh, in Srinagar or Colombo. In an essay 'Plural Dialogues' I had shown, how two states in South Asia, India and Pakistan, engaged in attrition for fifty years had used the language of dialogue as the language of control, and conversed between themselves so that they could fight; how they spoke of confidence building measures because they were confident of the power of words to sustain violence.[2] Their dialogue had been the background in which they had proceeded to war, creating warring artefacts, and then staged the next round of dialogue where they could pause to reflect on the preceding phase of violence. Dialogue was in such case a part of violence. There was, therefore I argued, a politics in dialogue, and a politics of dialogue, in much the same way a theorist had said, a 'politics of friendship'.[3] I also showed how a master discourse of arms control had erased all other aspects of the relations, and a plural dialogue had been subsumed in an overwhelming narrative of a standardized conversation that had transformed the former into a monolithic, monothematic and monolingual conversation. The received history of dialogue had overwhelmed all other histories of dialogue and peace.[4]

On the basis of what we have seen as features of a dialogic situation where peace accords arrive through an asymmetrical conversation between state and the rebels, we can now push our inquiry further into the impediments for a plural process of dialogue by investigating how states conduct the politics of ceasefire – the most concentrated form of dialogue in an armed conflict situation. We shall discuss here the dialogic situation obtaining between a state engaged in war and the armed adversary. Once again, the purpose is not to write another essay on conflict-management. The relational arguments in this inquiry into dialogues on ceasefire are more in the nature of parenthesis that aims to move to some kind of a thesis on plural dialogue. I turn once again first to the experiences of the northeast. The invitation to the readers is to read it along with the story narrated in the previous chapter – but this time the chronicle concentrates on the ceasefire between the state and the armed rebels in Nagaland.

The overwhelming desire for peace among the people of a land marked by a fifty year insurgency led to the declaration of third ceasefire in Nagaland on 1 August 1997, this time agreed upon between the Government of India and the main Naga rebel force, the National Socialist Council of Nagaland (NSCN).[5] The Khaplang group of the NSCN also declared a 'good-will ceasefire' before the year ended on 15 December. Though inter-community clashes such as between the Nagas and the Kukis continued with less intensity than what it was few years before, with ceasefire and the commencement of political dialogue between the Government and the NSCN, there was hope in the public mind that the violent past would be buried in quest for peace. The ceasefire was the result of political talks started between the government of India, particularly under the then Prime Minister Mr. Deve Gowda, and the rebel leadership. The Naga Hoho (council summit) also declared the willingness of all sections of the Nagas to live in peace.

What was less noticed amidst the widespread sense of relief ranging from cautious optimism to euphoria was that along with this had started the old game of creating and propping up in form of the Nagaland chief minister, S.C. Jamir, an opposition to the rebellion and undermining the political legitimacy of the rebellion, more particularly the rebels. Just seven months before the declaration of ceasefire, Mr. Jamir had declared on 28 December 1996 his intention to seek Interpol aid to secure the extradition of the rebel leadership.[6] This was an indication that the failed history of ceasefire of 1964 would be repeated. At that time, the participation of A.Z. Phizo, the unquestioned leader of the NNC, in the ceasefire talks had been opposed on the ground that he was in exile. The year of 1964 is in any case worth remembering. Having created a new administrative unit called the Naga Hills Tuensang Area and placing it under the Ministry of External Affairs, Pandit Nehru had entered into a 16-point agreement with a section of the

Nagas in 1960 to ultimately create the state of Nagaland within the union in December 1963. That this policy of grant of statehood was a measure of *realpolitik*, dividing the Nagas into moderates and rebels, and deepening the chasm between the propped up local government and the society, and within the society and therefore was not the assured road to peace was probably known to the government also. Six rounds of talks between the government and the rebels were held between 1964-67. Not surprisingly nothing substantive emerged, though the rebels among whom there was a young man later to become the chairman of the NSCN, Isak Chisi Swu, had participated in the dialogue with remarkable patience. The peace mission's efforts had brought nothing except the fruitless ceasefire agreement of 1964.

Just as the ceasefire agreement of 1964 signified nothing in terms of achieving a substantive political dialogue, the possibility appeared in 1997 again that this ceasefire too would become sterile. The conditions agreed upon by the two sides namely, unconditional talks, talks at highest level, and venue of talks to be outside India, proved almost woefully inadequate, for meanwhile, threats against the rebels continued, divisive tactics went on, and the substantive part of discussion continued to be repeatedly postponed. Typically Jamir commented in 1997, and this was indicative of government line of thinking, that the Naga Students Federation's theme of rapprochement among various sections of the Naga society was not enough, and that without spending too much time and energy over what happened in the past to the Nagas, Nagas should have started thinking of rapprochement in terms of 'bridging the gap between past, present and future'.[7]

Meanwhile talks continued with no substantive progress. Both Mr. Jamir and The Baptist Church of Nagaland wanted to become players of peace, as if the pathway to peace was not enough complicated. Predictably the rebels, wise after the experience of the sixties, refused to accept the entry of Jamir, who in their eyes was the manipulator of the agenda of peace talks, in the World Baptist Conference at Atlanta in 1997.[8] However the ceasefire initially agreed by the NSCN for three months continued. Other problems besides the Jamir factor that symbolized 'the sins of the past' and the 'manipulative politics of peace talks of the present' soon raised their heads if the government was not already aware of these. After all there was every reason to remember, as the rebels lost no time to remind anyone willing to listen to them, that the earlier two ceasefires in 1964 and 1974 were either preceded by or followed a split in the rebel ranks and a political grant of limited autonomy, which meant to them only surrender. The ceasefire in 1964 had been preceded by statehood, and that of 1974 had been followed by the Shillong accord – both now recognized as disasters in the chronicle of peace management and peace politics of the State in South Asia.

Even in 1964 the ceasefire had extended to the rebel operations in the neighbouring area of Manipur. But this time the ceasefire was for long confined only to Nagaland areas to deny the stand of the NSCN that Naga areas had all along included parts of what was to be known later as upper Assam and Manipur, and that an eventual political dialogue would have to deal with not only the administrative (in the Naga eyes 'artificial') entity created by the Indian State, called the Nagaland, but the Naga nation. The dove, Mr. I.K. Gujral the successor to Deve Gowda, declared in Parliament that fratricidal confrontations between Naga groups and others, also between different Naga rebel groups had seriously disturbed public order, and ceasefire in Nagaland would now usher in progress.[9] Inspired reports emanated to the effect that peace talks would lead to the break-up of Manipur. The Manipur Chief minister Rishang Keishing went on his own to declare in the state Assembly that the union home ministry had assured him that 'there was no question of disturbing Manipur's territorial integrity in view of the peace talks with Naga rebels'.[10] The union home secretary Mr. K. Padmanabhaiah declared that the agreement would be valid only in the state and a panel would be formed to supervise the agreement.[11] But while certain ground rules were framed and decided upon such as halt to ambush, no aggressive patrolling inside the hills and jungles, joint monitoring, formation of a panel, etc., political aggression by the union government continued. It was at best a 'moth-eaten peace and autonomy cheque' being offered by the government.

Meanwhile the Armed Forces (Assam and Manipur) Special Powers Act was implemented as vigorously as ever. Section 4 (a) of the Act confers power on the officials of the security forces to 'shoot to kill', a legal abomination permitted by Chapter III of the Indian Constitution. Section 7 of the Act (along with Sections 197 and 45 of the Code of Criminal procedure, and the Army Act of 1950), in a grotesque celebration of impunity, protects officers of the armed forces and other officials of the State from criminal prosecution irrespective of their crimes.[12] On the basis of this, raid and search operations continued. Threats of detention without trial, raid, search and seizure, shooting to death, public control measures, confiscation, and declaration of public assemblies as illegal, remained as ever. Nagaland still now after four years of ceasefire remains out of bound of the operation of Articles 19 and 21. The ceasefire monitoring office of the NSCN in Dimapur was suspended for sometime on grounds of illegal activities and was allowed to function only recently. In the Naga areas of Manipur, land alienation of the Nagas through extension of land acts such as Sardar Hills District, Manipur Land Revenue and Land Reforms Act or New Land Act for Hill Areas continued. The ceasefire did not bring a halt to the operation of the old colonial laws such as Land Acquisition Act of 1884 and Forest Laws. Yanganpokopi was taken from Maring Nagas for a national park. One hundred thousand acres of land were taken from Tangkhul Nagas. Loktak hydroelectricity project resulted in land alienation among the Zeliangrong Nagas. Similarly the political talks were interrupted by new demands and questions by bureaucrats to the extent that the NSCN leader Mr. T. Muivah was forced to declare, 'We are not toys to be played around with'.[13] The government of course had the last laugh when Mr. Muivah was arrested in Thailand early 2000 on grounds of illegal entry and stay, a development that showed in its wake, according to the rebels, the reach of arms of the state. So, though the ceasefire was extended in 2001 with effect from 1 August for another year, ominous rings were already in the air.

And all this was not a product of a malice driven campaign. Mr. Swaraj Kaushal, the government interlocutor in 1998-99, spilled the beans at the time of his fall from grace, when he remarked in his widely quoted report, 'We agreed to the word 'ceasefire' whereas this should have been "cessation of hostilities".[14] Mr. Kaushal of course did not explain what he meant by that. But a public politics of peace can make its own meaning. Cessation of hostilities, contrary to what some mean by it (armies remain in combat position but do not engage in hostile acts), is a broader term. It would have implied not only ceasefire, but also withdrawal and cessation of all measures considered as *hostile*. It would have meant in the first place withdrawal of the Armed Forces Special Powers Act, the scourge of the people there. It would have inaugurated a gradual but concrete plan of demilitarization including withdrawal of army to the barracks, stop to physical checks, continuous insults and harassments. It would have implied cessation of menacing patrolling, full restoration of civil liberties, and above all the withdrawal of ban on the NSCN. Also it would have meant cessation of hostilities in Manipur and upper Assam. In short, with a broad-spectrum step, this would have signalled the agenda of creating trust – essential for peace dialogue.

Instead what happened? On the alleged ground of inter-Naga clashes, the ceasefire-monitoring cell office of the NSCN in Dimapur was closed down. The mutual charges of violation of ground rules were allowed to continue. On 6 April 2000, Major Peter Parik of the Indian army reportedly asked the NSCN to vacate the designated camp at Munglaimukh. The complaint about this went unheeded. Before that on 22 March Captain Jouhal of the Indian army had issued reportedly a similar 'request' for movement of the armed forces personnel in certain rebel-controlled areas. Again this complaint went unheeded. Mr. Lotha, a member of the monitoring group of the NSCN was arrested early 2000 and held without trial in Diphu jail. Several other cadres were detained under the National Security Act during the ceasefire period. The houses of some members of the rebel army were raided. On 25 July 2000 Mr. Momba Moa, a rebel army sergeant major, was arrested at Jingkieng Nongshiliang in Meghalaya. Some of the outposts of the rebel monitoring arrangement were forcibly closed down. For six months the government took no initiative to hold a meeting of the joint ceasefire-monitoring group. The Indian government's argument was that these outposts were being misused for rebel activities and regrouping. The army's argument was that while there was a ceasefire, its normal activity under the Armed Forces Special Powers Act should and would continue. The added irony to the already twisted scenario was that the NSCN was an illegal organization. It had been allowed to function within 'strict limits' by the government in the interest of peace, and it wanted the NSCN to realize that. No wonder, the NSCN thought that ceasefire had only enabled the government to continue its policy of suppression under the garb of cessation of fire. It complained of 149 incidents of violation of ceasefire by the government side. Equally not surprisingly the government following the murder of Mr. Mungro, a senior functionary of the Khaplang group, and the attempted assassination of Mr. Jamir thought that the ceasefire was for the rebels a 'license to kill'. Ceasefire became a banana joke. The Nagas argued if no substantive dialogue was to take place, and injustice and denial of basic rights continued, what worth was this ceasefire? If the Indian constitution could be amended for eighty times, why could it not be done one more time for bringing peace? If legal-bureaucratic process were to be allowed to overtake the political process, would not this ceasefire embitter the situation already sullied between two unequal parties?[15]

Ceasefire the world over in the crooked history of peace politics has been shown to be a part of the politics of war. That is why the Liberation of the Tamil Eelam (LTTE) for a long time did not take government offers in Sri Lanka on ceasefire seriously. Ceasefire collapsed there again and again. In Kashmir the Jammu Kashmir Liberation Front (JKLF) declared ceasefire; it led to nothing. The ceasefire declared by the *Hejbul Mujahideen* in Kashmir in 2000 collapsed again on account of lack of trust and the (deliberate or otherwise) lack of a policy of the Government to carry forward the political dialogue.[16] Ceasefire has been often attempted to wear out the foe, to create a stalemate, to exasperate him, to choke the agenda of reconciliation and peace with which the adversary has come forward due to fatigue or desire for reconciliation.

Ceasefire de-linked from the agenda of democratic peace, that is to say, the participatory route of wide range dialogues, is typical of a state-centric peace strategy, where the state thinks that by talking to the leader of the adversarial camp, a tête-à-tête between two leaders, peace is achieved. It is a thinking that divorces the condition and feelings of people, their desire for justice, and their activism for peace from the process of peace.[17] The experience of recent Palestinian experience is not that unique and should be enough to make us realize the futility of the much talked of peace process in the post-colonial regions of the world. Such a singular process ignores the alternative ethic of truth and reconciliation, which is built on the path of *transforming retributive justice to restorative justice*. Peace constituencies in a conflict land, as the Palestinian peace activist Hanan Ashrawi has demonstrated so tellingly in her memoir, do not live by diplomacy, but on the basis of popular participation and social-political initiatives that make the dialogue process democratic.[18] The ceasefire in Nagaland was fast crumbling, and people who would enjoy such situation were those who lived off guns. The only way for peace politics to salvage the peace process in Nagaland was to involve the wide sections of people there in the social dialogue on peace – the Naga Students Federation, the Naga People's Movement for Human Rights, the Naga Mothers' Front, the Naga Hoho, the Church, and various other solidarities who had sympathy for the cause for which the rebels had taken up arms, but who desired justice and peace.[19] Though, by large the official dialogue remained confined between the state and the rebels, the unofficial dialogues between constituencies of society prospered, deepening and enriching the politics of peace, though unfortunately neither the government nor even the rebels who were supporting the peace constituencies in Nagaland had shown the vision to argue for a democratization of the process in terms of broadening its agenda and structure of participation in peace dialogues.

The ceasefire was ultimately renewed on 17 June 2001, this time extended to Manipur, Arunachal Pradesh and Assam; and then large-scale violence broke out in Manipur, particularly its capital Imphal, immediately in its wake. From 1997, the NSCN (I-M) had been demanding the extension of the coverage of the area of ceasefire to all Naga inhabited areas, of particular significance to Manipur, where many Nagas live in the hills (while non-Nagas live in the valley), and parts of which had been traditional Naga areas before the colonial administration, local chieftains and independent India's government had carved out a separate state called Manipur with a dominant non-Naga population. To the non-Nagas in Assam, Arunachal Pradesh and Manipur, the extension of ceasefire appeared as a signal that the central government would now agree to Nagalim ('greater Nagaland' to some, and historical vindication of Naga claims to some others), mass demonstrations erupted against the extension of ceasefire in Manipur, the assembly building and several government offices were set to fire and gutted down, ten died in police firing and sixty others mortally injured.

The AMSU (All Manipur Students Union) called for a three days strike in Manipur to protest against the extension of the ceasefire. The popular agitation, fiercest in Imphal, became dreadfully violent. The electronic media kept the Indian people riveted to the unrest in Manipur. On the third day of the strike, mobs stormed the government buildings and torched the State Assembly Hall after lynching several legislators and former Ministers. The police had to open fire to stop the unruly crowds from invading the house of Radhabinod Koijam, the last Chief Minister of the State before the proclamation of the President's rule. Manipur came under a curfew, but violent protests continued. The government promised to review the extension of the ceasefire.[20] So severe was the protest and violence over declaration of peace, that many Naga public organizations were forced to declare that they were for amity and reconciliation. The Naga organizations including the All Naga Students' Association of Manipur, (ANSAM), the Naga Women's Union of Manipur (NWUM) and the Naga Peoples Movement for Human Rights (NPMHR) condoled the 'loss of 13 young lives' in the firing of 18 June. A statement issued by them appealed to all 'Meitei brethren' and the Naga population in the State to 'maintain communal harmony at this critical juncture'. 'The need of the hour is peace and dialogue among ourselves', the statement emphasized.[21] On 24 June, Thingaleng Muivah himself told an Indian press correspondent in Bangkok that the agreement on the ceasefire with the government of India 'deals only with the ceasefire, nothing beyond that. The declaration of a ceasefire is not against anyone and it does not deal with territorial limits.' He explained: 'We have agreed not to fight in any part of India or Asia or anywhere. There is no reason for anyone else to interpret it more than that.'[22] The Naga HoHo, the apex body of all Naga tribal councils issued a statement on 17 July to the effect that the violent protests in the Imphal Valley 'revealed the extremely complicated and sensitive challenges the Nagas face to evolve a workable solution with our neighbours. Ultimately, we need a settlement that will ensure security, stability and peace for ourselves and for our neighbours. Our actions will decide whether we develop and survive together in a changing world or drag one another down to a common hell of perpetual animosity and violence, which can only be destructive to all.' The statement called on the Nagas not to 'ignore or treat lightly the intensity and depth of our neighbours' insecurity about their future', and went on to suggest that the leaders of both the communities should sit together and find a common approach to deal with the controversy.[23] These statements from the Naga organizations did not mollify the protesters. Violent agitation continued. Thousands of Nagas settled in the valley of Imphal fled back to the hills for safety.[24] On the 27 July, the Home Minister of India told the reporters that the government had decided to rescind the Bangkok agreement and to once again limit the ceasefire to the State of Nagaland.[25] The agreement on extension of ceasefire to areas outside Nagaland was rescinded. Ebullient crowds in the valley of Imphal greeted the announcement with boisterous celebrations on the streets.[26]

The ethnic politics of the post-colonial state had thus played itself out once more and the poison fruits were there for all to see. This was the classic situation, what Foucault called *governmentality*, where everything became a matter of administration, governance, mediation by power, and relation of power and rule. The call for reconciliation came, late, how late that only future can tell. Here is one Naga response in form of a dialogic manifesto. It is characteristic of the new situation of mix – of national aspirations, the ethnic brew in which the nationalist nectar prepares, a tired assertion of what one party hold to be true, coupled with weak admissions on the need to dialogue with others, reasons of state and consequently governmental compulsion, sense of reconciliation, realization of the presence of the other, in other words, a situation of dialogue. The statement after describing how Naga areas were taken away from the Nagas, concluded with this:

> 1. Rather than attempting to dissect other people's lands for a greater Nagaland, we had instead been victims of territorial dissections by foreigners for two centuries. Presently we had been dissected into Burma and the Indian states of Manipur, Assam, Arunachal Pradesh and Nagaland. The actual fact is one where some of our neighbours have become geographically bigger at our expense.
> 2. These dissections were done without our approval or consent and even information, though on our part, we had furnished all necessary information followed by actions to the effect that we are a nation with a concrete geographical land different and distinct from all other nations on earth.
> 3. We hold the British, Indian and Burmese governments responsible for this horrible mutilation of our ancestral lands. We also hold them responsible for the millions of Nagas whom they had butchered over a span of two centuries through aggression, invasion and interference.
> 4. While ardently desiring to maintain the best of neighbourly relationship with all our neighbours, we also declare, that we will not be bound by any artificial boundary lines drawn by foreigners without our consent.
> 5. We also further reiterate that in demanding a greater Nagaland, we are not demanding even an inch of ancestral Meitei lands or anybody else land. Our simple plea and appeal to all our neighbours and the world is that they recognize the fact that there is indeed a race and nation called Nagas with a distinct geographical land called Nagaland.
> 6. For a final and amicable settlement we extend our hand of friendship to all our neighbours to settle this dispute in a peaceful manner. We feel that this is the right approach, because the lands under dispute do not belong to the Prime minister of India or a bygone foreign imperial power.
> 7. We firmly believe that this is a must, because the verdict of history is there to prove that no nation on earth exists without a concrete geographical land and that the land of a people is inalienable and unchangeable.
>
> God had gifted these lands, under dispute because of foreign interference, to our forefathers for our habitation. *Tomorrow it will belong to our children and future generations. Let us rise up and settle the disputes amicably so that future generations will live in cordial relationships for all times to come. And in this attempt, rather than arguing over the tiny area of cease-fire coverage, let us even extend the cease-fire to the whole North Eastern region and even the whole world so that nations will learn to live in peace with one another.*[27] (Italics mine)

The Naga leader Muivah's response to the outbreak of violence in Manipur, briefly mentioned earlier, was cryptic and indicative of a situation where dialogic needs, possibilities, strategies, and governing techniques were dreadfully enmeshed. Ceasefire as we can see focalized all dimensions of peace by trans-valuing them; it turned the politics of cessation of hostilities into a field of 'governing ethnic relations by the state'. In that interview he said,

> ...There is no secret (deal) between the Government of India and us. The declaration of cease-fire is more than three years old. If anybody is saying there was (a deal), it must be their own creation. The cease-fire issue has been open to all. So, there cannot be any secret motive in it. The purpose of the cease-fire was to create a suitable atmosphere to discuss the Indo-Naga issue. It was time we examined or explored a possible peaceful solution to it...We have come forward to stop the killings...They (Government of India) and their generals openly declared that a military solution to the problem was not the right approach. But there should be a political solution to it. So we came out and talked. We are committed to the peace process. When the Government of India asked for a cease-fire, we agreed because a suitable atmosphere must prevail. Without declaring a cease-fire, it was meaningless to say we were ready for a peaceful solution. The Government of India and the NSCN now have an understanding that we (the Indian armed forces and the NSCN forces) should not fight anywhere. So, we have come forward to stop the killing and the fighting. This (the intention of both parties) needs to be appreciated by every sensible person.[28]

To the question at to what his answers were to claims that the extension of the cease-fire to other states also involved constitutional issues and serious law and order conflicts, and therefore needed to be mediated, Muivah went on,

> In that case, the Nagas will not respect us for accepting the Indian Constitution. It is not possible. Our people have fought for their freedom. If anything is going to be imposed on us, we will not accept that...The agreement is very clear. It deals only with the cease-fire, nothing beyond that. The declaration of a cease-fire is not against anyone and it does not deal with territorial limits. It is between two separate entities (representatives of Government of India and NSCN, which is me) in which we have agreed not to fight in any part of India or Asia or anywhere. There is no reason for anyone else to interpret it more than that. It is a surprise to us that some of our neighbours are taking an exception to it (the cease-fire extension). We can go and meet and tell them there is no danger to anyone...The Naga Hoho (Naga assembly) has told us that they will make an approach to the other states. The cease-fire does not harm them. If they have interpreted more than that, we are sorry. We not only want a peaceful atmosphere with the Government of India, but also with people around us (the other Northeastern states). How will the NSCN (I-M) discuss the issue of a greater Nagaland with the neighbouring states? That may come up later during the process of negotiations. At the moment it is difficult to say because the issue hasn't come up yet. However, it may not be easy. It is here to stay and it will be the government. Without that our people will start questioning us, 'What is your assurance? They'll ask us.' We have access to many neighbouring nations and states. We are able to maintain that because we don't do anything against the laws of the country or detrimental to the society. We know Thailand, Myanmar, and China. We know how to respect our neighbours. This is a very essential part of our political policy.[29]

> Only future can tell, how the rebels will be able to wade through the murky waters of dialogue taking along with them their movement, rebellion, and their people, to the shores of reconciliation and justice. The ceasefire has of course been still holding amidst various and sometimes violent reactions from different quarters, divisions among communities in the region, occasional exchanges of fire and rampages by the army, the continuing terror-tactic, and war-cries against peace. The state politics of ethnic management, perched now on the need for cessation of hostilities, will probably see this out. Such a conundrum of 'rescued and endangered' ceasefire has five lessons for the politics of dialogue.

One, the overwhelming desire for justice and peace saw the extension of ceasefire in Nagaland amidst all signs of a crumbling arrangement. Thus all the organizations had some role or other to play in advocacy of ceasefire. Dialogue made possible the politics of peace.

Two, this account is one of the most decisively current history – this is the history of a war-peace continuum in which we live, where the present slips into history as swiftly as imaginable, and can be thus recorded only as past so that it can enter the narrative of peace politics.

Three, ceasefire in a post-colonial scenario is mostly a declaration of the *continuum*, rather than being an usher of 'cessation of hostilities' and peace. Low-key war makes low-level peace, which means continuity of other hostilities associated with war.

Four, land remains one of the most important issues in conflict and any settlement, and yet the most ignored by peacemakers. At times, land is the core of what is known in academic language as 'geopolitics'. Yet, we must not forget that, this geo-politics is in a way politics of the people, because the overriding concern around land stems often from the ways people are settled in that land. It is in this combination that we find the inflammable intersection of geo-politics and ethno-politics.

And, finally, the plurality of a context of peace makes the event of ceasefire singular by leaving all its marks on it, yet hiding itself in the light of the singularity of the event.

The Singular Event and Its Multiple Traces

When in a statement dated 21 December 2000 that spoke of 'space for peace', the LTTE had all of a sudden declared a month long unilateral ceasefire effective from the Christmas, the government of Sri Lanka had no immediate comment to offer. A Sri Lanka Broadcasting Corporation radio bulletin said that the President Chandrika Kumaratunga then travelling in Europe Radio would comment later. The Foreign Minister Kadirmargar ruled out any immediate positive response, and the opposition leader Ranil Wickemesinghe welcomed the announcement and urged the government 'to seize the opportunity'. A Reuters report by Scott MacDonald datelined Colombo said, '...The announcement by the LTTE is the biggest breakthrough in efforts to end war since 1995...and the Government has been forced on a diplomatic defensive since Prabhakaran called three weeks ago for unconditional talks and dropped previous demands for a truce and troop withdrawal first, both politically impossible for the government to meet'. The Norwegian interlocutor Erik Solheim spoke of 'still a gap to be bridged'. The Vice President of the main Tamil constitutional party, the TULF (Tamil United Liberation Front), Anandasangaree spoke of so many lives already lost in war and therefore no motive should be attributed to the announcement of a unilateral ceasefire, which 'should be welcomed by both Tamil and Sinhalese people. We must take for granted that they are serious and the government must respond positively'.[30] Erik Solheim's visit to the jungles of the north seemed to have paid off. After the massacres of Chemmani and Bindunuwewa and the over all deteriorating human rights situation, the international donor community also reminded the government of Sri Lanka of the financial burden of the war-campaign by declining to pledge a single yen, dollar, euro or pound in its annual meeting in Paris in 2000. This was at a time when defence expenditure had shot up by 40 per cent, and flight of foreign capital became 9 times higher in the first ten months of 2000 – 3.2 billion Sri Lankan rupees as compared to 475 million rupees in 1999,[31] and the interim budget for January-April 2001 had already decided to allocate 26 billion rupees (US $ 455 million), about 20 per cent of the total spending during this time. In any case defence expenditure went up dramatically last year and rose to 6 per cent of country's GDP.[32]

In such embattled time and zones with nearly 70,000 people killed in the 17 year long war, the minority community of the Tamils said that a 'space for peace' had been created by the Tigers' initiative, while the majority community said that such offer of peace was not according to the rules of the game. The Sinhalese press decried the offer of peace by saying that everything for peace including ceasefire had to be negotiated to achieve; nothing would come of its own. There was nothing called a 'unilateral' ceasefire in the lexicon of peace. Kadirgamar demanded that the LTTE should be proscribed in the UK, its office in London should be disbanded, and the hoax of peace politics be called off. He said, talks with LTTE should continue, so should continue the war till talks succeeded.[33] Meanwhile Solheim added that the great power of South Asia 'India is very helpful. We are also keeping India informed about developments in the peace process'.[34] The pro-Government Tamil group, the EPRLF, widely seen by majority of the Tamils as having been co-opted by the state, predictably decried the unilateral ceasefire saying, 'Tamils should not take LTTE ceasefire seriously'.[35]

War, and the politics of peace inscribed in war, thus pose the question: What can be the agenda of a plural dialogue in a situation so singularly defined by parties, that even unilateral gestures of peace politics do not create dialogic possibilities, and both adversaries pose the question of '*permanent* peace', defined by each in own way, against a *temporary* ceasefire (because ceasefires have been mostly temporary in its history in the region and even the non-engaged armies of India and Pakistan remain in a state of high alert on the LOC in Jammu and Kashmir and keep exchanging artillery shells)?[36] In other words, how is one to understand the dialogic structure of a paradoxical situation, in which talks on ceasefire are being consumed by talks of war and peace? *In this question we have the clue to the main difficulty in our inquiry. It is this, if a situation is empty or near empty of dialogic procedures and contents, how can dialogue fill it up with itself? How can a situation extricate itself from its own self?* The universe of dialogue is marked by the circularity of acts, at times by the closed nature of the acts. All the more therefore we cannot be content with a liberal theory of dialogue. Dialogic acts call for a critical perspective, which means taking account the context of contest, conflict, war, and peace – the multiple traces that make dialogue possible in one situation, difficult in another.

To the State and the majority community, as has been with victorious powers in wars, the crucial issue along the pathway to 'permanent peace' is always the de-commissioning of arms of the adversary. One organ of the Sinhalese opinion was outraged at the offer of ceasefire for *it violated peace*, and indignantly commented, 'The approach adopted in negotiating peace accords in Kosovo, Sierra Leone, and other regions of conflict, was to negotiate with the rebel leaders. The price paid was rewarding the rebels by first recognizing them, and then granting them amnesty and placing them in positions of authority. There was no discussion about surrender of arms or the specifics of the punitive action for non-compliance of the agreements. Consequently, the peace negotiated was fragile...'[37] In this received knowledge of peace process and peace dialogues in Sri Lanka that included an intense interest about the Good Friday agreement, it was said that, 'the Government should borrow a page from the Good Friday Agreement and insist that de-commissioning should take place from the commencement of negotiations'.[38] Indeed visions of peace vary widely in the society, as various as the peace constituencies. A sane observer in an otherwise intensely indignant nationalist press in Sri Lanka skirted the issue of hard choices in South Asian peace politics so dramatically brought to light by the sudden announcement of unilateral ceasefire by the rebel group, and mused, probably dialogue (in his word 'engagement') would begin towards 'further dialogue' on matters related to fighting such as return of captives, and dead bodies, protection of refugees, opening up of civil supply route, and then abruptly commented, 'It is simple: *even a ceasefire must be negotiated.* All previous attempts at ceasefires (there have been three) have failed. The LTTE has been the side that has abrogated the previous ceasefires. The Sri Lankan State, on its part, has been accused of doing much to undermine the conditions of ceasefires of the past. Even the previous PA government has been accused of not doing enough to sustain the last ceasefire. (Emphasis mine)[39] The confused moderate voice of the commentator Lakshman Gunasekara went further and wondered in a tone that almost echoed the voice of the hawks, 'The party that unilaterally abrogated previous ceasefires is now once again calling for a 'ceasefire' at the very moment it was supposed to be preparing to enter negotiations without any conditions. In fact, if other factors are also brought into consideration, the LTTE's 'ceasefire' offer becomes even more suspect'.[40] One should leave aside the wild speculations in the nationalist mind about why the rebels wanted peace and why at that time it wanted so, as one newspaper conjectured, 'LTTE prefers to have no ceasefire during the talks' and then within four days told in a bewildered tone, 'Ceasefire offer triggers ceaseless pressure.'[41] LTTE, amidst all this, extended the ceasefire by another month from 24 January 2001.

This is once again a classic situation where a single act of declaring unilateral ceasefire consolidates all differential traces of a situation and becomes a congealed act. It is still too early to say definitively as to how the practices of peace politics will respond to such acts/events without getting trapped in the latter's singularity and succeed in re-inscribing the plural traces of peace politics in dialogue. In the case of ceasefire in Nagaland we saw that the vagueness of political discussions was counterbalanced by the concrete demand for ceasefire, which has held till date, for besides several other factors, the demand for a public politics of peace linked the plural vision of human rights and justice with peace process. In Sri Lanka, as the peace question remained de-linked from any broad vision, even a unilateral offer of ceasefire could be rejected by the political class that thought and still thinks that simply agreed declarations of vague political principles could/can ensure peace. The Palestinians among all rebels, engaged in more than decade old peace dialogues, and have learnt the nature of the fruits of a poisoned tree. A memorandum released by the Palestinian Authority on the eve of the Clinton's departure from US presidency criticized the style of US mediation and admitted,

"As a result of vaguely worded arguments...both parties to the conflict have mistakenly assumed, at different times, that either the Israelis had accepted to end the occupation, or that the Palestinians had agreed to forgo some of their fundamental rights. Such ambiguity made it possible for both sides to sign agreements that they could interpret diametrically opposed manners to their domestic constituencies...and has very little implementation."[42]

If in the mind of the political class in Colombo the revocation of ceasefire by the LTTE on earlier occasion weighed heavily, it meant that it was not ready to appreciate the possibility of any other perspective to the earlier story of ceasefire, that victims could have their own perspective, and they might have their views of why and when they needed peace, ceasefire, and could be exasperated with the peace politics of the rulers who saw peace politics as element of war strategy and would rarely take concrete measures towards settlement. The victim's view, euphemistically called the 'northern view' in the political parlance of Sri Lanka, was brought out clearly by S.J. Emmanuel in an article, 'Kumaratunga-Prabhakaran Talks – A Northern View', wherein he raised the question of 'a cold war situation within peace talks' where the issue was 'food or political talks'.[43] The victim wanted the easing of cordon first, restoration of supply of food items (even baby food items were being controlled), the withdrawal of draconian measures imposed by the Vavuniya military authorities, and measures to ameliorate the deprivation of thousands of farmers and fisher-folk, in short the withdrawal of the weapon of blackmail and threat with which the government wanted to bring the adversary into submission. The government said that the victim (that is the Tamil of the north) was grumbling too much and demanding too much food, which left 'the Tamils wandering how a government which was so reluctant to give sufficient food and medicine and other basic requirements for life could give political rights and freedom'.[44] The LTTE argued in the negotiations that, 'the initial stages of negotiations should give primacy to the immediate and urgent problems forced by (our) people'.[45]

In the peace talks, as the chroniclers noted, more than the interests of the peace constituencies, the interests of bureaucracy, military establishment, factional party leaders, and a singularly monolithic leadership of the armed adversary were active. There was no representation of other groups, victims' organizations, or movements. It was more a military staff exchange of views than a State acknowledging that it was talking to a political leadership opposed to the State. Ironically therefore while there was no real accommodation of the demands of the adversary, the lack was covered up by euphoria and gestures in ample measure. The composition of the government team left much to be desired. There was reluctance to move from a cessation of hostilities to ceasefire from the government side, and we must remember that earlier too the LTTE had declared a unilateral ceasefire on Chandrika's assumption of office as President. The monitoring committees delayed abysmally in functioning. There was equal reluctance to further lift the economic embargo and the ban on fishing. There was reluctance in opening a military-free passage at Pooneryn. And, finally, there was reluctance of the army establishment in implementing the decisions taken. The situation reflected almost the characteristics of the situation in Nagaland where too the army and bureaucracy ensured that the control-measures should continue, and not much importance should be given to the monitoring mechanisms for observance of ground rules, while exchange of fire was held in abeyance.

This tradition of wearing out the adversary with gestures and no substantive accommodation, talk and encirclement, effecting splits in the opposition ranks, and erasing the manifold traces of the desire for peace by various methods, has a history in Sri Lankan politics. In the Thimpu talks also the same had happened, and the talks had collapsed after just two rounds of discussions. Fundamental positions clashed, and talks were effectively killed. The 'proximity talks'[46] also arranged by India ignored the basic humanitarian desire of the people, particularly the victims. It was no accident therefore, given the structure of the dialogue, that the accommodation between President Premadasa and LTTE (1989-90) was to collapse, when again the state demanded that the adversary had to withdraw all capacities to mount hostile acts, in return conceded little (but more than what was to be given later), retained control measures, and de-linked peace politics from the issue of human rights and dignity, which in this case implied in the first place an acknowledgement of the humanitarian dimensions, in other words recognition of the victim's perspective.[47]

Once again, a ceasefire has been arranged between the government of Sri Lanka and the LTTE with the mediation of the mission from Norway. And once again, we must look very closely into the politics of dialogue. The signing of the memorandum of understanding by both Sri Lanka government and the LTTE, took place at a specific conjuncture in the world political situation, when an alternative policy for a negotiated political settlement was permeating the thinking of the LTTE leadership. The world powers had turned their attention from the traditional centres of protest and dissent to the hostilities of the emerging Islamic radicalism and other political rebel groups throughout the world. The US with its allies had not been involved in violations of civil and political, but had gone to the extent of infiltration into other territories and subduing the sovereignty of poor nations. The US and its allies within a short period of recent history had violated the sovereignty of ex-Yugoslavia for the sake of defending the Kosovar-Albanian national community against the Serbian majority and recently subdued Afghanistan as a retaliation to the 11 September attack on the Word Trade Centre and the Pentagon. In such situation the PA government and its Foreign Minister Kadiragmar became successful in their campaign for the total proscription of the LTTE by the US, Britain and the EU countries. India, of course, had already proscribed the LTTE. The proscription had hurt the LTTE whose main stakes lay not mainly with the Chandrika and Ranil Wickramasinhge governments, but with the international community, which housed and influenced specifically the Tamil Diaspora, and which within the context of the global market and capital, was increasingly disapproving of the concepts of separatism, partition or a formation of a total independent nation-state. Against the ideology of globalization, and in the background of global politics, nations could now exist under the umbrella of international guarantee only by partly accommodating to some extent the idea of co-existence and cultural diversity. Thus, the 'third dimension' of a conflict suddenly became crucial. The role of the Bush administration and its Western allies became critical in the dialogue for ceasefire. In the MOU on ceasefire, the third party facilitator, the Royal Norwegian Government, eventually turned out to be an arbitrator who could even take final decisions that affected both parties to the extent of making the Monitoring Mission the crucial authority in the work of demarcation of the lands in the North-East for the Tamil speaking people – and this in the background of the fact that the national question of the Tamil people in Sri Lanka would remain only half solved, if the Sri Lanka government refused to acknowledge that the issue of land (consequently territorial configuration) was the core-problem of the present dispute, and that it might make partition of Sri Lanka a virtual reality.[48]

The Third Dimension

If the issue of land overwhelms many dialogues on war and peace, the other feature of the dialogic context in Sri Lanka is the 'lurking presence' of the *third dimension* in a dialogue. Too often dialogue is considered as a bilateral act; what we forget is the silent supplement – the 'third presence' in conversation, whose role relates to the impasse, who symbolizes the other traces, seemingly effaced by a bilateral dialogue conducted on singular issue/s. The intermittent and alternating appearance of the third dimension that symbolizes the other traces of a dialogic situation besides the one represented by the dialogic actors and the act, is the sign of a process where other signs are being erased, but the erasure is never complete. The third dimension appears in many ways, in many intervals, and many scenarios. Thus, in the midst two stands, two viewpoints, two adversaries, and two themes (of war and peace) – on which and between whom the dialogue on ceasefire had been held in Sri Lanka – the issue of human rights appeared as the *third presence* symbolizing other concerns, constituencies, and themes. It also raises the significant question for peace politics, When does the agenda of human rights emerge in a dialogue on peace? When does a dialogist or a 'third party' push forward such an agenda in the ongoing talk, select the moment for the push, after which the dialogue may irretrievably shut the door on all such 'extraneous' concerns? It is also significant to note, as the Sri Lankan scenario bears out, the third dimension may not at times present itself formally; but it is there in the form of daily chorus on these 'extraneous' issues (such as, medicine, food, water, administration of camps, journey passes) on which negotiations go on at the public and governmental level.

That is where peace politics stands. It is a conundrum. Antonio Gramsci spoke of catastrophic equilibrium – equilibrium of forces moving towards catastrophe. The state and an equal armed adversary engaged in politics of peace, no one capitulating or conceding where to its mind concession is capitulation, but both forced to talk by the social as well as global pressure for peace, wide sentiments for justice, own exhaustion, and the sheer necessity to keep the structure of on-going politics intact – in such situation ceasefire appears as the singular act of a destiny called the 'war-peace continuum'. In this continuum where ceasefire is the dramatic form of a closed circle, we have the fault-lines appearing tantalizingly before a public politics of peace. These fault-lines suggest ways of prying open the closure. The fault-lines appear in form of the third presence, wide reactions, desires, engagements in peace politics away from the Camp David style meetings, which for no other better term I call as the public politics of peace where the multifarious and multitudinous participation in peace talks will be the sign of the day. It will imply democratization of themes, structures, actors, and consequences. The only way as I see it is to effect a reversal of theology – turning the humanitarian into most political, not necessarily in the sense that Kofi Annan suggests, but in terms of democratic politics which among others makes humanitarianism a part of human rights. That is where the middle space (often appearing as the third dimension) is to be invoked again to make the rules of Geneva Convention not what they are, that is to say making war humanitarian, but what they are not, which is to say, making humanitarian conditions the condition of making peace – the route to banish war by making peace constituencies compulsory participants in the future round tables whether in Paris, or Thimpu, London, or Dublin, or even Camp David. In Sri Lanka, in the uncleared areas, the government has given away the responsibility of relief supply to the army, which denies some of the basic supply items and thus prevents the war-ravaged population from fulfilling basic needs. The framework of humanitarian assistance cannot but be in the context of human rights, which means that representatives of the people from the affected areas will have a say in determining priority and process. And this can be done when humanitarian assistance is seen as a part of the human rights of the war-affected people. Without such an agenda and programme, how can democracy encourage a spirit of reconciliation from a wronged person and a suffering people who have lost trust in civil institutions, and as a victimized community lives in its own world, who knows that 'humanitarian activities and humanitarian intervention' never take into account the normative costs, that is, the fundamental standards, the human costs, economic costs, social costs and the physical costs of war, but on the contrary imposes more costs by intervention? If one takes a close look at the very detailed table (Table 7.1) given below, which maps the victims' voices for humanitarian space and peace in Sri Lanka, the argument for placing desire (for peace) as counter to power (of war), context (of multiplicity) as counter to singularity (of an act), and responsibility (for violence) as counter to unilateral way (to subordinate), at the heart of dialogic politics will become clear. Only such counter-positioning will make dialogues for peace possible – this is, in one phrase, *the politics of plural dialogues*.

Table 7.1 is drawn from the report of a two-day workshop (Colombo, 3-4 April 2000) organized by the Consortium of Humanitarian Agencies on the profile of humanitarian needs of the population in various districts in Sri Lanka. The aim of the workshop was to create a platform for a cross-regional and inter-ethnic dialogue between different district representatives, create an opportunity to clarify and prioritize issues, and make recommendations together with the identification of the organizations that will be potentially responsible for the main components of the framework of humanitarian assistance.[49]

Table 7.1 Prioritized issues along with recommendations for each district

AMPARA DISTRICT

Categories as per framework	Prioritized Issues	Causes	Recommendations (what)	Responsibility (who)
▪ Program Priorities	▪ Why does agriculture fisheries not livestock bring enough income?	▪ Restrictions on the people's movement. ▪ Lack of investment. ▪ Govt's economic policy	▪ Stop war ▪ Divert funds away from war ▪ Change the policy towards peoples aspirations	▪ Govt., LTTE & other Militant groups ▪ Govt. ▪ Govt.
	▪ Why are the women affected?	▪ Loss of the bread winner ▪ Lack of awareness/soci al & cultural setup ▪ No proper plans for women dev.	▪ Stop war ▪ Awareness programs ▪ A national program for women ▪ District level programs	▪ Govt., LTTE & other Militant groups ▪ NGOs & CBOs ▪ Govt. & NGO Coordination ▪ Govt. & NGO Coordination
	▪ Why is there degradation in Health & Sanitation?	▪ War & Economy ▪ Living without basic requirements	▪ Stop the war ▪ Awareness programs ▪ Allocate more funds	▪ Govt. & other Militants ▪ Govt. & NGOs ▪ Govt. & NGOs
▪ Peace & Reconciliation	▪ How can we avoid discrimination?	▪ No equal treatment ▪ The parties concerned do not understand the reality	▪ No partiality (avoid) ▪ Awareness programs ▪ Awareness programs	▪ Implementing agency ▪ Implementing agency ▪ Implementing agency

		▪ Communalism	▪ By enacting & implementing new laws ▪ Selecting dedicated officers	▪ Govt. ▪ Implementing agency
	▪ How does language influence peace building?	▪ Lack of knowledge of other's language ▪ Unwillingness to learn other's language ▪ Lack of proper policy making in regard to education	▪ Encouraging people to learn other's languages ▪ Providing opportunities to learn other's languages ▪ Awareness programs ▪ Compulsory from yr. 1 ▪ Provide needed resources (teachers) ▪ Proficiency bar for language teachers (Vs.)	▪ Govt. & NGOs ▪ Govt. & NGOs ▪ NGOs ▪ Govt. ▪ Govt. ▪ Govt.
	▪ What are the proper methods for peace building?	▪ Non awareness in understand-ing the reality ▪ No peace ideas within us	▪ Awareness ▪ Visiting different areas ▪ Awareness ▪ Meditation ▪ Religious activities ▪ Social activities	▪ Govt. & NGOs ▪ NGOs ▪ NGOs & Civil society & religious groups ▪ NGOs & Civil society & religious groups NGOs & Civil society & religious groups ▪ NGOs & Civil society & religious groups

JAFFNA DISTRICT

Categories as per framework	Prioritized Issues	Causes	Recommendations (what)	Responsibility (who)
▪ Program Priorities	▪ Persons who have lost their livelihood	▪ Civil war ▪ Ethnic violence ▪ Disappearing persons	▪ Stop war ▪ Bring peace ▪ Bring peace	▪ Govt., LTTE, Third party ▪ Govt., LTTE, Third party ▪ Govt., LTTE, Third party

	▪ Unemployment	▪ Lack of industries ▪ Vacancies not filled ▪ Restricted areas for fishing/ agriculture activities	▪ Establish industrial estates ▪ Establish free trade zones ▪ Fill all vacancies ▪ Remove restricted zones	▪ Govt. ▪ Govt. ▪ Govt. ▪ Govt.
	▪ Education and vocational training	▪ Lack of laboratory/ computer facilities in schools ▪ Lack of vocational training institute and some courses ▪ Restricted Youth Service Council activities	▪ Supply chemicals, glasswares, equipments, computers to schools and vocational training centres	▪ Govt. & NGO
	▪ Agriculture, fisheries, livestock farmers etc.	▪ Restricted areas ▪ Restricted time of activities ▪ Lack of marketing facilities ▪ High costs of inputs	▪ Allow free movement in all the areas ▪ Transport facilities to be improved between Jaffna and Colombo	▪ Govt. ▪ Govt. & NGOs
	▪ Transport: passenger commuting and transport of goods	▪ No transport facilities by road/rail ▪ No regular transport services by sea/air	▪ Stop war	▪ All parties
	▪ Land mines	▪ Civil war	▪ Stop war	▪ All parties
▪ Peace & Reconciliation	▪ Discrimination	▪ In almost everything, such as employment, promotion, price of essential items, fund allocation	▪ Impartial implementation of govt. policies	▪ Govt.
	▪ Language		▪ Implementation of govt. policies	▪ Govt.

	▪ War as a means of income	▪ Commission paid for several middlemen in business, politics, govt. officials in obtaining services, goods, etc.	▪ Remove restriction on transport of passengers, goods, etc.	▪ Govt.
	▪ Peace building efforts	▪ Civil war	▪ End war ▪ Peace negotiations ▪ Supply of items, services comparable to Colombo prices	▪ Govt.

TRINCOMALEE DISTRICT

Categories as per framework	Prioritized Issues	Causes	Recommendations (what)	Responsibility (who)
▪ Programme priorities	▪ Persons who have lost their livelihood	▪ The ethnic violence after 1983 ▪ Employment discrimination ▪ Discrimination in education	▪ Restore peace by negotiations ▪ Treat all alike, purely on merit	▪ Govt., militants, international agencies ▪ Govt. and private sector ▪ Govt.
	▪ Gender based issues	▪ Prevailing civil war in NEP ▪ Male domination in society ▪ Culture based ▪ Customary behaviour	▪ Restore peace early ▪ Develop mutual understanding by discussing problems ▪ Awareness programme about civilized societies	▪ Husband and wife ▪ Religious institutions and NGOs
	▪ Unemployment	▪ Improper planning by state sectors in investment for industries ▪ Lack of organizational facility to train youth in specialized fields ▪ Displacement caused by	▪ Proper planning by state sector to have even distribution among districts ▪ Youth to be trained in demanding fields by state and NGOs ▪ Restore normalcy early by ending war by negotiations	▪ Govt. & donor forum ▪ Govt. & NGOs ▪ Govt., militants, international bodies

		prevailing war situation and restriction imposed by security forces		
	▪ Health and sanitation	▪ Lack of facilities in the state run medical institutions, especially in uncleared areas	▪ State run Institutions to be equipped with staff and drugs. Restoration of medical facilities in uncleared areas.	▪ Govt. & NGOs ▪ Govt.
		▪ Displacement of people as a result of ethnic violence causes overcrowding of welfare centres and camps	▪ Early resettlement after restoration of infrastructure	▪ Govt. & NGOs
		▪ Unplanned activities permitted by authorities (industrial pollution, drainage problem, etc.)	▪ Regulations to be observed by authorities in granting permission for activities	▪ Govt. & NGOs
	▪ Education and vocational training	▪ Denial of education for orphans and children in uncleared areas where father and mother go for odd jobs	▪ Institutional support to be provided for orphans and poor children in uncleared areas.	▪ Govt. & NGOs
		▪ Step motherly treatment by state sector (free books and clothes)	▪ State sector to treat all alike by distributing free books and clothes at the same time	▪ Govt.
		▪ Unplanned activities of the state sector as well as private sector to cater for the drop outs of schools	▪ State and private sector to have training programmes in demanding fields	▪ Govt. & NGOs

		▪ Lack of training facilities in the traditional employment fields (fisheries, agriculture) to modernize their techniques	▪ State sector and NGOs to go in for training programmes in these fields	▪ Govt. & NGOs
▪ Peace & Reconciliation	▪ Language	▪ Though Tamil also was made an official language the so called Govt. administrators failed to entertain same to date. ▪ The Govt. has failed to honour and implement Tamil language provision in the constitution through administrative machinery. ▪ Lack of understanding between communities because of language barrier.	▪ Govt. to take firm action to honour and implement same immediately. ▪ Adequate facilitate to be extended to all communities to study other's language	▪ Govt. ▪ Govt. & NGOs
	▪ Discrimination	▪ Employment and higher education non availability of higher educational institutions. ▪ Harassment by security forces. ▪ Lack of state investment in NEP.	▪ Equal opportunity for all in employment and higher education. ▪ Strict regulations for security forces, proper training by International institutes like ICRC. ▪ State investments equally distributed among districts.	▪ Govt. & Private sector ▪ Govt. & International bodies. ▪ Govt. & Foreign investors.
	▪ Peace building efforts.	▪ Lack of peace building efforts in the south where as in the	▪ Awareness workshops and seminars to be conducted in the	▪ Govt, NGOs & Religious bodies. ▪ Govt. Militants

		North people have suffered enough and are looking for peace. ▪ Huge amount of money spent for war machinery depriving development activities. ▪ Degree of uncertainty is high all over Sri Lanka.	South with demonstrations especially through religious bodies. ▪ Stop war by negotiation.	and International bodies
	▪ Encouraging inter-ethnicities.	▪ Misunderstanding among communities created by Politics. ▪ Denial of equal opportunities in all aspects by the state. ▪ Not understanding the inner feelings of each other.	▪ Bring communities together for inter cultural exchange programmes. ▪ All opportunities to be given purely on merit basis in state sector. ▪ Cultural exchange programmes and get togethers at religious institutions.	▪ Govt, NGOs ▪ Govt ▪ Govt. Sector, NGOs & Religious Bodies.
	▪ War as a means of income.	▪ Not for all but for a few. ▪ 2000 year old rich culture is being ruined. ▪ Law abiding is lost.	▪ Restore peace by negotiations. ▪ Stop war with negotiations. ▪ Restore peace through negotiations and enforce law and order. Using Govt. Machinery.	▪ Govt, Militants and International bodies ▪ -do-
▪ Development	▪ Rehabilitation alone is not enough to create a balanced situation in the country. ▪ To pave a sustainable peace situation in the future.	▪ Prevailing situation in NEP has prevented new investment or industries. ▪ For last 15 years. Same time existing industries also were destroyed.	▪ Development activities in infrastructure facilities & new investment programmes to be boosted in NEP to bring a balanced situation among regions in the country.	▪ Govt. and International Donor Forum.

VAVUNIYA DISTRICT

Categories as per framework	Prioritized Issues	Causes	Recommendations (what)	Responsibility (who)
▪ Program priorities	▪ Socio-economic deterioration	▪ Inter racial harmony disturbed ▪ Breakdown of traditional employment's ▪ The downfall of the Vanni Market	▪ Political, Administrative and religious leaders to create awareness of the inter racial, social and cultural and religious values and also promote correct attitudes.	▪ The state and the LTTE. ▪ Different politicians, religious leaders and NGOs.
	▪ Education and Vocational Training	▪ Schools are not held regularly. ▪ Lack of facilities for the displaced students. ▪ Irregularity in allocation of resources.	▪ The state to provide school facilities, buildings, trained teachers according to the need. ▪ To provide transport to the students.	▪ Central Govt. & the Provincial council.
	▪ The Loss of means of earning.	▪ Agricultural inputs for the farmers in the unprotected areas. ▪ The difficulty of farmers living in protected areas to go to unprotected areas for cultivation. ▪ The exodus of the business community from the town.	▪ Permit the transport of un prohibited goods in required quantities. ▪ All un official payments collected by different organizations to cease.	▪ Security forces. ▪ The state.
	▪ Health and Sanitation.	▪ No specialist consultation. ▪ No facilities for additional patients. ▪ The deficiencies in preventive program.	▪ Provide specialist consultant doctors, officials and also with the necessary instruments and materials.	▪ Central Govt. and Provincial Council.

	▪ Gender	▪ Cultural deterioration. ▪ Right to marriage being curtailed. ▪ The identity of young girls being lost.	▪ The state and the NGOs to take necessary measures to develop the economy.	▪ Govt., NGOs, Religious organizations.
▪ Peace and Reconciliation	▪ Language	▪ Both the groups lack the knowledge of other's language. ▪ No facilities for adult education. ▪ Both the groups tend to depend on English.	▪ Appoint security personnel with adequate knowledge of Tamil language ▪ Orient children from pre school stage to bi- lingual practice.	▪ Security forces. ▪ Central Govt. ▪ Provincial council.
	▪ Discrimination	▪ Mutual disrespect for each other's culture. ▪ False promises of the rulers. ▪ State policies not accommodat-ing.	▪ To launch a program for the appreciation of cultural values by the inter cultural district committees.	▪ The district secretary.
	▪ War as a means of income.	▪ The livelihood by the security services. ▪ Transport of prohibited goods to the un- protected areas and issuing licenses to unauthorized persons. ▪ Sending prohibited goods to the unprotected areas at a higher price.	▪ To establish an Institutional framework to regularize transport of goods, passengers, trade, and other official duties and implement the regulations.	▪ The State, Security forces, and peoples organizations.

▪ Peace and Reconciliation	▪ Breach of peace	▪ Unbalanced allocation of resources. ▪ Building mistrust among the ethnic groups. ▪ Wrong political promises.	▪ To make the allocation and distribution of aid and resources transparent.	▪ The State and Donors.
	▪ Unbinding of the Inter-racial binding.	▪ Partial political outlook. ▪ Misunderstanding among the ethnic groups ▪ Terrorist activities.	▪ In organizing development and other relevant activities to make it inter-racial and ethnically heterogeneous.	▪ The state, NGOs, Multi racial organizations.

POLONNARUWA DISTRICT

Categories as per framework	Prioritized Issues	Causes	Recommendations (what)	Responsibility (who)
▪ Programme priorities	▪ The decrease of income from agriculture, Fisheries and Livestock activities.	▪ Houses, Paddy lands, livestock lost and no new enterprises. ▪ Because of the open market economy, the rice mill owners and the buyers are lowering the prices. ▪ The scarcity of labour and the higher wages for labour.	▪ Security operations to be minimized during the paddy cultivation seasons. ▪ Fixing fair prices and implemented by the state for the produce. ▪ Introduce new technologies at affordable prices for the farmers. ▪ Revive the labour sharing practices. ▪ Ensure security for the immigrant labour from other areas.	▪ The District secretary, Responsible security personnel, Civil security committees. ▪ State sector, farmer organizations, rice mill owners, NGOs, Credit supplying Institutions. ▪ Voluntary Organizations Brokers, Relatives.
	▪ Socio-economic deterioration	▪ Unsuitability in the life style due to the war psychosis. ▪ Inability to take the cows to the jungle for	▪ To encourage the development of the personality and the capacity building within the context of the present reality.	▪ Govt. & NGOs. ▪ District secy, Security chiefs, Civil security committees. ▪ Community leaders.

		feeding, the loss of paddy fields, decrease in income. ▪ Demotivation for work due to psychological drawbacks.	▪ Formalizing the security situation and to make the civil security committee to function without disruption.	
	▪ The deteriorating health condition of the communities.	▪ Inability to have a balance diet. ▪ Inaccessibility to health services. ▪ Not taking preventive measures to prevent diseases, lack of education.	▪ To prepare an awareness programme on preparing a balance diet with the food resources available in the area. ▪ To activate the inefficient health institutions and appoint officers. ▪ Provide health education. ▪ Promote indigenous medical system.	▪ Govt. & NGOs. ▪ -do-
	▪ The spread of alcoholism among the people.	▪ The feeling of insecurity and unsuitability created by the war situation. ▪ Escapism from problems. ▪ Free availability of illicit liquor and state sponsoring of liquor.	▪ Openings for a new approach towards life and encourage for the same. ▪ To help them to get motivated to become aware of the reality of the problems by providing education. ▪ Controlling of illicit liquor, minimizing the state sponsorship for liquor.	▪ The religious leaders counselling service, women's Organizations ▪ -do-. ▪ Police, excise dept., Women's organizations. ▪ Politicians, The religious leaders and NGOs.
	▪ Child and women abuse and the situation of insecurity.	▪ The breakdown of the institution of family due to the prevailing war situation. ▪ Emigration of father/mother.	▪ To ensure the identity of the family. ▪ Provide facilities. Ensure security. ▪ Controlling of mother of under 16 children	▪ The village leaders, Religious leaders, Govt., NGOs, ▪ The department of probation and child care.

		▪ Death due to war of father/mother. ▪ Suicide mother/father. ▪ Child labour.	migrating for foreign employment introducing new projects. To have a long-term programme for the widows and the orphans of the war. ▪ To take legal action against those who employ child labour.	▪ The children's secretariat. ▪ Voluntary Organisations ▪ The department of labour.
▪ Peace and Reconciliation	▪ Breach of peace.	▪ Prevailing political party system. ▪ Economic inequalities. ▪ Strengthening of LTTE. organization.	▪ Strengthen non-partisan peoples organizations for promotion of politics at the village level. ▪ To improve the market to guarantee the price for products. ▪ The Tamil community and their organizations to pressurize the LTTE to come for peace negotiations. ▪ Invite them for such dialogues. ▪ To encourage peace efforts through International organizations.	▪ Peoples organisations, Sarvodaya. ▪ The state sector, private sector, efforts integrate institutions. ▪ UNO, University, Teachers associations of the North and the East. ▪ Voluntary organisations.
	▪ The ethnic disunity	▪ LTTE attacks ▪ Displacements. ▪ Attitudes of some security officers.	▪ Not to arouse racial feeling by terrorist attacks. ▪ To change the perception of some security officers regarding the Tamil people.	▪ Govt, NGOs, Security forces.
	▪ War as a means of living.	▪ Unable to engage in cultivation. ▪ No other	▪ Provide security measures to the paddy lands. ▪ Introduce new	▪ Security forces, Govt. Agencies, NGOs. and the

		employment opportunities. ▪ Easy access to join the Armed forces and terrorist organizations.	projects. ▪ Create employment opportunities.	private sector.
	▪ Increased use of arms in the area.	▪ Martial and terrorist activities. ▪ Issuing guns to the villagers. ▪ Illegal ownership of arms.	▪ Minimize the terrorist activities ▪ To give training on firearms and make them responsible through awareness. ▪ To arrest those who possess illegal arms.	▪ Security forces. ▪ -do-
	▪ The language problem among the ethnic groups.	▪ The Sinhalese language act. ▪ Inability to use one's mother tongue to get things done. ▪ Deficiency in Communication	▪ Both groups to learn the languages. ▪ To appoint translators to do things in one's own language.	▪ The Department of Education. ▪ The State.

BATTICALOA DISTRICT

Categories as per framework	Prioritized Issues	Causes	Recommendations (what)	Responsibility (who)
▪ Programme priorities	▪ How could the ill effects of environmental degradation be improved in Batticaloa District.	▪ Indiscriminate felling of trees and mangroves for security and other reasons. ▪ Unplanned construction of infrastructures in urban and town areas in the district. ▪ No sewerage system in the town.	▪ Create awareness among security forces and public about the detrimental effect to the whole Island. ▪ Formation of special planning committees comprising Go/NGO (local) in district level. ▪ Major scheme for sewerage system.	▪ Go/NGO/ Defence authorities. ▪ -do-
	▪ How far education is affected in the district.	▪ Inadequate attention to fulfil the needs of school	▪ Bring this issue to the notice of Hon. Minister of education and	▪ Parliamentar-ians, local heads of Departments.

		buildings, staff and basic facilities in un-cleared areas. ▪ Government policies in acquiring school buildings, to use the army personnel. ▪ Political and ethnic discrimination allocating funds for development activities.	Donor agencies. ▪ Change in Govt. policies to evacuate army personnel. ▪ Formation of multi ethnic advisory committees to advise the allocating authorities.	▪ Defence officials. ▪ Govt. Agent/ NGOs.
	▪ What factors could be attributed to social and economic deterioration.	▪ On-going conflicts. ▪ Economic embargo to the cleared and un-cleared areas. ▪ Denial of funds for the development activities in un-cleared areas.	▪ Changing present policies. ▪ Lifting restrictions. ▪ Formation of advisory /monitoring groups comprising local leaders, NGOs and Govt. Officials (Ds.s)	▪ Govt./Parliam-entarians/ Security personnel. ▪ NGOs.
	▪ What are the social impacts brought upon the society due to mentally and physically disabled persons.	▪ Indiscriminate arrests, torture, sexual harassment, and detention. ▪ Indiscriminate shelling shooting and presence of pressure / landmines in whole area. ▪ Cordon operations restricted movements and lack of recreational opportunities.	▪ Create awareness among army and paramilitary personnel. ▪ Give awareness of the legal education and setting up of unit within Ministry of Justice. ▪ Peace committee members to get involved in such instances. ▪ Provide recreational opportunities to disabled.	▪ GO/NGO/ Peace committees
▪ Peace and Reconciliation	▪ How does language affect peace and	▪ Implementation of language policy.	▪ Implement programmes according to the	▪ Govt. Ministry of Education. Local

	reconciliation	▪ Distorted history of Tamils in text books. ▪ No opportunit-ies for children multi ethnic group to understand each other.	constitution. Tamil education lists to be in the editorial board. ▪ Arrange picnics, Sports, get together among children of all ethnic groups.	authorities.
	▪ Why war has become a means of income.	▪ Power hunger. ▪ Selfishness. ▪ Fishing in troubled waters.	▪ Power hungry people to be chased away. ▪ Earning motivation to be discouraged. ▪ Highlight them in Mass Media.	▪ Govt; ▪ NGOs. ▪ Peoples Organisations.
	▪ How political policies affect peace.	▪ Recruitment methods. ▪ Decisions / plans taken at top level. ▪ Party politics.	▪ Bring back competitive examinations instead of quota system. ▪ Decentralize programme. ▪ Keep away from party politics.	▪ State ▪ -do-

ANURADHAPURA DISTRICT

Categories as per framework	Prioritized Issues	Causes	Recommendations (what)	Responsibility (who)
▪ Programme priorities	▪ Agriculture and other related activities as a source of living has been lost.	▪ Agricultural activities of the border village farmers have been affected. ▪ Farmers leaving border villages due to the terrorist activities. ▪ Being limited to the refugee camps farmers have become dependents.	▪ The war to be ended.	▪ All the stake-holders of war, Govt., People in the civil society.

	▪ Children orphaned and victims of abuse.	▪ Parents lost due to war. ▪ Children in the refugee camps. ▪ Loss of parental care.	▪ Facilities to be made available for the protection of children.	▪ All the stake-holders of war, Govt., People in the civil society.
	▪ Education and vocational training.	▪ Educational resources lost due to war. ▪ Intellectuals are less attractive to the area. ▪ Increase of poverty.	▪ Start poverty alleviate programme. ▪ Provide facilities to upgrade education and vocational training.	▪ Govt. & NGOs.
	▪ Drugs and alcohol.	▪ Becoming addicted to liquor due to the stress created by the war. ▪ The laws not enforced. ▪ Easy availability of liquor due to state sponsorship.	▪ Create awareness through animation programmes. ▪ Govt. to take action to enforce the law.	▪ Govt., NGOs, Religious institutes.
▪ Peace and Reconciliation	▪ Discrimination.	▪ Unequal treatment based on ethnicity. ▪ Officials influenced by political interferences. ▪ The corrupt practices of officials.	▪ Equal and just treatment to all the ethnic groups.	▪ Govt., People.
	▪ The language.	▪ The languages not given equal opportunity. ▪ Even at a critical situation, opposition against the equal opportunity for the languages. ▪ The protest	▪ To protect the human rights by providing equal language opportunities and thereby ensure right implementation.	▪ The State, Peoples Organizations.

		against equal language opportunities arousing radicalism.		
	▪ War as a means of living.	▪ A majority being engaged in earning large income through improper means due to war. ▪ The socially abused victims created as a by-product of war. ▪ Through this another group becomes rich.	▪ End to the war and create peace.	▪ Stakeholders of war, people.
	▪ Peace building efforts.	▪ Those related to the problem have not explicitly stated about their ▪ Willingness or commitment ▪ Failure to implement practically the peace effort. ▪ Some groups using the peace efforts to realize their selfish motives.	▪ Peace efforts to be made transparent and genuine.	▪ All the stakeholders of the war.
	▪ Encouraging the inter racial relationship.	▪ The influence of the extremist racists. ▪ Mass media not providing the necessary corporation. ▪ The members of the two ethnic groups not knowing each other's languages.	▪ To create human rights awareness. ▪ Make the media to provide equal services to all the ethnic groups. ▪ Learning each other's language.	▪ Govt., NGOs, Politicians of the two ethnic groups.

	▪ Participatory methodologies.	▪ Being limited to books and lectures. ▪ Lack of transparency in the project implementation. ▪ In the preparation of the estimates. ▪ The beneficiaries are not made aware of the above. ▪ Difficulty in implementing the project due to ignorance.	▪ Make all those who participate in the participatory methodologies aware of the programmes.	▪ NGOs, Beneficiaries, Donors.
	▪ Continuous monitoring and evaluation.	▪ In practice not really being implemented. ▪ Loss of confidence due to lack of transparency in financial transactions. ▪ Not being made aware of the relevant matter.	▪ To have a continuous monitoring and evaluation.	▪ Govt. & NGOs. ▪ Beneficiaries.

In the results of this collective exercise details of which are before us in these preceding fifteen pages, we can witness what I have termed as the dynamics of plural dialogue. In the next chapter I propose to discuss the implications of this in details in terms of a politics of friendship. But we can note here at least this much, that in this dialogic exercise, the plurality of the context makes up the text because it does not hide itself here, it represents itself with more fidelity; because in the desire for peace the dimensions of justice, fear, poverty, gender concerns, inequalities, anxiety for security and peace, distrust for administration, revulsion at widespread corruption, longing (to communicate with the outside – other victims, other districts, non-affected areas) and the belonging, and the aspiration for dignity – all play themselves out. The victim population of each affected district marks out its prioritized issues, the causes it thinks to be behind those issues, its recommendation, and its configuration of responsibility.

Interested readers can do a statistical exercise to bring out the relative importance of the issues charted out in this long table; but my intention is not to show their relative incidence, but their concatenation, the inter-textual nature in this exercise that goes by the name *context*, but for us which is the text of dialogue, of peace. This is different from the mono-dimensional concept of peace that we visited in chapter 2 and in the two accounts of ceasefire negotiations described in this chapter. The anthropology of a plural communication is present here in its pure form.[50] Therefore is the question, how far do dialogues on ceasefire take into account these desires, wants, scarcities, attitudes, expectations, and their interconnections, to firm up cessation of hostilities leading to durable and just peace? How is it that, ceasefire negotiation like a monster eats up all anxieties, aspirations, dimensions, and expectations, and leaves almost no trace around?

The memories of My Lai, in the Quang Ngai province of South Vietnam, that witnessed a massacre of 347 unarmed villagers by the US troops on 16 March 1968, the circumstances in which the horrible episode was exposed more than a year later, the attendant court martial of 17 officers and the eventual conviction of Lieutenant William Cassey, pardoned by Richard Nixon in 1974, may have been effaced from the popular memory outside Vietnam no sooner than peace returned. I do not deny that peace often returns in this way. Obviously I am not arguing here against the return of peace, so that more My Lais can happen, nor I am arguing against the negotiations on ceasefire, which everyone knows is the first step towards return of peace. That is absurd. But circumstances that make such atrocities happen, then hush them for long periods, and when exposed, inspire vehement denials by the establishment and all out attempts to stymie inquiries are always present around dialogues for justice and peace. South Asia has seen its own My Lais on the peripheries of its own democracies, in the northeast, in Punjab, in Kashmir, in Jaffna Peninsula, and in remote rugged lands of Balochistan. These shape dialogue in South Asia in deep ways.

Cessation of Hostilities

If ceasefire is understood as cessation of hostilities, we must reflect on what will constitute a violation, a renewal of hostilities, in other words an aggression. If aggression is understood as simply an initiation of hostilities (a1), then it is clear that all circumstantial considerations are irrelevant, what is relevant is who initiated the renewal. But there may be situation, where circumstantial considerations become critical such as some promises are not being kept, other conditions are being enforced, or follow-up measures are not be implemented, one party may think of renewing hostilities, interpreted as committing aggression (a2). There may be yet another situation, one party may find out that ceasefire is being been used by the enemy to launch a surprise attack in near future, so that to forestall it in face of such impending disaster, the party renews hostility (a3). Though we are discussing here dialogues on ceasefires between states and what is considered its 'non-state enemies', yet the import is the same as that of the failed history of stopping aggression by the League Covenant, the Kellog-Briand Pact, the London Charter, and the UN Charter. The point is, negotiations on cessation of hostilities show the almost compulsive progression from a1 to a3. Ceasefire negotiations concentrates on who should not fire the first shot, in other words you cannot be the one to fire the first shot – the brutal simplicity leading from a1 to a3. How to dissolve the threat, and make the question of intention irrelevant amidst the sea of issues relating to peace and justice clamouring for attention, is beyond the modalities of ceasefire negotiations, because in such negotiations the institutional legitimacy of the other is not recognized, and the relations between legitimacy, justice and peace are deliberately ignored. I am aware that in recent times, measures have been explored and devised ranging from confidence building measures, and legitimizing the opponent simultaneously while negotiating with the opponent, to instituting third party or multilateral supervision. But surely the high rate of failures of such negotiations and devices (leading to restoration of peace and democracy) must lead us into questioning the presumed basics of such process, and consequently appreciating the strength of adopting a relational stand in the contentious politics of our time.

Or take this report on the Agra dialogue between India and Pakistan (14-16 July 2001) on restoring peace in their mutual relation, particularly in Kashmir. Though this was not a ceasefire negotiation, the situation had all the characteristics of one such. The two supreme leaders of two countries were discussing how to stop the Kashmir war, also the undeclared war across the LOC and Siachen. The report in the *New York Times* said,

> The leaders of India and Pakistan, nuclear-armed adversaries, abruptly broke off two days of marathon negotiations on Monday night. They failed to make any substantive progress on their intractable conflict over Kashmir, a land claimed by both nations, or on reducing the risks of a nuclear exchange.
>
> There were also ominous signs that the summit meeting may have deepened the Pakistanis' already deep mistrust of the Indians. The talks between Pakistan's military ruler, Gen. Pervez Musharraf, and India's prime minister, Atal Behari Vajpayee, ended without any joint statement after repeated extensions of the talks through Monday. The general cancelled a visit to a famed Sufi shrine in Ajmer to make time for more negotiations, but left empty-handed at midnight.
>
> Two Pakistani journalists, interviewed separately, said General Musharraf's chief of staff, Lt. Gen. Ghulam Ahmed, had told them on the record, insisting that he be quoted, that the Indians had backed away three times from agreed joint statements. Nasim Zehra, a columnist for an English-language newspaper in Islamabad, The News, said General Ahmed had asked her to pass on his account to the rest of the press corps. And Hamid Mir, editor of an Urdu daily in Islamabad, Ausaf, said, 'There was a sense of betrayal that a number of times things were agreed to and gone back on.'
>
> The Indians and Pakistanis did not answer questions on Monday night or this morning about the summit meeting or give any official explanations of what went wrong. They said they would hold separate news conferences later today...Many Indians blame the general for a Pakistani incursion into the Indian state of Jammu and Kashmir that brought the two nations to the brink of full-scale war in the summer of 1999 and ruptured high-level talks between them until this weekend.

Despite the lack of progress, the general and the prime minister are expected to keep talking: Mr. Vajpayee accepted the general's invitation to visit Pakistan. In a brief statement, India's spokeswoman, Nirupama Rao, said this morning, 'I am disappointed to inform all of you that although the commencement of the process and the beginning of a journey has taken place, the destination of an agreed joint statement has not been reached.' The two sides' longstanding conflicting views on Kashmir bedeviled their attempts at compromise. In blunt remarks Monday morning at a breakfast for senior Indian journalists, General Musharraf insisted that there could be no real progress until the conflict over Kashmir was openly acknowledged – a statement he has made many times before.

'This is what we've killed each other for,' he said. 'I know it much more. I've fought in the two wars.' The Indians were riled that he publicly suggested that they were not ready to face reality on Kashmir while negotiations were still going on. Several hours after his lengthy comments were broadcast on a private television channel, Indian officials took the highly unusual step of releasing a statement the prime minister had made privately when the talks with Pakistan began. In it, Mr. Vajpayee said any negotiations on Kashmir would have to include what India refers to as cross-border terrorism. India maintains that the militant groups supported by Pakistan and based there and fighting a protracted anti-India insurgency in Kashmir are terrorists who are crossing the border into India. Pakistan calls these same militants freedom fighters.

India defines Kashmir as an inalienable part of its nation of a billion people, while Pakistan, with a population of 150 million, has long taken the position that the Kashmiris should be allowed to vote to join Pakistan – a stance the general hinted he might soften if ever talks really got under way. Kashmir's status has been debated since British India was divided in 1947, when mostly Muslim Kashmir was assigned to mostly Hindu India. Pakistani journalists said the general needed to keep his focus on Kashmir to avoid angering hard-line elements of the military, as well as Islamic fundamentalists who believe that Kashmir should be part of Islamic Pakistan. General Musharraf himself said today that he might as well settle down in the big house in Old Delhi where he lived as a child rather than go home to Pakistan, where his family moved in 1947, if he were seen as selling out on Kashmir. 'There are definitely compulsions on my part that I must talk about it,' he said. The scene of the blunt-talking general breaking the diplomatic china was just part of a day full of surprises, the most troubling of which occurred after the talks broke off.

More than a hundred journalists rushed to the hotel where General Musharraf had gone to bid Mr. Vajpayee farewell. For an hour and a half, they waited outside the gates. The Islamabad journalist, Ms. Zehra, who had come to the meeting as a guest commentator for the state television station, spoke intensely on a mobile phone to the general's chief of staff, who was inside the gates, then circulated with his account of what he called the Indians' inconstancy. She said the chief of staff had also told her that the general had wanted to hold a news conference, but that the Indians had refused on security grounds – which left some Pakistani journalists simmering with anger at the insult. Just before midnight, the general's entourage of 33 vehicles, illuminated by flashing red and blue lights, streaked past the waiting journalists, who then crammed into cars to return to the hotel where they were based to wait for an Indian statement. After Ms. Rao read out the single paragraph, outraged Pakistani journalists began shouting out questions about why democratic India had denied the General a place to hold a news conference. She refused to answer any questions, but they chased her and four members of her

> staff into the lobby. An Indian official who witnessed the incident said India's press team literally ran away. One journalist hit Ms. Rao on the shoulder, while others pushed and shoved the staff. 'How can they do that to a lady?' Sanjay Verma, a member of Ms. Rao's staff, asked. The shouting was still going on when more journalists poured out of the Indians' news conference. Khalid Hanif, chief reporter for the *Daily Jang* in Islamabad, was emotionally and loudly demanding to know why the general had not been allowed to hold a news conference.
> It was a rancorous ending to a meeting that was supposed to bring India and Pakistan together. As she left the hotel where the reporters had been waiting for the general, Ms. Zehra sighed and said: 'What a pity. It's finished. Let's go home.'[51]

Therefore the poser raised earlier needs repeating, namely, *how is it that, ceasefire negotiation like a monster eats up all anxieties, aspirations, dimensions, and expectations, and leaves almost no trace of these around?* As if the clue of the murder is left everywhere, that is to say, the clue is left nowhere. All this becomes part of a game of peace. With ceasefire negotiation, war begins again leading onto to new search for cessation of war. In this alterable act, there is focalization – peace trapped in its alterability. Here is then the irony. War must lead to cessation of hostilities. And cessation of hostilities must end as *cessation*, and must re-form as peace. But cessation being the copula makes peace its hostage. And each time there is cessation, not only war but peace too dies with it. That death we shall say is necessary, because we cannot afford to allow many interests and concerns play themselves out in the dialogue on ceasefire, if we want peace. Therefore peace must die in order for the cessation of hostilities to re-form as peace. In this repeated return of death, peace faces its gravest challenge, namely *how will it negotiate its mediation, and how will it look back at the time when it was being re-formed*? In a way then, this is a distinct politics, the politics of no war-no peace, of the middle ground, of giving ground, of propinquity, of the third ground, of contending with possibilities emerging in future – the politics of treating the interregnum as a determinant of how much progress will have been made.

Such mapping of concerns will allow us glimpses of the way understanding emerges – understanding as the core of dialogic politics, as an exercise in spatial imagination – however, more important, understanding as an exercise in poetics. It is a theme that I try to address in the next chapter. But at least we can say, that perhaps with more such mapping of concerns of the victims living in the long night of solitude with no dialogue with anyone outside, the experiences of victim-hood will effect the transgression, so direly needed, from dialogic politics as element of state politics to dialogic politics as part of justice and peace. In this anthropological exercise we see the limits of the current politics of dialogue, which is tied to the death of desire and to a void that we for lack of no other word can express by saying, whistle, talk, dialogue, peace.

Notes

1 I attempt the same investigation in the context of nationalist struggles in East Pakistan/ Bangladesh in my essay, 'Leaders and Legacies – Stories in the Time of Transition', *Indian Economic and Social History Review*, 37 (4) September-December 2000.

2 'Plural Dialogues' in R. Samaddar (ed.), *Space, Territory and the State – New Readings in International Politics* (Hyderabad: Orient Longman, 2002), Chapter 7, pp. 151-165.

3 Jacques Derrida, *The Politics of Friendship* (London: Verso, 1997).

4 In 'Plural Dialogues' I write in details about how the events of nuclear politics has completely overwhelmed a plural dialogic situation between Pakistan and India.

5 One of the notable accounts of such desire is Kaka D. Iralu, *Nagaland and India, Blood and the Tears – A Historical Account of the 52 Years Indo-Naga War and the Story of Those Who were Never Allowed to Tell It* (no publisher's name, Kohima, 2000). In the introduction to the 542-page account of the war, violence and the army atrocities, Iralu writes, 'In a war it is not the mightier nation that triumphs but evil that gets the ultimate victory. This victory of evil however can be stopped when the warring parties sit down together and thrash out the factors and issues that have led to war in the first place. And as far as the author is concerned, this fifty two year Indo-Naga war is a war that should never have been fought'.

6 'Jamir May Seek Interpol Aid to Extradite Rebels', *The Telegraph*, 29 December 1996.

7 'Jamir Remark Irks Naga Students', *The Telegraph*, 30 April 1997.

8 'NSCN (I-M) Not To Attend Atlanta Meet', *The Hindu*, 16 July 1997.

9 'Gujral To Make Sure That Demand For 'Greater Nagaland' Is Not Entertained – Centre-NSCN Talks Next Month', *The Telegraph*, 26 July 1997.

10 Ibid. 'Keishing Upset – Feels Sidelined', *The Asian Age* (section, The North East Age), 27 July 1997; also *The Sentinel*, 29 July 1997.

11 'Panel To Supervise Centre-NSCN (IM) Truce', *Assam Tribune*, 31 July 1997.

12 On the impunity given by the Act, see, Sebastian M. Mongray Vs Union of India, *AIR 1984*; *People's Union for Civil Liberties Vs Union of India,* (1997) 2 JT 311; Nungshitombi Devi Vs. Rishang Keishang, (1982) *I GLR 756*; Luithukla Vs. Rishang Keishing, (1988) *2 GLR 159;* People's Union for Democratic Rights Vs. Union of India, (1991) 2 *GLR*; on the Act, see Gautam Navlakha.

13 'We Are Not Toys To Be Played Around', Statement by Muivah, *Hindustan Times*, 7 September 1997.

14 'This PM Has No Respect For His Word', Statement by Swaraj Kaushal, *Times of India*, 25 July 2000.

15 Information relating to allegations of violations of ceasefire agreement, Press Statements issued by Information Service of the Government of the People's Republic of Nagalim, 14 & 15 July 2000; also *Northeast Vigil*, 1 August 2000, http://nevigil.freeservers.com.

16 On this fiasco, Sudhanshu Ranade, 'The Kashmir Gambit', *Economic and Political Weekly*, 35 (52-53), December 30, 2000, pp. 4599-4601.

17 The chief US negotiator in the Balkan peace diplomacy writes in details of how the US encouraged two out of three parties in the conflict to go on an all out offensive before the NATO-UN forces would enforce ceasefire in the Balkans. See Richard Holbrooke, *To End a War* (New York: The Modern Library, 1998), pp. 62-63, & 191. Holbrooke takes pride in the doublespeak that he along with other US negotiators employed in advancing interests that they thought fair in using the imperative of ceasefire in the broad plan of a political strategy. This is a classic account of how *not to end a war*. The extent to which doublespeak can go in making ceasefire a part of war, it is necessary to study Holbrooke's account that, for instance, says,

'I want to be very frank in the privacy of this room', I replied. 'We asked them not to take Banja Luka, but we did not give the Croatians and the Bosnians any other 'red lights'. On the contrary our team made no effort to discourage them from taking Prijedor and Sanski Most and other terrain...It would help the negotiations greatly if these towns fell.

'I am very concerned that we will be blamed publicly for encouraging more fighting and more bloodshed', Tony (Holebrooke's colleague) said. 'We should emphasize peace. This may not be your view, but you should say it in a way that doesn't exacerbate differences on other fronts...

Christopher (Christopher Warren, the US Secretary of State) said he agreed with Tony as far as public statements went. I had no problem with that; the negotiating team had been careful in public, so much so that journalists in the region continued to think and report, that we were trying to stop the entire offensive. But I refused to stop the offensive. (p. 172)

Ambassador Holbrooke sums up all these by saying in his account, 'I explained our strategy to a reporter at the time: 'If we can get a ceasefire, we'll take that. If we can get some more constitutional principles, we'll take them. If we can settle Sarajevo, we'll do it. *We're inventing peace as we go*' (italics mine)'. (p. 175).

18 Hanan Ashrawi, *This Side of Peace* (New York: Simon & Schuster, 1995).

19 To know these unrecorded voices of peace in Nagaland and their impact, see Jehan Perera (ed.), *Peace Process in Nagaland and Chittagong Hill Tracts – An Audit Report*, SAFHR Paper Series 5, Kathmandu, 1999; also see the report (mimeo) of the 'Civil Society Meeting on Human Rights, Justice and Naga Peace Process', Kohima, 18-19 March 2001; also *Naga Resistance and Peace Process*, a compendium of documents relating to peace process in Nagaland (Delhi: Other Media Communications, 2001).

20 '11 killed; Protesters Burn Govt. Offices, Manipur Assembly', 'Centre Rushes Additional Forces', *The Times of India*, 19 June 2001, p. 1; Harish Khare, 'Centre Finds itself in a Bind', *The Hindu*, 19 June 2001, p. 1; also *Reuters*, 19 June 2001, 'Eleven shot dead as mob torches State Legislature', www.online.reuters.com.

21 A press statement from Imphal, 21 June 2001, circulated by the Naga campaign network by Akum Longchari.

22 www.rediff.com/news/2001/jun/25inter.htm.

23 Naga HoHo's Press statement from Kohima, dated 17 July 2001, circulated by the Naga campaign network by Akum Longchari.

24 www.rediff.com/news/2001/jun/24naga.htm.

25 Reuters, 27 July 2001, 'India backtracks on peace deal with tribal rebels' – www.online.reuters.com.

26 Ibid. I am also grateful to the unpublished report on the situation by Ram Narayan Kumar and Laxmi Murthy (2001).

27 Kaka Iralu, 'Nagas and Two Centuries of Territorial Dissections', 20 June 2001, circulated by the Naga campaign network by Akum Longchari,

28 Interview of NSCN (I-M) General Secretary, Thuingaleng Muivah, by M.J. Vijayan, www.rediffmail.com, June 26, 2001.

29 Ibid.

30 'LTTE Outflanks Government?', *Sunday Leader*, 24 December 2000, p. 10; see also the foreign minister's interview 'LTTE's Ceasefire Offer Opportunistic' to *Daily News*, 5 January 2001, p. 9.

31 K. Ratnayake, 'Sri Lankan Government and LTTE Inch Toward Negotiations', *Daily Mirror*, 25 December 2000, p. 9.

32 'Sri Lanka Peace Charade Goes On', *The Island*, 14 January 2001, p. 12.

33 'LTTE Can Be Proscribed in UK – Kadirgamar', *Sunday Observer*, 21 January 2001; *Sunday Times*, 7 January 2001.

34 'Solheim Lauds India's Role in Lankan Peace Process', *Sunday Times*, 7 January 2001, p. 13.
35 'Tamils Should Not Take LTTE Ceasefire Seriously', *The Island*, 1 January 2001, p. 7.
36 'A Permanent Peace vs. Temporary Ceasefire', *Daily News*, 13 January 2001.
37 Neville Ladduwahetty, 'Negotiations and Peace Accords', *The Island*, 28 January 2001, p. 8.
38 Ibid. see also, 'Of Ceasefires and Clowning', *The Island*, 24 December 2000, p. 10.
39 Lakshman Gunasekara, 'What happened to the 'Talks'?', *Sunday Observer*, 31 December 2000, p. 7.
40 Ibid.
41 *Daily Mirror*, 21 December 2000, p. 8; *Daily Mirror*, 26 December 2000, p. 8.
42 'Palestinian Report Discredits US Mediation', news report, *Hindu*, 24 January 2001, p. 16.
43 S.J. Emmanuel, 'Kumaratunga-Prabhakaran Talks – A Northern View' in Kumar Rupesinghe (ed.), *Negotiating Peace in Sri Lanka – Efforts, Failures and Lessons* (London: International Alert, 1998), p. 279.
44 Ibid. p. 280.
45 Prabhakaran to Ratwate, letter dated 8 December 1994, cited in 'Kumaratunga-Prabhakaran Talks', p. 280.
46 The US diplomats used the tactic of 'proximity talks' extensively in the nineties of the last century in the Palestinian problem. For a graphic account, see *This Side of Peace*, chapter 12, pp. 219-222. On proximity-talks in the Balkan negotiations during shuttle diplomacy and in Dayton, see Richard Holbrooke, *To End a War*. Holbrooke defines 'proximity talks' in this way, 'I phoned President Carter and listened in fascination as he described how he had tried without success to get Sadat and Begin to talk directly to each other. He had then reverted to 'proximity talks', a diplomatic technique originating in the mid-east negotiations held in the 1940s at the UN, in which the mediator moves between the two parties, who rarely meet one another face to face – a sort of 'shuttle diplomacy by foot'. (p. 205).
47 On the Premadasa-LTTE negotiations, see Bradman Weerakoon, 'Government of Sri Lanka (President Premadasa) LTTE Peace Negotiations 1989-1990' (pp. 145-159) and A.J.V. Chandrakanthan, 'The negotiations Between President Premadasa and the LTTE – A Retrospective Assessment' (pp. 161-171) in *Negotiating Peace in Sri Lanka – Efforts, Failures and Lessons*.
On the victims' story of the humanitarian disaster in Sri Lanka, there are several accounts and reports. Didier Bertrand provides an exhaustive bibliography of English, French and Tamil sources, a selected version of which appears at the end of an equally important report by Didier Bertrand, 'Going Back, but Where? Forced Repatriation of Tamil Asylum Seekers from Europe', *Refugee Watch*, 12, December 2000, pp. 9-19 & 37-38.
48 My attention to the third dimension in this context has been drawn by Linus Jayatilake's circular of 8 March 2002, Colombo.
49 I am grateful to Jeevan Thiagarajah for providing me access to the report. The report was subsequently placed before the second peace audit exercise organized by the South Asia Forum for Human Rights (SAFHR) in Bangkok on 30 August-2 September 2000, and made available in Shahid Fiaz and Ranabir Samaddar, 'Peace Audit report 2 – Peace process in Sri Lanka', SAFHR Paper 8, Kathmandu, April 2001; information on the report also available on www.safhr.org.
50 See particularly section 2 of 'Peace Audit report 2'.
51 'Kashmir and A-Arms Defeat Leaders of India and Pakistan' by Celia W. Dugger, www.NYTimes.com, 17 July 2001.

Chapter 8

Friends, Foes, and Understanding

The Need to Understand is not anything Special, It is a Matter of Prudence

The politics of dialogue then proposes for us an agenda of understanding. The task of dialogue is political, but the political carries within it its supplement, the *ethics and poetics of dialogue*, which we can express through the word 'understanding'. Therefore, in beginning to grapple with the core of dialogic politics, we must visit some of the commonest notions inhabiting the dialogic universe, often ignored in political theory. Jurgen Habermas had commented in *Theory and Practice*, 'On the road to science, social science has lost what politics formerly was capable of providing as prudence'.[1] Probably more than theory, prudence or common sense, from whose history relating to dialogue I shall try to pick out certain fragments, will enable us to write of friendship, trust and understanding, of the politics of understanding, in the process something about our understanding of the politics of dialogue and democracy. It revolves round the fundamental issue: how do we understand each other? How do we appeal? How do we dialogue among ourselves? In other words, what is the politics of appeal, the politics of familiarity, the phenomenological form of familial proximity? Do nations and peoples of this region feel the need of loving each other, that is to say, do they feel that it advisable to love to be loved, to understand in order to love, in short, to love so that they can understand? In trying to make sense of these questions we shall see that what Habermas indicates as prudence contains within it the practices reflecting on our historically formed notions of territoriality, belonging, and conversation, practices rationalized as the geo-politics and ethno-politics of friendship.

These questions refer to a politics of understanding, also a model of that understanding. It is obvious that nations of this region fail miserably if we take a maximalist model of accommodation. Such a model will tell that it has to be a maximum of friendship, the friendship has to cover all issues of existence, and that such friendship cannot permit the loved and the beloved to be incommensurable in any way. Such a model of mutual understanding disallows asymmetry, because the risks of asymmetry complicate the egalitarian schema of friendship; disallows any difference between the effective and the virtual, between a fragile friendship and death. A maximal model does not admit of the possibility that sympathetic understanding may erupt from a loss of union, mourning may give rise to renewed partnership, and anguish may inspire understanding and tolerance. In such judgement, no instability in friendship is allowed. Stability is crucial. Brotherhood, and from brotherhood complete identity of life is primary, primal.

It is apparent that nations of South Asia, particularly the three nations of the erstwhile-united land, do not follow the maximal model, and by that judgement cannot be friends. Yet, do they have no understanding of each other? If the maximal model is not practised here, must that mean that there are no discernible laws of understanding in this region? Before, however, we answer that question, we must probe into the reasons of the impossibility of a maximum alliance, understanding and friendship, which was sought to be internalized during the cold war era as the normal model of friendship.

Behind friendship lies the architecture of a desire – the desire to live, to achieve liveability, to prefer, and to love; in short there is morphology of accommodation. The impossibility of absolute friendship shows the impossibility of the laws of equality, and indicates the inevitable attempts to make that impossible possible, a possibility of periodic upheavals, and of recurring testimonies to the gain and loss of friendship. Alliance is a weight, sometimes an overwhelming weight, of the attempt to ally, to reconcile, to make others conform, and to constitute the incompatible values in compatibility. This also suggests a truth. The truth is about the nature of the proximity lying at the heart of a friendship, and the differences that this proximity produces – proximity's absoluteness, therefore without resemblance, without attraction, without dissociation, and without the possibility of significant or reasonable preferences. There are all kinds of partners who are partisans: born partners, jealous partners, partners of solitude, and partners of differential understanding. Often the partners invite each other to be members of a singular community, an anchored community – another name of secret society. They are like the owls in the full light of the day. Such partnership is in the long run the premonitory sign of a breakdown. In maximizing the essence of partnership and the mutual obligation to answer, such partisanship takes the notion of common good to the point of ridicule. We shall then say, what can be common has ever but little value, common good is a contradiction, and at the end of the day it can be found by only conjuring up secrecy. Belonging to the kingdom of a spiritual and therefore a hyper-political community, it assures only currency, presence – a contemporaneous life. Belonging to contemporaneous life, it experiences harrowing tremors of existence.

A maximal friendship (like cold war friendship) depends not only on maximum unity, but also on maximum enmity, maximum hostility. In its imperatives, we find reconstituted enmities, re-constitutive enmities (like post cold war enmities). The logic of total friendship is ineluctable. It kills understanding. Its testament becomes an epitaph on accommodation, an elliptical justification of hostility. It begins with democracy, but soon turns again it, stabs it with the dagger of absoluteness, totality.

There is another model of understanding whose key is difference. Accommodating difference, reconciling to the permanence of difference bespeak a strategy whose stakes are limitless when placed in the history of the relations in the region. In short we speak of a dialogic order, whose basis is a different *politics of understanding*.[2] In arguing for a politics of understanding as the core of dialogic politics today, it is obvious that I have begun with Derrida's argument about politics of friendship, but I want to carry it to a different direction in the context of a possible politics of democracy and dialogue for reconciliation and justice.

Reflexivity and an Alternative Argument

In opposition to the maximal idea stated above, I wish to present here briefly an alternative argument of partnership and friendship relating to the history of post-colonial south Asia. It not only contradicts the model discussed above, but denies and defies the very wisdom of that history of claiming total union that led nations of this region to desire total friendship, demanding total surrender of the self to the other, or vice versa. I may confess that this alternative argument may seem a kind of mythography.

I shall give the example of the politics of understanding Bangladesh. The fifty-three years of 'our' independence and thirty years of 'their' independence are historical facts. These are two different countries, sovereign neighbours. India is a Hindu majority country. Bangladesh is Muslim majority. India is more industrialized. Bangladesh is less. India had never a military rule, Bangladesh, for fifteen years, was ruled by the army. In our attempt to understand Bangladesh, we are to give due respect to these facts. In other words, we accept 'South Asia' as the *given* region, composed of *given* nation-states, and we accord respect to the state-system *given* to us. We are to hold our 'contra-history' in rein so that the desire to belong to our past does not oppose history. But alas, in understanding a neighbour this is often not possible. The relations of past continue to structure the present and this is at the core of the problem of reflexivity, and what I term as 'contra-history' that the present history has to carry within. I hope to make clear in the following pages, how these two elements – reflexivity and contra-history – make a possible politics of dialogue.

What is reflexivity? The OED meaning of the word suggests that, 'the action of the verb is performed on its subject'. In other words, we may say, it implies the dynamics of mutuality in space(s), word(s), time(s), and text(s) – with knowledge and existence, both caught in that dynamics. It is a principle that tells how the object and the subject are perennially reversed in their relation. It speaks of how the self is thereby constituted, that is constituted through being re-constituted. If one definition of nationalism is that it is a system of cultural signification, this signification thrives on reflexivity – in other words, a nation does not by itself only, determines its signs. The signs are determined by elements of otherness that helps in configuring these signs. Nations thus rarely avoid the problem of reflexivity. Even if we stop short of generalization, we can see at the least how the nation in India constitutes itself by reflecting on its neighbours, and how the existence of a neighbour, an intimate knowledge of her, becomes crucial for determining India, in other words India-ness. Reflexivity features in the strategy of finding similarities and differences.

Linguistic nationalism has been turned into a problem by religion in this subcontinent. Anti-colonial nationalism was similarly problematized by religion. A nationalist strategy that seeks identity in common factors (like language, geography, anti colonial resistance), at the same time probing into differences (like their language/our language, their religion/our religion) thus becomes inevitable. The dynamics of understanding thrives on the chiaroscuro of identity and difference; they always raise the possibilities of contra history, that is to say, the history of the desire to go against history, only to deny its own presence immediately after. Given this context, in understanding Bangladesh, the Indian has to appreciate this psychology of the margin, known to India as Bangladesh. Besides, many crucial cultural and social-psychological artefacts such as autobiographies thrive on margin – margins of fact and memory, history and literature, identity and difference, nation and the person, conflict and accommodation, and desire and resolution. It is impossible therefore to deny these artefacts such as autobiographical writings and the memorial accounts a place of utmost importance in appreciating the dynamics of understanding in South Asia.

I shall confine myself to few examples here. The late, famous essayist Qazi Abdul Wadud's *Saswata Banga* spells out in evocative terms the pathos of an 'eternal Bengal'. The memoirs of Abul Fazal (*Rekha Chitra*) or Abul Zafar Shamsuddin (*Atmasmriti*) and the surrealistic memorial accounts in Sunil Gangopadhyay's *Purba Paschim* and *Jochankumari* invite critical reading into the dynamics of nationalist reflexivity, for they show how memory becomes increasingly stronger to ultimately create the eternity called Bengal. Social science researches in South Asia on political relations have not paid due attention to the role memory has played in determining our self, in the making of the ambivalence that we know today as South Asia. Our Vangiya Sahitya Parishad, their Bangla Academy, our *Sahitya Sadhak Charitmala*, their *Jibani Bichitra*,[3] our Tagore, their Nazrul and our Sarala Devi Chaudhurani, their Rokeya Sakhawat Hossain – these are all not merely historical parallels, but a play of memories too.

The second way in which we find reflexivity problematizing our understanding of Bangladesh is the multiplicity of our selves. I am a Bengali who has suffered partition. I am an Indian who has gained independence. I am a South Asian who is aware of other South Asian countries. Multiplicity of selves is, of course, not enough, for though all these three selves determine my attitude to Bangladesh, it may be that at a critical moment one self may triumph. The others are left to the wilderness, eventually as haunted houses for ghosts. The triumph of a one-dimensional self should not astonish us, given the imperatives of modern nationalism. The riots of 1950, the Enemy Property Act, the controversy over Tagore centenary in East Pakistan, the dispute over the waters of the Ganga, the reverberations over the destruction of the Babri Masjid – all these and many others remind us of the history that 'multiplicity of selves' presents. Yet, and my argument here is that, precisely these suggest a contra-history, they show how we have not reconciled ourselves to partition, they too have not. Indeed, contra-history is the refusal to accept the history of the modern craze for territorial integrity; this is the dogged refusal, the intransigence of a self to allow itself to be cut into totally separated two selves. The project of forging friendship in South Asia has remained wilfully blind to the multiplicity of self, though the everyday balancing act in our sensitivities and our political conduct regarding Bangladesh shows that we only fool ourselves.

One can argue in the vein of post-colonial reasoning, such as of Ashis Nandy, that it is the intimacy of relations that creates multiplicity as well as the rigid one dimensionality that then submerges multiplicity.[4] There is some merit in this observation. Ashis Nandy of course speaks of intimacy in a different context, where colonialism so much mixes with or engulfs the native self that the latter is lost and has to be first recovered before it is free from imperialism. Speaking in our context, we may see that intimacy can also bring in its wake a certain kind of reflexivity which does not sink our 'multiple self' into oblivion, but makes it sensitive and reciprocal, responsive to the stirrings of mind, matter and heart. Polar ends such as the South Asian universal versus the parochial, the nation-state centric material versus the spiritual, the achieving or the competitive versus the non-achieving, and the rational versus the emotional, do indeed meet in as much as they are created because of our intimacy with our neighbours. It is this reflexive relation born out of intimacy that always raises the possibility of contra-history, contra-factuality – the ground of an alternative appearing as a mythology.

This point needs a little elaboration. We of this Bengal know them so much. They too know us. We know our mutual betrayals, hopes, and despair. Thousands of people cross border and ferry news from here to there and there to here. Our rivers, ecological-agricultural regions, transportation networks, clan ties, affinitive links, folk regions, remain spread across the entire land. Above all, we know, what they write, how they think, they too know of us. To deceive ourselves of the pangs of proximity, we have changed categories into opposites: migration is infiltration, border trade is smuggling, empathy is interference, area study is policy prescription, and neighbourhood is 'near abroad'. But while this metamorphosis shows a rigidity of mind born of intimacy, our memory, literature and our thousand and one responses speak of the multiplicity of selves too. It is not enough to say that this is a remainder of partition. Of course, social sciences, journalism, state policies and other elements of the language of power seek to impose our standards on Bangladesh, completely misreading her self (selves). But beyond this universe of unilateral acts, the reality of alternative presences finds out other histories, seeks other possibilities, and seizes the signs of multiplicity born out of intimacy, dialogue, and reflexivity. After all, each history leaves in its trail a remainder, a different present; each historical act produces other possibilities besides the one actualized. Reflexivity thrives on that. The question of open choices producing reflexivity therefore turns past as present, then to a fractured present, to a remade past, onto to a new past – a lingering hope, almost a despairing hope of a different future.

We are, it seems then, hopelessly caught in a bind. For, to define ourselves, we have to now define our neighbours. Our liberal present has to produce a not so liberal neighbour, our secularism has to find a fundamentalism there, our modernity faces someone's traditionalism, and our economic progress is counter-posed to her poverty. And this we do, not only to define our triumphant self, we do it to resolve the quarrel and tensions in our many voices. In order to westernize our present, we must non-westernize others, just as in order to challenge the given present, we must re-open a settled past. The first produces history out of half-truth, the second counter-poses possibility to history in order to drag history into the region of mythography. Our attempt to understand our neighbour cannot settle for either. More than the scholarly articles of international relations and South Asian relations, the literary accounts we have earlier referred to reflect that fluidity. We can think again of the account of Abul Hashim, written much after the forties and fifties. He writes of the Muslim League, of the impending partition, yet he writes of the entire Bengal that might not have been partitioned. In speaking of the partitioned Muslim land, he in fact speaks of the whole land. Or we can take the famous journalist and novelist Abul Fazal's *Rekha Chitra*. He writes of Chittagong, how he repeatedly went back there to recover his self – yet this Chittagong is not of Bangladesh, it is the Chittagong of eternal Bengal. Interestingly, these accounts written much after the partitioned independence never make partition an inevitable outcome. Partition is an accident, a play of power, of fate. The fluidity of the narrative is almost impossible to catch in the nationalist language of friendship and enmity, the language of international relations and area studies. The innocence expresses itself in many ways in these narratives, and raises its head again and again in face of a settled history of friendship and enmity.

Reflexivity therefore calls for a politics of new language of understanding and dialoguing with a neighbour. The received language of international politics is inappropriate in this task. It is very difficult to shake off the image of the 'Bangladeshi Mussalman' from Bengali or the Assamese upper caste Hindu minds, which think of the former as a primal force, representing un-tempered and un-mediated primitive impulses. This image will cease to be legitimate only when we stop offering primacy to the military metaphors of security and fear in our understanding of Bangladesh, metaphors that pass off as politics. This politics is after all only current history, desire and contra-factuality have no place in it. It implies that the language of understanding our neighbour cannot be primarily of fear and anxiety. In fact, the language of empathy exists only by (almost wilfully) neglecting the political primacy given to the metaphors fear and security – a point that I laboured on in *The Marginal Nation*. The language of empathy relives a failed past, an act often termed as cultural, as a strategy to re-deploy the fluidity. Our own novels such as Tagore's *Gora*, or their poems such as those of Shahid Kadri or Shamsur Rahman tease the limits of politics and subvert them with new ethical questions. In a way, they often poetize where politicizing would have failed.[5]

The evolution of a dialogic language bases itself on the exercise of coping with the problematic of friendship, fear and anxiety of a security-centric universe in which we are born and spend our life. Thanks to our political concerns about state system in South Asia, centres of Bangladesh Studies have grown up in India – in Delhi, Jaipur, and Calcutta. Dissertations are written and books are published. The student in Bangladesh studies has to read about Bangladesh, then do field work, best if he/she can stay there for a longer period and be a participant-observer and then he/she comes back and writes (a thesis or a book). Writing certainly appears as the culmination point of this process – though from the point of view of Bangladesh, it must be an occulted process to be crowned by the rite of writing and being written. In this corpus of written texts on Bangladesh, Bangladesh is the sick child of South Asia – haunted by floods, famine, coups, assassinations, instability, obscurantism and unrest. The neighbour becomes a neighbour by being 'anthropologized'.[6] And, it is not that these writings do not contain fragments of truth. Because they contain such fragments, they represent a kind of ethnographic authority that presumably the early colonial ethnographers in India also had. Slowly, imperceptibly almost, vernacular references are pushed out, indigenous voices silenced, the 'I was there' or the 'I know' factor accorded primacy and the dialogic part of the encounter pushed to the margin through the *act of writing*. Paul Rabinow in 'Representations are Social Facts' draws our attention to Clifford Geertz's famous paper on cockfight, where the cognitive process is separated from the texts it generates, the actuality of discursive situations marked by interlocutors is filtered out. The anthropologist establishes that s/he was there and then disappears from the text. His/her authority is thus established.[7]

Yet to say this is not enough. For we cannot say with confidence, and we do not know as yet, how to ensure that the authoritarian text can be changed into dialogic text, and the dialogic dimension of our interactions with our neighbour can be brought to bear in our texts. Do we need new forms of writing – experiential and interpretative? A writing that almost resembles talking, conversing? Charming as it sounds, that too cannot erase the factor of power or authority from our language, as our newspaper and other media accounts of Bangladesh and other neighbours supposedly experiential, show. All these, at the end, are guided by the meta-narrative of fear, anxiety, and authority. Dialogue, a critically existential encounter and an ethical attitude are ruled out.

Any day we can see a number of newspapers and magazines in India, engaged in culturally producing Bangladesh through their representations of/on Tasleema, Nurjahan, imams, parliament boycotts, civil strife, massive emigration, floods and starvation. The sociology of this cultural production shows two things. First, this culturally produced neighbour becomes a part of our self, our habit. Second, this cultural production is part of a specific politics of neighbourhood, and it is very difficult to avoid such politics in a South Asian interface, because the bottom line reveals the politics of cultural representations. In seeking ways to understand a neighbour, we must therefore go beyond the anthropological truth of production of authoritarian texts, and find out how in the world of contentious politics, the themes of dialogue, trust and accommodation work themselves out. I propose to make a hesitant start towards that task in the next two sections of this chapter. I begin with an attempt to examine the elements that lie beneath the political dynamics of dialogue constituting friendship, accommodation, and understanding. And from that I want to make an attempt to understand the rules of formation of the language of dialogue.

Elements of Dialogue for Understanding

There are two ways in which my argument of pitching friendship and understanding (indeed as will be soon clear, I am here arguing for a democratic inquiry) on the theme of dialogue may be construed. These two ways are ironically contradictory. First, it may look like an ethical or a pious tract. Second, it may seem a leaf straight out of a book of enlightened self-interest. Let me place my response to these two assumptions at the outset, for these clarifications will put the submissions that follow in this chapter in a clearer light. First of all, I am neither contesting here the ethical underpinning of what is argued here, nor I am denying the factor of interest. But in pointing out the ethical dimension of understanding, and the process of friendship formation and accommodation of interests, I am only suggesting a historical course. It is a course through which the desire for friendship refuses to be satisfied with historical verdicts, and furrows long tunnels imperceptibly within this history to be able to face it with what I have called mythography. I am also suggesting that the accommodation of varying interests, reconciliation, and partnership are best understood as historically achieved relations.

In trust, dialogue and understanding, we have therefore the working of history that contains within it the desire to step beyond the historically set limits of accommodation. These are the 'truth games' that make relations unstable; but they encourage in an assuring manner a politics of trust, and trust in building up an ethics of friendship. To clarify this point of historicity of relations, on which trust and understanding build up, I want to make a move here by way of picking out certain contemporary themes in inter-state and inter-nation relations in this region. I wish to hold them up in the mirror of the politics of accommodation to show how a historically predicated argument of relations that admits the nuances of contra-history is a better tool for understanding the dynamics of dialogue.

Trust and Dialogue

I want to begin with the crucial theme of trust. Trust (or, the idea that the logic of trust is operational only within networks of 'traditional', that is already formed, relations constituting a network,[8] and therefore by logic is not operational where this tradition is not evinced to naked eyes) occupies a critical place in military metaphors also. Any day we hear from men of strategic circles that that X country is not to be trusted, is untrustworthy, and hence can be trusted at one's own peril. This argument is then extended inward from inter-state and inter-nation relations in this region to classify certain groups of people in the country as trustworthy and untrustworthy. We can hear any day that Pakistan committed 'treachery', that LTTE went back on cease-fire and broke trust, or that X country cannot be trusted in view of her behaviour. Likewise we are told that that Chinese assurances can never be taken at face value in view of their persistence in their own design, or, X community is essentially untrustworthy for their allegiance to Z country, or, Y people at the frontier do not have faith in the nation. In *The Marginal Nation* (1999) in my discussions on trans-border migration in South Asia I showed how the issue of trust that expressed itself in a philosophy of not trusting acts as a pillar in the security architecture of India.

But this raises the question, why do people trust? There are three answers. One, A can trust B because A by trusting B gains. Two, A trusts B because B trusts A and therefore A's gains can be calculated in terms of the accumulated gains of A and B whose mutual trust increases the stock of dividend of trust. Three, A has no other option than trusting B, for B in relation to A invokes trust so much that A has to engage with B in this game of trust to find out the trajectory of relation and relational gain. True as all these explanations are, it is clear these three explanations are built on an exchange-scenario.[9] Even when the explanation is that people trust each other *within* an entity (network or collective such as family, locality, community) because that is how they live or because only that is how they can live, in the process increasing the total volume of social capital, such explanation cannot avoid the fundamental argument of mutual exchange.[10] In a desperate attempt to escape the hold of a neo-classical bind, it has been argued that, 'Trust is a key by-product of the cooperative social norms that constitute social capital' or 'Trust, it should be recalled, is not in itself a moral virtue, but rather the by-product of virtue; it arises when people share norms of honesty and reciprocity and hence are able to cooperate with one another.'[11]

This neo-Weberian scenario does not explain three things. First, it has to depend on a circularity of arguments. For it may be asked, if trust is a by-product of virtue, what begets virtue? If the answer is, trust in norms that constitute social capital begets virtue, then the question would be, if trust is born from social capital, what begets social capital? If the answer is cooperative social norms, that is to say trust, what begets trust? It seems that unlike physical capital (land, machine, money, stock) social capital does not have to reproduce itself on an expanded scale, so that we can continue depending on the holy trinity of virtue-trust-social capital for continuity of 'social goods'. Second, it does not explain why trust breaks down, except by resorting to long term structural explanations of the kind Fukuyama undertakes in *The Great Disruption*, and yet this explanation cannot deny that moments of increase and collapse of trust are historically noticeable political periods. Third, what is the purpose in which trust is put to use? Is it to run day-to-day life, or, to increase prosperity, or, to tide over adversity? In other words, in trying to avoid the compulsive scenario of exchange, the discourse on trust enters into a bind. The result is because it avoids a simple historical-political truth that trust augments in *actions*, *moments of historical actions*, that in these moments people start trusting, increase their capacity to trust, realize the need to trust, and develop the willingness to trust in times of contention, as they live in the midst of contentious history. The only other way is to admit that trust results from a realization of the profitability of exchange. That is to say, that the game is played out in the mirror house of market. While, as I admit, trust may be induced by the notion of gain, the point to note is that such a mirror-arrangement that makes perceptions of the dividends of trust possible is faulty on two counts. It cannot take into account the huge stock of trust available in society, because not all stocks can be represented in the mirror of exchange. Secondly it has no way of explaining why, if trust benefits from exchange, the total stock of trust does not always increase, as in many cases, through exchange-activity. In other words, the game of gains scenario even when based on a dynamic format cannot explain why people trust each other, why people should trust each other.

My plea, therefore, is that, in order to develop a theory of dialogue, we have to search for the historically conditioned roots of trust; and the plea is for a search engine that can negotiate the galaxy inhabited by networks, ties, relations of affinity, histories of mutually conditioned entities, histories of political actions, and of course the complex of morals that marks this galaxy.

This is a more fruitful way. We shall then open our eyes to acts of altruism that may occur in pursuit of morals, or may happen driven by an appreciation that such altruist act can bring in a net increase of the fortunes of a group/community/nation, or simply because such acts entail 'satisfaction' . But in all these increases we shall find that the net stock of trust has also increased. History presents before us ample evidences of all these three reasons. Charity, compassion, altruism, trust – all these are based on historically evolved ways of understanding, accommodating, reconciling.[12] Since values are collectively produced, these become the collective goods of society where goodness and volunteerism become social actions beyond the comprehension of the rational action theorists. Indeed it can be shown in a detailed analysis how co-operation and contention have not been mutually exclusive in popular history. In what Charles Tilly calls 'contentious politics', we find trust playing critical role in building up solidarities of various kinds.[13] Mutiny, war, revolution, strike, have produced great moments of trust; similarly exodus and disasters have also produced trust, which have then resolved into great social residue in various forms and ties, ultimately increasing the net social stock of trust. In case of South Asia historians of Naga politics have wondered, why the Naga rebels refused to take advantage of the crisis in 1971in cornering India and securing gains.[14] Similarly every student of India-Pakistan relations knows that, Pakistan declined to take advantage of the situation in 1962 that India against India. Similarly, China did not push for any advantage in 1999 when the border war between India and Pakistan was being waged, even after the bitterness between the two countries following the Pokhran II nuclear test by India. India does not take advantage of the Tibetan issue in a big way in spite of mutual irritants between the two countries over Tibet – an issue that should have permanently clouded the fundamental norm of co-existence.

Or, let us think of the following questions posed by some chapters from the contentious history of India.

What leads the host population to offer shelter to the immigrants who may be aliens? All accounts, nationalist and cosmopolitan, agree that without the massive quartering of the aliens by the host population the aliens cannot survive so much so that nationalists and patriots say that the country's borders are shrinking inwards.[15] Again, how does explain from the standpoint of the rational theory of trust the genesis of the Khilafat alliance in 1919-21in the nationalist times in India, before that the Lucknow Pact of 1916 between Gandhi and Jinnah, and after that the C.R. Das-Fazlul Haq pact, better known as the Bengal pact?[16] Or, one has to take note of the complete political character of the steps of the West Bengal Government into initiating the recently concluded pact on Farakka (1998) between India and Bangladesh?[17] Conversely, we can also note the extreme disillusion among Muslim nationalist mind, when Congress rejected the Bengal Pact, or, refused to leave any ground for the Muslim League in the ministries in 1936-37, or when it rejected the Cabinet Mission plan.[18] Again a historical study of the post-colonial Naga political history will tell us the extreme disillusion of the Nagas when the Government of India broke the Akbar Hydari accord with them, or of the

Kashmiris, when the Government went back on promises built on Article 370 with respect to Jammu and Kashmir?[19] Or, how shall we explain the 'strange' occurrence when the Prime Minister of Nepal, B.P. Koirala and Chairman Mao Ze Dong met in Henchow in March 1960 during Koirala's visit to China, and they resolved during their conversation not to aggravate the India-China tension over borders that were already building up.[20] Trust, and/or then a sudden cataclysmic loss of trust is the answer in all these cases. Evidently what are considered as fairness, justice, commitment, are important. They insulate trust networks. Conversely a lack of these jeopardize the trust-stock. The fortunes of democracy depend on how insulated these networks are, and how the total stock is able to increase on the basis of familiarity that trust induces. This holds true of the dialogic situation not only within a country, but in relations among peoples, nations and states also. In both we can observe the growth of trust-stock through the growth of trust networks. As Tilly shows, the growth can be gleaned in the observable shift from individual to the network, separate beliefs to shared stories of fortune and woe, from distinct and separate cognition to conversation, from unilateral action to interaction, and from deliberation to everyday talk. The transformed locus and basis of trust points out, that the strategic height in the politics of understanding is not the individual participant, but the collective process that generates trust. The task for a critical theory of dialogue is to understand the process – the mechanisms, sequences, outcomes, reproductions, and then fathom and fashion an appropriate politics of conversation and dialogue.[21] In Palestinian peace politics this is called *amanah*, which means trust reposed in someone who has to keep it – a responsibility and a treasure. In fact, in the Palestinian peace activist Hanan Ashrawi's autobiographical chronicle of the public politics of dialogue, *This Side of Peace*, the concept of trust comes back again and again as leitmotif in her tale of dialogue and peace politics.[22]

Inequality and Dialogue

Discussion of trust leads us to the theme of inequality, another critical factor predicating trust and dialogue. Though the issue of in/equality itself is of significance in discussions of social relations, my concern here specifically relates to the issue of how inequality affects trust, thereby political accommodation. Apart from the basic fact that inequalities are produced from the ways all kinds of social goods and capital are produced and traded, and reinforce the power structure or create new power structure, inequality is a product of various classificatory modes. These are the classificatory modes in which we observe, gaze, exchange, relate, correspond, converse – in short, we make our appearance. Again I am speaking here of individuals and groups as well as individual states and groups of states engaged in relational acts.

It should not be difficult to see how our ways of taxonomy induce inequalities. The arrival of modern economy in South Asia, as probably elsewhere, has ensured that individuals have to pass 'scanners' to land up in various slots with determined attributes and formats, in the process joining people having similar characteristics. In an ironic way therefore taxonomy skews social mobility (and as we shall see soon, and this is more significant in our context, it plays significant part in determining poles of social conversation, its structure, limits, and the capacity of dialogue to contest inequality). Efforts, inheritance, connections, learning, do not make much difference in this space, which is like a segmented market, based on classifications of various kinds. Classification produces hierarchy and from that, inequality. Markets serving as scanners, sort out individuals according to the marginal increases in productivity caused by incremental inputs of the resources they embody or control. These scanning monitors can classify individuals and groups according to indices of attributes, which in modern economy are financial capital, machines, land, and now with increasing importance, information and knowledge. But if that is how the social sensors work, why do not societies break up in unworkable boxes deaf to each other?

The point is that as against these sensors classifying and slotting individuals and groups, we have also individuals and groups conversing with each other. Assuming that inequality will remain in some amount, and has always remained, we need to study the dynamics that temper inequalities produced and exacerbated by markets. I refer to the mechanics of dialogue. Dialogists as interlocutors build on previous experience, improvise, and bring new ideas, values, and goods through intercourse. The collective production of ideas and identities does not necessarily produce symmetry in relations, but the mechanics compel continuous engagement and prevent a party from running away with all the goods. The 'social colloquy' that Charles Tilly speaks of is a dynamics for engaging with inequalities. Thus while inequalities may result from economic domination, patterns of interlocking that produce unequal networks, asymmetrical control over resources, unequally paired categories, opportunity-hoarding, and categorical distinctions, and we cannot claim that dialogues undo any of these, yet dialogue increases the strength to negotiate inequalities. The practice of dialogue is significant for three reasons.

First, certain specific modes of dialogue reactivate the residue of morals, which are then transformed into moral resource in fighting inequalities.

Second, dialogue by locating the fault-lines in structures of coercion weakens organized centres of coercion, which have always thrived on inequalities.

Third, certain forms of dialogue augment trust and trust builds up solidarity against inequalities.

Indeed, it will be worthwhile to see how the relations between the three elements I am referring to here – inequality, dialogue, and trust – work in a dynamic format. Grossly we can say that, increase of inequality weakens trust. Increase of dialogue strengthens trust and weakens inequality. Decrease of dialogue shows increase of inequalities and a decrease of trust. The scenario holds greater relevance in the current global context, also in our own regional context. For in this context, the three factors, finance capital (more potential as it becomes more volatile), control and management of information, and control of science and technology reinforce inequalities and have the capacity to twist the mechanics of dialogue to make the latter compatible with the former. Ironically, civil wars, genocide, proliferation of arms, and mercenary activities in state and non-state forms have created a situation, which invokes the need for dialogue with increasing frequency. But these are the situations, which invoke dialogues without necessarily reducing inequalities. Here we have the most intriguing element in the dynamic scenario just sketched above. An increase of dialogue may not necessarily decrease inequality or correspond to decreasing inequality, while increasing inequalities will certainly affect the total stock of trust.[23]

Yet there is more to the issue of the relation between dialogue and inequality that bring us to one more aspect of this relation besides the three mentioned above. What we have not said explicitly so far is the issue of inequality of power, the asymmetries of position. The question hence arises, if trust breaks down completely and the asymmetries of power assert themselves to an extent that dialogue collapses, how will dialogue be restored without which democracy cannot proceed? One answer is that, force is needed to restore dialogue. Democracy comes by the powerful (authoritarian) route; 'we must break the egg' to bring in place the right sort of institutions, set the governmental order, and discipline anarchic social forces. That was the logic of Samuel Huntington, when he appreciated the political legitimacy of modern rule in Turkey in his thesis on 'political order in changing societies'. That is again the logic of the way Russia has been democratized; this process of democratization of Russia after 1991 is again a process of 'breaking the egg', when hundreds of thousands Russians had to perish, sink under poverty, and had to endure the emergence of the mafia at all levels. The parliamentarians had to be gunned down, the parliament house had to be bombarded, and the political society had to accept the humiliation of forced participation in repeatedly rigged electoral order – all this so that Russians could become subjects of democracy. This is again the current US strategy of resorting to force in order to restore conversation in society (the most recent example is Afghanistan where constitutional deliberations began after the US and its allies demolished any opposition to their plan). Apart from the moral question inherent in 'coercing someone to dialogue', the problem with the metaphor, as Alexander Motyl, one of the most astute observers of Soviet and East European politics points out, is that we do not know how many eggs have to be broken before the omelette can be served.[24] In any case the social cost of ruining trust and dialogue is overwhelming for democracy. The result is in fact what Antonio Gramsci termed as 'passive revolution', whereby the state emerges victorious over democracy, and authoritarianism permanently inscribes a conversational order. We are, then, left with the other route – the route of dialogue, familiarity, and rebuilding trust.

But how can that route be operational in a situation of complete breakdown of trust and dialogue? We are in fact faced with several such situations in post-colonial regions today, situations worsened by the incapacity of the liberal order to understand others. To say the least, this incapacity is marked by a paradox, akin to almost the liar's paradox. For example, John Rawls at the end of his celebrated essay, 'The Law of Peoples' (1993), asks and answers,

> If it should be asked whether liberal societies are, morally speaking, better than hierarchical societies, and therefore whether the world would be better place if all societies were liberal, those holding a comprehensive liberal view could think it would be. But that opinion would not support a claim to rid the world of non-

> liberal regimes...Political liberalism holds that comprehensive doctrines have but a restricted place in liberal democratic politics in this sense. [25]

In other words, liberalism cannot become comprehensively liberal without negating its essence. Surprisingly, Rawls has not seen the logical fallacy of this position that arises when a doctrine of democracy is not built on a theory of dialogue. What is termed as barbarity and total loss of civility, is often the total breakdown of trust between communities, and hence of dialogue. As much as this is true between communities, is also true of the relation between countries, peoples.

How can trust be restored?[26] How and why will the weak, the victims of asymmetries of power, agree to dialogue or communities agree to restore their relation to some form of co-existence or familiarity? How will dialogue become the key to restoration of trust and thus reconciliation? In other words, how can the reconciling practice, that is restoration of trust and dialogue, inhere justice, and can conceptualize minimal justice as the path to peace and democracy?[27] How can the victims, say of genocide, regain their dignity in situations where regimes, both at the national and international levels, have failed to address the question adequately due to their own divisions, interest groups and 'high politics'? Or, how can dignity be restored where memories and trauma imprison their lives and history, because justice has not been given the opportunity to address the crime of genocide, and perpetrators of murders have been politically rehabilitated in positions that have the capacity of influencing the course and conduct of national and international politics? Or, haunted by memory and the desire to escape and occasionally take revenge, the victims turn into permanent escapees? This is a major question reflecting on the issue of the relation between truth telling and dialogue, which calls for separate discussion. But one can say at least few things here, before we proceed further.

Plural Truths and Plural Dialogues

If truth about injustices is about ending judicial immunity/impunity of the perpetrators of crimes, and helping a society to put the past behind it, it requires nothing short of full knowledge about atrocities committed (on both sides). We still do not know how exactly truth telling heals or succeeds in bringing about dialogue and reconciliation. But the right and the ability to tell the truth restore pride and dignity in the abused. Since truth telling can also become banal, as it is fast becoming a staple in the diet of transitional peacekeeping, truth itself becomes a victim of the daily truths of manoeuvre and manufacture of silence under new conditions. Truth telling is often de-linked from the process of justice. Lawmakers and judges ignore the vast amount of truth about injustices and abuses revealed by truth commissions. As a result reconciliation becomes another unequal act. It also shows the almost bottomless difficulties in reconciling truth with office, state, and justice. Truth commissions are only commissions. In South Asia, citizens' and state-recognized human rights commissions sometime engage in dialogues with the victims; they also organize occasionally dialogues of the victims and bring out the truth about injustices. But rarely these commissions have the enforcing powers of the courts, though they practise a broader mandate in dialogues with the abused. Truth commissions are different from international tribunals also, such as the International Criminal Tribunal for the former Yugoslavia and the International Criminal Tribunal for Rwanda. These are established in response to massive state violence and may function with powers and purpose of a court, but they are singularly incapable of restoring trust and dignity, for beneath the international hype there is very little scope for the truth to come up and make the victim stand up in confidence to negotiate the transition to a less violent and constitutional politics. And, as far as country-specific cases are concerned, in country after country, commission reports are prohibited from having judicial effect as in Guatemala. On the other hand if as in South Africa where the commission has the power to pardon if the perpetrator of abuse come out with truth, and nothing short of full truth with specified time, truth may have some effect on the process of justice marked by less unequal dialogues.

Here is the role of dialogue. As victims press hard and testimonies mount, truth if not resulting in direct justice in courts, contributes to making justice possible in the long run by incorporating the truth about injustice into public politics, and making it a part of democratic politics. This is what happened after the publication of the report of the Shah Commission, which inquired into the excesses committed in India when the national emergency was declared in 1975. Revelation of truth leads to few trials, but it establishes before all a broad pattern from among countless events of crimes, impunity, and injustices. Thus it is important to inquire, how is the dialogue for truth organized? How is past dealt in public sphere? What are the relationships between the adversaries when truth is spoken? How many (versions of) truths are coming out? Does truth telling encourage the process of justice? Does it get acknowledgment by broad sections of society? What is the process through which society encourages truth? Is truth telling in front of a commission instituted through national legislation, number of drafts, closely linked to dialogues between parties and other political forces? In other words, we are once again pushing for a relational account – truth, yes; but relating to what, related to what, relates to what? We are again emphasizing here the need to study the institutional practices of dialogue, dialoguing for truth that combines demands, desires, and responses.[28] At least this much is clear:

First, the mechanics of dialogue for minimal justice would have to be such that dialogue is not subsumed under the asymmetries of power.

Second, the middle space (often called the humanitarian space) will be able to organize the mode/s of dialogue so that truth about (how populations became) victims, inequalities, and injustice comes out to reinforce the point that reconciliation and justice are inseparable.

Third, truth and reconciliation become the reason of dialogue.

Fourth, the truth about the destruction of trust, and about barbarity cannot be single. Dialogue becomes what I termed earlier – 'plural dialogues'.

In having many themes and many audiences the dialogists widen the question of justice. Hanan Ashrawi writes,

> Once I was asked how many audiences I had in mind when I gave public briefings or responded to official questions. I began counting. The strata of listeners became a host of internal ghosts fighting for my soul and speech. Was I expected them to keep them separate, to identify each one and address in its own parlance? Would I speak to our own constituency, a people under occupation, the same I would address Israeli public opinion? Did I exercise self-censorship when targeting the ears of the leadership in Tunis? Which Arab constituency or government did I envisage when speaking the language of our torn world? As a woman was my discourse gender-based? Was I slipping into Western garb and an occidental mindset when I faced the foreign media? ... Suddenly I faced captive of the Word, which I had revered and severed for so long. The *logos*, formidable as an enemy and overwhelming as a friend, had become both master and slave. I began to resent the original question, how many audiences I had in mind, which had opened up such a Pandora's box of linguistic and existential self-consciousness.[29]

This is the point where the iron grip of state over the link between dialogue and political order becomes weak. Society repeatedly dying for reconciliation brings into life plural dialogues that not only reflect the plural process of truth and reconciliation, but they become a factor constantly predicating the existent inequalities. In other words, democracy becomes a condition of reducing inequality.

Situations often arise where instead of an outright victory of one party in the battle for justice (Bolshevik victory in Russian civil war, Allied victory in the Second World War, or the communist victory in China in 1949), there is a stalemate. A negotiated settlement ensues, there is a gradual shift from illegitimate authoritarianism to democracy, democracy is fragile, and popular unity remains episodic. The capacity of the outgoing regime is still enormous. But equally strong is the memory of all injustices, asymmetries, and the determination to prevent recurrence of such massive human rights violations and oppression. And finally, the spirit of the constitution, that is to say, the mood of framing new rules and abiding by them, marks the beginning of the new order. In short, in such situation, dialogue is not ensuing from the constitution. Rather the constitution bases itself on the dialogue about the truth of injustices, the desire to heal the wounds expressed in those dialogues, and the overwhelming urge to combine the desire to punish the guilty with the desire to restore justice through dialogue, procedure, and an agonizing inquiry. It is a long way from Nuremberg Trial and the present Hague trials for crimes against humanity. It shows the inadequacy (I admit the necessity) of these modes and the urgency of the politics of restoration of agency/autonomy to the victim. Dialogue to find out truth, to speak out the truth, to inflict shame on the perpetrators and accomplices of injustice, to voice the imperatives of accountability and to demand procedures to punish the guilty, cannot erase asymmetries of power. But it becomes a necessary mechanic in facilitating the transition from injustice and occasional retributive justice to restorative justice (restoration of dignity and the combination of justice with reconciliation) and recognition of past injustices.[30] Here is what one correspondent wrote of dialoguing for restoration of dignity by stressing on justice,

> The diminutive witness struggled unsteadily to gain her feet and then launched into a story that had been seared into her memory over the six decades since the events: Wan Aihua was eleven when Japanese soldiers stormed her North China village, killing her parents, raping her, and beginning her enslavement as a 'comfort woman' to Japanese forces. Under the repeated prodding of the tough young Chinese prosecutor, her story crescendoed to an emotional peak and she collapsed on the stage, requiring emergency hospitalization. Her testimony was among many, presented by 75 former comfort women to the Women's International War Crimes Tribunal on Japan's Military Sexual Slavery...The tribunal offered a unique perspective for posing issues pertaining not only to Japanese war crimes, but also to those committed by other nations.[31]

Democratization of Relations and Dialogue

Dialogue that speaks of reconciliation on the basis of truth, justice and restitution, thus restores the moral content of trust, brings out the fallacies of a 'rainbow nation', and thereby strengthens the political ethic of democracy. In other words, the practice of dialogue presents before us here the theme of democratization of relations, because as in the cases of trust, inequality, and the manifold nature of truth predicating the mechanics of understanding, the factor of democratization of relations also reflects on understanding and accommodation. Admitting the complexities in the three-way figure of inequality, dialogue and trust, it becomes necessary to see, first, how democracy becomes a casualty if understanding evaporates, and second, how democracy aids understanding and vice versa. It is clear that I am again taking a relational stand. We do not have to go into the controversies in democratic theory now, though what I am going to argue may indicate the way the issue of democracy can be looked at in our context.

The significant issue to note in the context of studies of democracy is that, democracy today arouses two kinds of discussions. One is the theme of transition, second is the theme of institutions. The moment policy-makers and analysts in South Asia sit together to mull over issues of democracy, the issues invariably are, is the state of affairs enough democratic? How can these be made democratic, or more democratic? What are the causal mechanisms for a democratic transition? What are the requisites? These questions belong to the theme of transition. Similarly they discuss, why are certain institutions not democratic? How to insulate democracy? What is the procedure and institutional practices of democracy? These questions belong to the theme of institutions. It is not difficult to see why democracy is perched today on these two themes – transition and institutions – clearly it is the post-restoration period beginning with 1989, the *annus mirabilis*. But this is significant, for with these themes being internalized in democratic theory, democratic theory creates democratic hierarchy, a ranking of more, less, and never democratic, and relegates to the back the issue of justice, equality and understanding. Inside the country of India, we have the stereotypical case of Bihar unendingly lampooned in the elite-controlled national media as lacking the inherent democratic strength. Outside, Pakistan is placed absolutely low in a similar ranking order; it cannot have democracy because democratic institutions are not sustainable there; and in any case Islam is unsuitable for democracy. In these examples and explanations, either the requisite mechanism for transition is held to be absent, or the institutions for democratic functioning are considered weak. In this reappearance of the structural-comparative school, we can locate the justification for democratic hierarchy, more accurately democracy producing its deficit.[32] The reason why democratic theory produces this is because the route is non-relational, static. As against this I am advocating a relational route that places dialogic relation at the heart of democracy. Only in this way we can remove from the core of democratic theory the metaphysical issue of transition and the required purity of institutions, and re-inscribe in it the themes of public rights, public claims to justice, mutual approval of the norms of justice, trust in the dialogic procedure, and an expansion of political society.[33] Democratization of relations becomes the issue at stake in democracy.

But how can we proceed towards that? Let us remember what was argued above on inequality and trust. The regime of relations existing in the region may promote inequality by making the weakening of trust networks its explicit goal. Such weakening may result in confrontation, colonization, civil war, and sometimes, outright annexation. We have ready instances of all these. Deficits if suddenly become too big, or shocks to democracy too strong, relations can become more asymmetrical, more disengaging, more dissociating. Trust-stock will plummet affecting democracy in a serious way. It is important therefore to critically analyse phases of sudden appearance of deficits and shocks (1946-48, 1974-77, 1991-93 in India, 1970-72, 74-76 in Bangladesh, 1995-2000 in Sri Lanka, 2000-2002 in Nepal) to understand why inadequate democratization of relations makes accommodation and understanding difficult.

One can refer in this context to a significant exchange of letters very recently in India. During one of the sporadic efforts of the RSS (Rashtriya Swayamsevak Sangh, the premier Hindu organization) to have an out of court settlement with the Muslims on the dispute around the Babri mosque in Ayodhya, its chief K.S. Sudarshan had written a letter to the convenor of the Babri Masjid Coordination Committee, Syed Shahabuddin on 22 September 2000. In it Sudarshan wrote,

> I have always looked at the Ram Janmabhoomi movement not as an attack on the pride and identity of the present and future generations of Indian Muslims, rather as a positive effort to induce them to dissociate themselves from the medieval mentality, religious intolerance, bigotry and vandalism, and help promote national integration and harmony. While adhering devoutly to their Islamic form of worship they as sons of the soil should also take pride in their pre-Islamic past...

To this Sahabuddin denied that the Babri mosque dispute was a part of centuries of problem, because Hindus and Muslims had 'participated creatively in the process of cultural synthesis, which was even extended to spiritual quests characterized by Sufism and Bhakti Marg'. He juxtaposed medieval vandalism with the Brahminic suppression of pre-Vedic religions, Hindu violence against Buddhists and Jains, and Shaivite intolerance towards Vaishnavites. And then he asked,

> Otherwise, how could they join each other in the First War of Independence against the British power in 1857, under an old decrepit...Bahadur Shah...? Why was the Ayodhya issue forgotten when Jan Sangh shared power in UP in 1967 and at the Centre in 1977?[34]

Sudarshan in his letter of 21 November 2000 had accused Sahabuddin of not 'looking at Ayodhya issue as part of centuries old bigger problem of reconciliation and harmony between Hindus and Muslims'. Sahabuddin replied that it was not a 'centuries old' problem of dispute and an issue of reconciliation, but of ending the communal divide that was of recent origin, putting an end to the use of history to rake up quarrels, and then use the argument of reconciliation in order to force unjust solutions. Sahabuddin replied his letter of 4 December 2000 with these words,

> History cannot be rolled and unrolled like a carpet. It cannot be rewritten by demolition-squads; they only write a fresh chapter, reeking of hatred and revenge. Nor should they judge the behaviour pattern of the past rulers and their armies, whatever their religion, on the basis of present standards. We have to come to terms with history.
>
> I appeal to you, Sudarshanji, to let the dead past bury its dead, and act in the living present for a glorious future of our homeland and all our people.[35]

The exchange is significant because it shows not only why and how dialogue is employed in modern politics, also because it shows at the same time the inadequate extent of dialogue in a democracy. It reveals the necessity to undertake a rigorous analysis of the acute lack of accommodation, understanding, and dialogue that marks today's democracies rendering them unequal, unstable and weak.[36]

Apart from communal conflict, one can look into other patterns of conflict also to find out how dialogue appears in a contentious field, relates to democracy, and the inadequacy of one reflects on the other. We may call it following Tilly 'contentious conversation'. In fact, we can see that both Sudarshan and Sahabuddin in the above exchange were appealing to the virtue of 'reconciliation', one arguing that the problem was rooted in the bigotry of one community, and was old, and therefore needed the relinquishing of claim by the said community. To Sahabuddin, such reconciliation was only a different name for injustice and misuse of past, and reconciliation lay in forgetting the past. They were indeed dialoguing, dialoguing was the form of contention; this is contentious dialogue. But even this was not all that the exchange signified. In political dialogue, time is a crucial reference. Sudarshan speaks of long past, Sahabuddin also refers to long past (Hindu bigotry against Buddhists in the early middle ages); but Sahabuddin refers mainly to immediate past when the long past has been 'misused' and communal discord and hatred have become crucial elements of politics. Though we cannot go here into the details of how time and the sense of time influence dialogue, Paul Virilio calls it 'chrono-politics', and we shall return to the issue of time in the politics of dialogue in chapter ten, at this point it is necessary to note that while democratic theory acknowledges the element of dialogue, it has yet to answer to the fact that contentious dialogue marks the perennial inadequacy of democracy. Dialogue in other words is not always the great harmonizer that liberal theory would have us believe, its operation is often contentious, and there is a dialectic in its relation with democracy. Contentious dialogue is the mark of democratic deficit and inadequate democratization of relations, in as much as dialogue builds trust and acts as a supplement.

Here I shall take a leaf out of the history of the civil liberty movement in India – from one of the Indian states marked by acute political violence. We have with us now, thanks to a group of human rights activists of Andhra Pradesh, the remarkable 'Report of the Committee of Concerned Citizens' issued in March 2000 on its efforts to dialogue with both government and the left wing armed guerrilla group engaged in a violent war against the state – the CPI-ML (PW) (Communist Party of India Marxist-Leninist – People's War).

The Committee of Concerned Citizens consisted of fourteen members. The members were S.R. Sankaran, a retired high civil servant known for integrity, honesty, and pro-people attitude in his official career, K.G. Kannabiran, a noted lawyer and the national President of the PUCL (People's Union for Civil Liberties), four well known university academics, a retired vice chancellor, a journalist, and an ex-legislator. The committee was an independent collective of individuals sharing concern on the pervasive climate of violence, brutality, and insensitivity in the context of the violent movement of the armed Left wing guerrillas in support of their demands, and ruthless state actions to suppress the movement. The Committee dialogued with both guerrillas and the state for more than three years (1997-2000) in what it called the 'search for democratic space'.[37] Its aim was to make a meaningful intervention to end violence in Andhra Pradesh, particularly in the Telengana region, and help bring about political consensus on certain fundamental reforms without which it admitted that violence would not end.[38]

The Report referred to the ideals of liberty, equality, and dignity inscribed in the Constitution, thereby reminding the government that the basic objectives of armed movement were in tune with those in the Constitution. It then raised specific issues, such as, lack of land reforms, extra-judicial killings, disappearances, the entire issue of impunity, semi-bonded labour, corruption, torture, rape of women, police brutality, bureaucratic indifference, unaccountability of legislators, and widespread corruption. It believed 'that it was possible to find a long term direction for a democratic restructuring of the society, which alone can completely address many of the issues being faced'.[39] It appealed the government to stop atrocities, killings by the police and the paramilitary forces, and engage in a public dialogue on constitutional ideals, the objectives of the armed guerrillas, and its own stated goals. It also 'unhesitatingly' condemned the 'arbitrary and mindless violence' of the guerrillas and 'their policy of extermination of individuals and destruction of public property'.[40] It issued open letters to both government and armed opposition, issued statements, met officials and leaders of both, organized public meetings and published correspondence in order to build opinion in favour of dialogue. The effort failed in bringing rapprochement. But from the point of practice and the theory of dialogue, the responses of the actors in the three-year conversation and the conversational sequence are significant.

The report ends with the publication of an internal circular of the CPI-ML (People's War). The circular refers to the agenda of dialogue followed by the Committee of Concerned Citizens, speaks of 'their liberal outlook, or vague notions like human rights and right to life', which lead them to 'project some of the mistakes that are occurring in our movement as human rights violations and to equate it with the repression of the exploitative government'.[41] Yet as the circular put it, the Party (of the rebels) was ready to correct those mistakes such as shortcomings in actions against informers and in the functioning of people's courts, which were marked by ruthlessness and arbitrariness, also shortcomings in form of narrow basis of justice committees, the use of physical punishment in resolving disputes among people, the indiscriminate destruction of public property, and attacks on police who were not related to the armed encounter between the security forces and the armed guerrillas. But while the circular admitted those mistakes and instructed the rank and file to correct them, it nonetheless reminded the dialogists of the Committee of Concerned Citizens to be aware of the fact that social change came not 'through superficial reports and recommendations without taking a clear stand on the problems of the toiling masses and the rights of those people, and without analysing it from the class struggle viewpoint'.[42] Then why was the Party interested in the dialogic agenda of the Committee? As the circular put it,

> The loss of life and property and suffering of the people in the war between repression and resistance has been causing anxiety to the democratically minded people. A section of such democrats have already come forward to protect civil liberties...They too have been laying down lives. But a new section, which has been completely away from the movement and is essentially liberal in character, has begun to come forward to join this struggle. Such people recently formed the Committee of Concerned Citizens to make an effort to reduce the 'violent struggle' between the Government and the Naxalites (read people). Despite its limitations this is a good development. It is favourable to the revolutionary situation when democrats feel the need to act even though without a direct role in the revolution. It only remains for them to decide which side they would be on – People or the Enemy. We have to make an effort to align them with the people. This task has acquired importance for us today.[43]

What was the response of the government? In a discussion with the Committee of the Concerned Citizens and the Cabinet Sub-Committee on 2 January 2000, the Home Minister admitted that the problem was political, it was one of justice and not law and order, and therefore it could not be left to security forces for solution. But in the words of the Government 'one thousand or two thousand people cannot force their views on the rest of the entire state', the 'Naxalites...should come to the mainstream'; and one Government spokesperson said that as 'Revenue Minster he had struggled hard to get land reforms implemented properly. But there were difficulties in implementation including intervention by Courts...Even if it was not easy...the Government will implement land reforms wherever possible. The Government considered itself fully accountable to people...(and) the dialogue on issues cannot end with just one meeting and...should continue'.[44] The meeting with the Chief Minister that had taken place nearly two years ago (10 April 1998) had been even more benign. The Chief Minister admitted lapses, said that he knew that land reforms had been inadequate, and that people had been facing unjust measures. He spoke of what he felt were the needed measures such as formation of self-help groups, more World Bank aid to control and ultimately remove poverty and diseases, more water and land development programmes, and assured that there would be no more fake encounters.[45]

Of course, fake encounters did not end, nor ended torture and summary killing, nor has ended poverty. The desperate appeals to the National Human Rights Commission (NHRC) were without result, so were without result the few directives of the NHRC to the Andhra Pradesh government. But in terms of contentious politics, the agenda of the Committee of the Concerned Citizens has been a landmark. Its report shows, how dialogue and democracy relate to each other; how dialogue is strategically employed in politics; how dialogue builds trust in one case and how it fends of pressure in another; how dialogue builds the middle section of society, and representing the middle section occupies at times the position of the centre; how dialogue reveals the way in which public actors in politics treat issues of politics as their private realm where (according to them) the public should not intervene,[46] and finally how dialogue by itself is a process of *contentious conversation*.[47]

The truth is that the amazing variety of ways in which democratization occurs depends on combinations and concatenations of three basic elements – expansion of trust-stock, negotiation with inequalities, and spurt in dialogues. This also explains why mobilization-repression-bargaining scenario (the nationalist period in India) or techniques of co-optation, containment, dissolution of coalitions, imposition, and categorization (the post-colonial period in Sri Lanka), cannot block the elemental democratic thrust. Indeed, in the relational scenario that I am sketching, democracy as a political order may have all these in order to keep itself steady. It may use the former strategy in order to control the forces of challenge. In the second scenario, its conduct may be marked by a strategy to co-opt these challengers to democracy. It may confine the contending forces, may dissolve old coalitions, may impose new rules, and may try to block the sudden surges of dialogue threatening democratic order by categorizing participants in various orders and demonizing some. Indian democracy adopted the last mode when confronting the challenge from the Left. It simply categorized them into civilized left and the non-civilized and unconstitutional left. But, thereby creating deficit in the stock of trust, reducing familiarity, and increasing inequality, democracy creates its own supplement – a dangerous supplement in form of further desire for democracy that may push the day to day existence of democracy to the point of possible nullity.

In short, the problem of accommodation and understanding is a political problem. And that is why, a theory of multiculturalism as a way to negotiate plurality is inadequate for critical engagements with the issue of a dialogic polity. The stakes are high here. The four themes – trust, inequality, plurality of truth, and democratization of relations – once we view them in the glass of a politics of accommodation, suggest serious intellectual inquiries into issues of friendship and enmity, war and peace in this region. The task of understanding is here irreducibly political. I do not deny that pursuit of understanding entails issues of culture; some wistfully say that we need a 'civic culture' of accommodation that strengthens trust and civilizes politics. But the practice of trust, dialogue, contending with inequalities, and pointing out the incessant production of democratic deficit, is a political task. And here I part company with those in search of the elusive civil society as the way to promote a dialogic order, or those in search of retrieving the collective memory of democracy as the way to rectification of democracy and restoration of friendship.

Rules of a Dialogic Act

By now I hope to have made my arguments clear as to where my fundamental critique of the military metaphors around friendship and enmity lie. All military metaphors are moved by the 'fear of the masses'. Fear of the masses in this age of delirium has not only evoked direct military metaphors leading to the emergence of notions of 'peacekeeping army', 'waging peace', and 'peace-building', fear of the masses has led to a virtual zeroing in on madness, distrust, quarrel, and restoring sanity. The same fear has led to finding out the rules of 'decomposition and re-composition of the multitude', in short rules of making democracy a fully functioning regime, which means making democracy work as its own mask. Thus we have strange bedfellows in post-colonial democracies that are so because they are paired as binaries – distrust/trust, insanity/sanity, anarchy/democracy, and collapse/reconciliation. These binaries make possible for democracy stricken with fear of the masses to co-exist with the various military metaphors.[48]

In showing the complex relations between democracy, dialogue, and understanding the intention was double.

First, I wanted to examine at least the possibility of a different logic of accommodation whose dialogic justification would not be grounded in the binary of security or morals. It is my belief that critical theorists of democracy should look for possibilities of understanding elsewhere – fundamentally, in the inadequacies and adequacies of democratic politics, and in the capacity of democratic politics at a given moment to encourage trust and understanding.[49] Auden had written in his poem, *New Year Letter*, '...true democracy begins / with free confessions of our sins'.

Second, I wanted to locate the roots of the inadequacy of the available language of accommodation. It is not enough to say that the rational actor approach cannot develop the language of accommodation. For, if behind accommodation lies understanding, we have to formulate its own politics. That political language may sometimes look like belonging to a self-interest theory or an ethical theory of friendship and of brotherhood/sisterhood. But let us make no mistake that such a politics will have its own language. The initial similarities, if there would be, are but tokens of the newness of the new formulations. Can we try to locate, therefore, at least some of the basics of the new language of the politics of understanding?

Here are some very tentative answers for colleagues in research on dialogues. At least the direction of my own work will be along these lines. Let us recall that in the beginning we had defined understanding as understanding difference. If things were identical, there would remain nothing to understand. But if trust, dialogue, the reflexive dynamics, and the ever-continuing process of democratization are based on understanding, what helps understanding? What are its rules? Of course we can say that understanding is based on trust, dialogue, and the ever-recurring process of democratization, and they facilitate in turn the former. There is nothing to disagree on this, except that causal theories always lead to chicken and egg explanations. Beyond an immediate point they do not help us to come to terms with a problematic. We must look into the rules of its formation, and not origin-accounts.

Dialogue to understand difference? This immediately leads us to ask how do we perceive things are different? We obviously look for signs, representations, and symbols. But our wisdom is compromised there, for in that act of locating difference we may be only looking at ourselves, a voyage of vindication of what we stand for in this world. Since symbols of difference are often cultural, we shall always lapse into what Edward Said had called in a different context 'orientalism'. The problem is therefore, how do we eschew the inevitable narcissistic acts in understanding difference? Some broad rules for such effort may be suggested, may be these are rules of thumb. (Therefore we shall allow many exceptions, revisions, and modifications.)

First, in lieu of looking for symbols, we take what strikes our gaze as elements of a different order. In other words, we do not rush into judging our neighbour who appears to us not as neighbour, but because of our act of perceiving differences, as persons of the 'faraway land'. This means in a fundamental sense, we take them as elements of an order, and not as signs of 'something', that something over the spiritual destiny of which we think we should preside.

Second, when something has caught our gaze, we can look into it only our history of obscurity, not manifested in our 'density of narcissism'. In other words, the 'different' is someone with whom we relate, because this 'difference' belies our truth, our god, our existence. What appears as different is so, because we have related, and difference is nothing but the mediation between my presence and my functions, acts.

Third, our dialogues becomes dialogues between what we can call 'orders of truths'.

Fourth, in the ensuing dialogue to understand (differences), we repose our trust in some residue of familiarity.

Finally, we locate the fault-lines in the relation, which allow us to deliberate and dialogue; our dialogues become acts in contending inequality.

These five principles not only suggest the possibility of a different politics of dialogue, they also tell us the limits of a liberal language of understanding. Tolerating difference will be nothing but just the high act of piety. The awareness of a different language of dialogue that takes into account the differences, the longing, belonging, and the relations lying in various repositories, requires a non-referential conception of language to be applied to the study of dialogues that incessantly take place in society, and among nations. Concretely it means exploring the systemic relationship between terms and propositions within the language rather than viewing these terms and propositions as only having some putative experiential reality of which they are assumed to be the expressions. In order to rewrite the political history of dialogues that will be able to indicate the dynamics of understanding in this region, we may have to therefore think of starting from the other side of the chain. An attention to the language of dialogue will imply discarding a simple notion of dialogue as some reflection of the degree of consciousness of dialogue among social beings and political entities; it will imply, as I tried very weakly to suggest, studying the huge corpus of dialogic experiences in this region very attentively, and the way in which the political language of dialogue is built, the strategy of dialogue is planned, and the language is conducted. Indeed the structure of the political language of dialogue, and the forms of accommodation and understanding conceive and define interests.

In political theory, dialogue has held a mysterious place. Political theory has not been as yet able to solve in its two thousand and four hundred years of history whether dialogue signifies a pedagogical enterprise, or a mode of production of knowledge, or a neutralizing discourse, a trope, or a matter of resolving identity through a critically existential encounter. In the *Great Dialogues of Plato*, for example, while it is clear that the dialogues serve as discursive ploy, it also appears as a way of life as Dion Chrysostom described of the Hellenic city of 70 AD. Long after Plato, Mao Ze Dong had said that dialogue was crucial for the correct handling of contradictions among the people, implicitly conveying the argument at the same time that only a certain political consciousness and a certain political leadership could ensure the culture of dialogue. Else, in a world of asymmetries of power, it became a dialogue of the deaf.[50] In the ideas of Antonio Gramsci, the place of dialogue remains complex, and is connected to his discussions on state and civil society. Gramsci had argued that it always happened that individuals belonged to more than one private association, and often to associations that were objectively in contradiction to one another. A totalitarian desire aimed at ensuring that, members of a particular party would find in that party all the satisfactions that they had formerly found in a multiplicity of organization, that is to say, such party aimed at breaking all the threads that bound these members to extraneous cultural organisms. It also aimed at destroying all other organizations or incorporating them into a system of which the party was the sole regulator. Gramsci saw in this desire a paradox – the parallel existence of progressive and regressive aspects. Such desire occurred when the given party was the bearer of a new culture, also when the given party wished to prevent another force, the bearer of a new culture, from materializing.[51] What Gramsci wrote of parties is no less true of nations.

Dialogue is communication. With the growth of institutions of communications we have come to accept the presence of something called as the 'public sphere'. The emergence of a public sphere belongs neither to the state nor to the individual citizen – a mediating sphere that mediates between society and state, in which the public organizes as the bearer of public opinion. Yet we must remember that this public sphere is often a degraded public life, where the public faces its various particularisms and long-term political disempowerment in the backdrop of short-term empowerment in the arena of the public. Thus the public sphere in the nation induces communication and dialogue only to retard it later. Dialogues thus turn out something to be managed. The institutions of dialogue such as parliamentary debates, diplomatic notes, exchanges and meets, state-community accords, the various social and political contracts, or say, trade union negotiations, become crucial to the science of government. The more power takes on what Foucault called 'capillary form', the greater is the importance of instituting and managing dialogues. Dialogue becomes a tool of governance; it becomes crucial in constituting the field that he termed as governmentality. It becomes the form of a transactional mode on which modern state and state relations base themselves. Faced with such reality, dialogue even though emanating from a more fundamental phenomenon of reflexivity remains tantalizingly perched on the ethic of the nation.

One fruitful way of understanding the tension around the concept of dialogue is to stress its character of mediation. In fact, the question of resolving or reaching position/identity through critically existential encounters, that is, through dialogues, belongs to the domain of mediation. Since there is an in-erasable asymmetry in social positions, we shall be forever searching for a 'subject supposed to liberate' positions from asymmetries to enable them to engage in an equal and fair dialogue, and then finding in place of such a subject, the existence of a constitutive lack. In simple words this means that a hegemonic power negotiating the asymmetries in society through instituting and managing conversations is fundamentally a short-term reality, while dialogues, that is, the mediating encounters, are the long-term phenomena of society. It is the lack that constitutes the reason of dialogue.

Two important corollaries follow from this. First, since the lack (of hegemonic power) constitutes itself as the subject, that is, such a lack promotes the mediating nature of dialogue, it implies a continuous process of dialogic acts building and predicating political relations. These signify the presence of a contentious communicative order, different from what Habermas describes as communicative rationality, because Habermas misses the factors of inequality, contention, and strategies in thinking of a communicatively rational world. Second, since the lack itself is the subject, making dialogue the reason of a communitarian and communicational existence, positions/identities are transient and precarious in a politically managed society. Identification always fails to achieve a full-fledged identity. Dialogue becomes thus the condition of the management of an incomplete society forever trying to be civilly and politically complete. In this new politics of knowledge the dominant mode still remains hierarchical, for the dialogic structure carries the elements of inequalities residing in society, and therefore represses constantly the other within. And hence whenever dialogue occasions friendship and increases trust, the dialogic situation is marked by an implicit recognition of equal right to interpretation, a rejection of the comparativist legacy – a willingness to open oneself up for remark and involvement.[52]

In the theatre of dialogues, if others are thus being constantly produced, we have to remember that 'they cannot be taken in, they cannot be expelled'. They give 'yes' to reconciliation and then they give it the slip. Dialogues are thus never finished. They do not produce harmony. They ensure the conditions of unstable stability, and see to it that the 'assured horizon of a hermeneutic question' remains impossibility.[53] Dialogues produce in short only understanding, which eschews absolute friendship. Two points need immediate attention on the basis of which we can develop our ideas little more. First, if neither identity nor difference is absolute, it means understanding is a function and product of what I have termed as 'plural dialogue'. Second, understanding would mean further, as I indicated in the chapter on peace accords, the development of a middle ground, not in the meaning of compromise but in the sense of a 'politics of communication' as well as 'a communication of politics', and the 'politics of democratization against the politics of democracy'. It is this aporia that will face us eternally. Therefore let us be sure that we are not speaking here of a rational field of communication where actors engage in communicating with each other in order to create an understanding on common values. We are speaking of the problematic of a production of consensus that lies at the heart of the issue of dialogue. Etienne Balibar writing of the politics of Spinoza posed the problem in this way, 'How can one produce a *consensus*, not just in the sense of the communication of pre-existing opinions, but above all as the condition of the creation of *communicable opinions* (that is, opinions which are not mutually exclusive)? And how can this consensus be produced since, as we have seen, the 'matter' of politics is constituted not of isolated individuals but of a *mass*, whose most frequent passion is fear, and to which everyone belongs, rulers and ruled alike? For a mass, in this sense, is a fearful thing, not only to those who govern, but even to itself.'[54] Yet in grappling with this problematic, we must remember that understanding being a relational practice, our engagements are marked by an unusual and therefore remarkable parsimony of self-gaze, contrary to what the theorists, in C.B. McPherson's words, of 'possessive individualism' would like us to believe. Dialogic imagination with its treasure of curious relations and appreciations promotes acts of understanding. In the practice of understanding we have therefore the transmutation – understanding becomes understanding of difference, and difference is then not only difference in form of understanding, but at times replaced by understanding. The breach in the unilateral act of defining, the sovereign act of the powerful, will be final.[55] As I propose to show in the following chapter with the help of an analysis of what goes on by the name of humanitarian affairs in present day globe, even the conduct of the humanitarian is most unilateral, and needs to be replaced urgently by a dialogic understanding of the paradoxes inherent in the humanitarian.

What will be the result of the inversion that I speak of above? Dialogue will become plural towards federalizing the political society. And, what will be the loss? The sacrifice will be the centred space – the space of a Delhi-centred India and India-centred South Asia. Un-centred (de-centred), the political space of accommodation marked by a variety of forms becomes irreversible. Turn, as Roland Barthes had once said, the gallery upside down, nothing would change except some inconsequential inversion of top, down or right, left.[56] For, we would have irretrievably dismissed political unilateral-ism and monologues, and ushered in a federal organization of the political space. We shall be able to live without the agonies of being chained to, and therefore carrying the anxiety of protecting, the centred space of which we are a part for we have related as parts to a centre.

Notes

1 Jurgen Habermas, *Theory and Practice*, trans. John Viertel (Cambridge: Polity Press, 1996), p. 44.

2 See for a critique of the modern politics of friendship, Jacques Derrida, *The Politics of Friendship*, (London: Verso, 1992).

3 These are the two respective literary series of publications of biographies published in West Bengal and Bangladesh – *Sahitya Sadhak Charitmala* (Calcutta, Bangiya Sahitya Parishad) and *Jibani Bichitra* (Dhaka, Bangla Academy).

4 Ashis Nandy, *The Intimate Enemy: Loss and Recovery of Self under Colonialism*, (Delhi: Oxford University Press, 1983). On reflexive nationalism, I have argued in details earlier in the introduction to Ranabir Samaddar (ed.), *Reflections on Partition in the East* (Delhi: Vikas Pub, 1997). On the notion of reflexivity, Pradip K. Bose, 'Text of Language, Language of Text – The Post-modern Dilemma', *Margins*, February 2000 (earlier published in Bengali in *Anushtup*, 27 (1), 1992).

5 I have discussed the interface of the two phenomena of territoriality and reflexivity in 'The Failed Dialectic of Territoriality and Security, and the Imperatives of Dialogue', *International Studies*, 35 (1), 1998.

6 For how the production of authorial knowledge of 'other' countries works out in education, see Volker R. Berghahn and Hanna Schissler (eds.), *Perceptions of History: An Analysis of School Textbooks* (Oxford: Oxford University Press, 1987).

7 Paul Rabinow, 'Representations are Social Facts – Modernity and Post-modernity in Anthropology' in Paul Rabinow, *Essays on the Anthropology of Reason* (Princeton: Princeton University Press, 1996), Chapter 2, pp. 28-58; earlier published in James Clifford and George E. Marcus (eds.), *Writing Culture: The Poetics and Politics of Ethnography* (Berkeley: University of California Press, 1986).

8 For such a view of trust in most succinct form, Francis Fukuyama, *The Great Disruption – Human Nature and the Reconstitution of Social Order* (New York: The Free Press, 1999), chapters 2 (particularly pp. 47-60) and 12 (pp. 194-211). Trust indeed in this logic is seen as social virtue residing in 'traditional goods' with historical impulses little to do with the building or maintaining these 'goods' or creating social 'bads'. The difference of my argument that rests on relations and historical events with this argument will be clear from the way I approach the issue. For the network-centric discussion, see also by Francis Fukuyama, *Trust – The Social Virtues and the Creation of Prosperity* (New York: Free Press 1995).

9 On this, Susan Cross and Robert Rosenthal, 'Three Models of Conflict Resolution – Effects on Inter-group Expectations and Attitudes', *Journal of Social Issues*, 55 (3), Fall, 1999, pp. 561-580. Cross and Rosenthal speak of distributive bargaining (hard, competition-based, agreement oriented approach with lose/zero-sum gain disputes), integrative bargaining (concession-making and search for mutually profitable alternatives), and interactive problem solving (parties recognize the issue to be a problem for them that calls for interaction).

10 Again I am summarizing Fukuyama's argument here. See also, Herbert Kelman, 'The Interdependence of Israeli and Palestinian National Identities – The Role of the Other in Existential Conflicts', *Journal of Social Issues*, 55 (3), Fall, 1999, pp. 581-600, where Kelman reminds us of what he terms as 'negative interdependence' in the zero-sum view of national identity and the 'existence in reality of positive interdependence between Israelis and Palestinians'. Readers can see how my argument here proceeds along a different direction on the question of dependence and its acknowledgement.

11 *The Great Disruption*, pp. 49-51; see also Diego Gambetta, *Trust – Making and Breaking Cooperative Relations* (Oxford: Blackwell, 1988).

12 Related to this, see Victoria Kahn, 'The Duty to Love – Passion and Objectivity in Early Modern Political Theory', *Representations*, 68, Fall, 1999, pp. 84-107.

13 On contentious history and politics Charles Tilly has written enormously. The reader can at least have an idea of Tilly's history and politics from his *The Contentious French – Four Centuries of Popular Struggle* (Cambridge, MA: Belknap Press, 1986).

14 On this see Subir Bhaumik, *India's Northeast – Insurgency Crossfire* (Delhi: Lancer Publishing House, 1996).

15 I write on that in detail in *The Marginal Nation – Trans-border Migration from Bangladesh to West Bengal* (Delhi: Sage Publications, 1999).

16 I have analysed the consequences of the disillusion among the Bengal Muslim leaders following the collapse of the Bengal Pact in terms of the evolution of a Muslim public in Bengal; see Ranabir Samaddar, 'Leaders and Legacies – Stories in the Time of Transition', *Indian Economic and Social History Review*, (4), December 2000.

17 How lack of trust played a crucial role in the deadlock over negotiations on Farakka waters becomes clear in the narrative of negotiations in Ben Crow (with Alan Lindquist and David Wilson), *Sharing the Ganges – The Politics and Technology of River Development* (Delhi: Sage Publications, 1995); Harry W. Blair, 'Democracy, Equity and Common Property Resource Management in the Indian Subcontinent', *Development and* Change, 27, 1996, pp. 475-499; also Ranabir Samaddar, 'Flowing Waters and the Nationalist Metaphors', *Studies in Conflict and Terrorism*, 20 (2), April-June, 1997.

18 On the reaction of Jinnah and his disappointment over the behaviour of Congress in 1930s-40s particularly during the cabinet Mission period, see the collection of essays in Mushirul Hasan (ed.), *India's Partition – Process, Strategy, and Mobilization* (Delhi: Oxford University Press, 1993).

19 The story of broken promises is described in details in chapter 7; see also chapter 9.

20 The complete transcript of the discussion between Mao Ze Dong and B.P. Koirala (1960) is now available with *Mao Zedong on Diplomacy* (Beijing: Foreign Language Press, 1998), reproduced in 'BP and Mao', *Nepal Times*, 25-31 May 2001, pp. 1 and 7. In the verbatim record, Mao is recorded as refusing to blame only India for the tension around India-China border, thanks Nepal for refusing to take any side in a dispute between India and China, speaks of resolution of issues between Nepal and China, and speaks repeatedly of the need for trust between developing nations. Similarly, B.P. Koirala speaks honestly of differences between Nepal and China, of his desire to remain neutral in disputes between India and China, and shares his views on ways to develop friendly relation between two countries.

21 The recent writings of Charles Tilly on trust and inequality and relation to democracy have seemed to me significant. See for example, Charles Tilly, *Durable Inequality* (Berkeley: University of California Press, 1998); 'Contentious Conversation', *Social Research*, 65, 1998, pp. 491-510; 'Power – Top Down and Bottom Up', *Journal of Political Philosophy*, 7, 1999, pp. 330-352; 'Where Do Rights Come From?' in Theda Skocpol (ed.), *Democracy, Revolution, and History* (Ithaca: Cornell University Press, 1998), pp. 55-72. I am very much grateful to Charles Tilly for suggesting me these essays and allowing me access to his recent writings. On the role of everyday talk, I am indebted to the writings of Jane Mansbridge; for example, 'Everyday Talk in a Deliberative System' in Stephen Macedo (ed.), *Deliberative Politics – Essays on Democracy and Disagreement* (New York: Oxford University Press, 1999), pp. 210-239; 'Feminism and Democracy', *American Prospect*, 1, pp. 127-136; and, 'Using Power/Fighting Power – The Polity' in Seyla Benhabib (ed.), *Democracy and Difference* (Princeton: Princeton University Press, 1995).

22 Hanan Ashrawi, *This Side of Peace* (New York: Simon Schuster, 1995), pp. 31, 142, 154-55, 229 and 262. Hanan describes how the women of Hamas come in a delegation to her house to dissuade her from joining peace talks in Madrid when so many husbands and brothers of Palestinian women have been illegally jailed and deported by the Israeli authorities and they remind her of her responsibility. At other places, in the internal quarrels in the Palestinian peace camp, each one is reminded of the trust that the people of the occupied territories have placed in them, and she writes that on one occasion how the entire citizens' delegation wanted to resign because of the irresponsible behaviour of the PLO and the Palestinian Council reminding the latter of the responsibility that was on their shoulders as they themselves were from the occupied territories.

23 To be clear I am indeed pushing here for a thoroughly relational account as opposed to a static account of a relational category. Donna Luff speaks of 'moments' of dialogue in her essay, 'Dialogues Across the Divides – Moments of Rapport', *Sociology*, 33 (4), November 1999, pp. 687-703.

24 Alexander Motyl in an unpublished comment in a seminar on 'Democracy and the Margins' organized by the University of the Negev, Beer Sheva, 19-22 May 2000.

25 John Rawls, 'The Law of Peoples' (1993) in See Samuel Freeman (ed.), *Collected Papers of John Rawls* (Cambridge, MA: Harvard University Press, 1999), p. 563.

26 For an instructive discussion that links reconciliation with justice, see, Rajeev Bhargava, 'The Moral Justification of Truth Commissions' in Charles Villa-Vicencio and Wilhelm Verwoerd, *Looking Back Reaching Forward – Reflections on the Truth and Reconciliation Commission of South Africa* (London and Cape Town: Zed Press and University of Cape Town Press, 2000), pp. 60-67 (hereafter *Looking Back Looking Forward*); a longer exposition of his arguments, 'Between Revenge and Reconciliation – The Significance of Truth Commissions', lecture delivered in the first peace studies orientation course of the South Asia Forum for Human Rights, Kathamndu, 7 February 2000 (mimeo).

27 N.J. Kitz (ed.), *Transitional Justice – How Emerging Democracies Reckon with Former Regimes*, Vol. 2 (Washington D.C.: US Institute of Peace Press, 1995); K. Asmal and others, *Reconciliation through Truth – A Reckoning of Apartheid's Criminal Governance* (Cape Town: David Philip Publishers, 1996); Donna Pankhurst, 'Issues of Justice and Reconciliation in Complex Political Emergencies – Conceptualising Reconciliation, Justice and Peace', *Third World Quarterly*, 20 (1), 1999, pp. 239-256.

28 On the need to take a concrete look into the practices of truth telling, see Priscilla B. Hayner, 'Fifteen Truth Commissions, 1974-1994 – A Comparative Study', *Human Rights Quarterly*, 16, 1994, pp. 597-655; also her, 'Commissioning the Truth – Further Research Questions', *Third World Quarterly*, 17 (1), 1996, pp. 19-29.

29 *This Side of Peace*, pp. 154-155.

30 For this, three other essays in *Looking Back Looking Forward* by Johnny De Lange, 'The Historical Context, Legal Origins and Philosophical Foundation of the South African Truth and Reconciliation Commission', pp. 14-31; Wilhelm Verwoerd, 'Towards the Recognition of Past Injustices', pp. 158-165; and Mxolisi Mgxashe, 'Reconciliation, A Call to Action', pp. 210-218.

31 Mark Selden, 'On Asian Wars, Reparations, Reconciliation', *Economic and Political Weekly*, 36 (1), 6 January 2001, p. 25; see also, Richard Falk, *Human Rights Horizons – The Pursuit of Justice* (London: Routledge, 2000), Ch 12, 'The Redress of Past Grievances – The Nanking Massacre', pp. 199-216. Yet, it will be wise to note, how these are termed as 'past', not 'present' – grievances that continue. In chapter 7, I show in the analysis on accords, how past grievances are matters of the present. The question is, how past does it have to be in order to be past?

32 See in this context, Umberto Eco, 'Tolerance and the Intolerable', *Index on Censorship*, 1 (2), May-June 1994, pp. 47-55.

33 In discussing the relation mentioned here, the writings of Robert Putnam are well known. See Robert D. Putnam, *Making Democracy Work – Civic Traditions in Modern Italy* (Princeton, NJ: Princeton University Press, 1993), 'The Strange Disappearance of Civic America', *American Prospect*, 24, 1996, pp. 34-48; and 'Bowling Alone – America's Declining Capital', *Journal of Democracy*, 6, 1995, pp. 65-78. However, it will be clear to the readers that the path I am taking here will differ significantly from Putnam's.

34 The exchange was reported in *The Times of India*, 15 January 2001, 'Ram an Excuse for Thrusting Hindutva' by Rajesh Ramachandran.

35 The entire correspondence is published in *Muslim India*, 219, March 2001, pp. 110-113.

36 See in the context of this debate, the article by the Indian Islamic theologist, Ashgar Ali Engineer, 'Da'wah or Dialogue?', *Islam and Modern Age*, June 2002.

37 *Report of the Committee of Concerned Citizens, 1997-2000* (hereafter Committee Report), published by S.R. Sankaran, Convenor of the Committee of Concerned Citizens, Hyderabad, March 2000, p. 101.

38 Ibid. 'Overview', pp. 1-8.

39 Ibid. p. 1.

40 Ibid. p. 3.

41 'Revolutionary Resistance – Liberal Democrats' Response – Our Tasks (CC Circular 2/1998)' in *Committee Report*, p. 83.

42 Ibid. pp. 82-96.

43 Ibid. pp. 82-83.

44 *Committee Report*, p. 36-37.

45 Ibid. pp. 71-81.

46 I am grateful to Arun Patnaik who has drawn my attention to the problematic of 'private freedom' in public politics which all want to enjoy – from right to left – which leads to their disinclination to engage in dialogue over what they think to be their domain. He mentions the mood of irritation among the armed left at the 'intervention' of the democrats in the politics of armed struggle; he also mentions the irritation of the constitutional left, referring to the objection of the erstwhile General Secretary of the Communist Party of India (Marxist) to the intervention by non-party left in left political matters. See Arun Patnaik's essay, 'The Concept of Individual Freedom', *Indian Journal of Human Rights*, 1 (1), 1997; also his unpublished essay made available to me, 'The Left and Civil Society – Towards a Critique of Private Freedom'.

47 The current liberal writings on deliberative democracy fail to take into account how the democratic procedure for negotiating disagreements is fraught with contentions of various kinds that I indicate here. See for example the celebrated work by Amy Gutmann and Dennis F. Thomson, *Democracy and Disagreement* (Cambridge: Belknap Press of the Harvard University Press, 1996) and some of the essays in the collection of commentaries on Gutmann and Thomson, Stephen Macedo (ed.), *Deliberative Politics – Essays on Democracy and Disagreement* (New York: Oxford University Press, 1999). However Michael Walzer's essay in that collection, 'Deliberation, and What Else?' (pp. 58-69), indicates some of the problems. Also see, Lynn Sanders, 'Against Deliberation', *Political Theory*, June 1997. My objections of course take a different route.

48 Etienne Balibar, 'Spinoza, The Anti-Orwell – The Fear of the Masses' in Etienne Balibar, *Classes, Masses, Ideas – Studies on Politics and Philosophy before and after Marx* (London: Routledge, 1993), pp. 3-37; also Warren Montag, 'The Pressure of the Street – Habermas' Fear of the Masses' in Mike Hill and Warren Montag (eds.),

Masses, Classes, and the Public Sphere (London, Verso, 2000), pp. 132-145. Michael Mann's discussions on what he says 'the dark side of democracy' are relevant in this context, they show how democracy has always been a companion of ethnic cleansing. In his forthcoming book, *Murderous Ethnic Cleansing*, he devotes considerable attention on 'the colonial dark side of democracy' with reference to the experiences of Mexico, Australia, United States, and particularly the genocide of Hereros, 1904-05 – www.theglobalsite.ac.uk.

49 Chantal Mouffe brings this out in *The Democratic Paradox* (London: Verso, 2000). She writes, '...the "constitutive outside" (is) not a content which would be asserted/negated by another content which would just be its dialectical opposite – which would be the case if we were simply saying that there is no "us" without a "them" – but a content which, by showing the radical undecidability of the tension of its constitution, makes its very positivity a function of the symbol of something exceeding it: the possibility/impossibility of positivity as such' (p. 12), and '...the condition of the possibility of a pluralist democracy is at the same time the condition of impossibility of its perfect implementation. Hence the importance of acknowledging its paradoxical nature.' (p. 16); See also her essay in this context, 'Deliberative Democracy or Agonistic Pluralism?', *Social Research*, 66 (1), Fall 1999, pp. 745-758. She writes, 'Awareness of the fact that difference allows us to constitute unity and totality while simultaneously providing essential limits is an agonistic approach that contributes in the subversion of the ever present temptation that exists in democratic societies to naturalize their frontiers and essentialize their identities.' (p. 758).

50 His essays, 'On Correct handling of Contradictions Among the People' and ' On Ten Major Relationships' in *Selected Works of Mao Ze Dong*, Volume IV (Beijing: Foreign Language Publishing House, 1989).

51 Antonio Gramsci, *Prison Notebooks*, Indian Edition, (Hyderabad: Orient Longman, 1996).

52 I am in agreement here with Ashis Nandy's thumb rules for an egalitarian dialogue, except that these almost look like constructing an ideal situation, 'east set up against west'. The point is, if all are equal in dignity, power, and an appreciative readiness, what is there to dialogue? See Ashis Nandy, 'Defining a New Cosmopolitanism – Towards a Dialogue of Asian Cultures', *Identity Culture & Politics – An Afro-Asian Dialogue*, 2 (1), January 2001, pp. 46-57.

53 Terence Ball, 'Deadly Hermeneutics; Or, Sinn and the Social Scientist' in Terence Ball (ed.), *Idioms of Inquiry – Critique and Renewal in Political Science* (Albany, N.Y.: State University of New York, 1987), pp. 95-110.

54 Etienne Balibar, *Spinoza and Politics*, trans. Peter Snowdown (London: Verso, 1998), p. 119.

55 This crucial question is brought out in a collection of essays, Anuradha Dingwaney and Carol Maier (eds.), *Between Languages and Cultures: Translation and Cross-cultural Texts* (Delhi: Oxford University Press, 1996).

56 Roland Barthes, *Empire of Signs*, trans. Richard Howards (New York: Hill and Wang, 1982), p. 109.

Chapter 9

The Non-dialogic World of the Humanitarian

Humanitarian Affairs and the Dialogic Agenda

Though the epics and ancient annals tell us that in the old days dialogues used to be regularly held for conduct of humanitarian affairs, such as mode of conducting wars, conducting peace, declaring truce, saving the innocents, preserving a town from demolition, or, rescuing the besieged women, and we need to only recall in this context the dialogues in the *Mahabharata*, or the dialogues between Saladin and the Crusaders, strangely today the conduct of the humanitarian affairs is taken as an unilateral act of governing and administering relief. And even though the norms of humanitarian affairs have behind them the deliberations of states, treaties, conventions, and declarations, the dialogic element has been divested from humanitarianism to the extent that the victims of abuses for whom humanitarian affairs are intended have not been allowed any power to engage in a dialogue on what need to be framed and undertaken for them. Issues of care for the needy have become linked to issues of power, and thereby they are now matters of governing, and not of dialogue.

How could such a transformation come about? How could one more area of politics be colonized by unilateral acts? In this chapter I take up the instance of refugee care and administration to show how one of the most celebrated acts belonging to the domain of humanitarian affairs in today's world is marked by a non-dialogic approach, and has only contributed to the growth of what Foucault called governmentality. I take up briefly issues of power, fear, and ethics – the three most significant elements in humanitarian affairs – to suggest why 'humanitarian' affairs must move away from the way the humanitarian is defined, and embrace dialogic politics in a dialectical way. My arguments will be complex, my instances are drawn from primarily South Asian experiences, and my suggestions will be depending lot on historical inferences, but I shall try to make the case as best possible.

I Hope that I Have Been Able to Put Forward My Case for Protection by the UNHCR

These words are from the letter of the sixty-six years old Thuingaleng Muivah, the General Secretary of the outlawed National Socialist Council of Nagalim (Isak-

Muivah) to the United Nations High Commissioner for Refugees (UNHCR),[1] after the UNHCR had rejected his application for the grant of refugee status to him. The Government of India and the NSCN (I-M) had been engaged in 'peace talks' since 1997, and Muivah who had been the chief negotiator of the NSCN (I-M) was on his way to The Hague via Bangkok on January 19, 2000 to attend the eighth round of the ongoing peace talks when he was arrested at the Bangkok international airport on the charge of travelling on a fake South Korean passport. The Thai immigration authority decided to prosecute him on the criminal charge of possessing a fake document and entering Thailand illegally. Mr Muivah was jailed for nearly eight months before he was released on bail.

After his arrest, Muivah had appealed to the UNHCR office in Bangkok for recognition as a refugee. In his application he had pointed out that at the time of India's independence the Naga indigenous people had opposed a British decision to include their homeland into the territories of India and Myanmar (then Burma). When the departing colonial power, the British and the Indian leaders refused to accept their legitimate demand, on 14 August 1947, the Nagas had declared independence and the establishment of an independent Nagaland comprising all the Naga inhabited areas in India and Burma. The Nagas had informed all major governments in the world, and the United Nations of their declaration of independence. In his letter, Muivah also informed the UNHCR that the Indian Government did not recognize the Naga people's demand for self-determination as a political demand, and regarded it as a secessionist movement to be suppressed with force. He further wrote in his letter that for nearly five decades the Naga freedom movement had been suppressed by the members of the Indian armed forces who had been given immunity under the Armed Forces Special Powers Act enacted by the Indian Parliament in the mid-fifties. He informed the UNHCR that as a matter of policy the India government did not grant passport to persons who challenged Indian authority and that as a Naga nationalist who had rejected Indian hold over his homeland, he could not have requested the Indian government for a passport. He had therefore appealed to the UNHCR to recognize him as a refugee under the 1951 UN Convention on Refugees.

The UNHCR rejected his application under Article 1F of the 1951 Convention, which meant that he was alleged to have committed offences, which could be characterized as 'war crimes and serious non-political crimes'.[2] The rejection letter issued by the UNHCR did not give any details of these crimes allegedly committed by Muivah or any evidence the UNHCR had of such crime committed by him. The NSCN protested against the decision of the UNHCR. They claimed that they had never been accused of terrorist crimes, that they were the only armed rebellion group on the Indian subcontinent who had never threatened civilian population, and that eminent peace personalities such as Jayprakash Narayan, the respected Gandhian leader of India and who had led a peace mission to Nagaland in the sixties of the last century, or generals such as Lt. General F.A. Vyas of the Indian Army who was in charge of the counter insurgency operations in Nagaland, and security experts like Murkot Ramuny and Chaman Lal had admitted that the Naga army had not killed civilians, or adopted terror tactics, and was a mass-based national movement.[3]

Thuingaleng Muivah's appeal to the UNHCR for grant of refugee status on *political* grounds (as distinct from *humanitarian*, whose significance we shall in course see), and the rejection of the appeal by the UNHCR pose three problems for a dialogic politics of justice and humanitarianism.

First, we have here the problem posed by the connection between grant of refugee status to a person declared an outlaw by a state, the need for his immunity from the draconian law of that state, and the viability of a peace process.

Second, we face here the problem posed by the location of the power to define, determine, and exclude from the status of the refugee.

Third, we are required to take into account the problematic of the political ethic of care and reconciliation in determining the humanitarian nature of an international arrangement.

As we shall see, these three problems sum up the unilateral nature of authority engaged in conducting humanitarian affairs, which have been divested of any dialogic role for the victims. Let us study the problematic of unilateral nature of humanitarianism by going into the details of the three issues mentioned above.

Connection between the Humanitarian Affairs and Human Rights and Peace Politics

Let me begin with the first, the immediate and obvious one, relating to the connection between the humanitarian and the human rights and peace politics. The appeal of the NSCN (I-M) that Muivah's deportation and arrest in India would put unbearable strains on the ceasefire and peace process apart from placing him at great personal risk, and that therefore Muivah should be given asylum in a third country has grounds. The UNHCR is the de-facto 'gateway' for third country asylum, being the only UN agency empowered to 'recognize' or 'exclude' persons from the definition of a refugee under the 1951 UN Convention Regarding Refugees. Earlier the NSCN (I-M) had been admitted to The Hague based international body, the Un-represented Nations and Peoples Organisation (UNPO). The Indian Government had objected to the application of the NSCN to the UNPO on the ground that it was a terrorist organization. The executive committee of the UNPO had rejected the submission of the Indian government for lack of evidence. The decision of the UNHCR not only amounted to denying Muivah the right to public hearing and the right to effective remedy, rights guaranteed by the Universal Declaration of Human Rights and the International Covenant on Civil and Political Rights, it unknowingly became a part of a policy of hostility instead of being a part of humanitarian settlement and reconciliation.

As recounted in previous pages, after several rounds of preliminary discussions between the leadership of NSCN (I-M) led by Thuingaleng Muivah and three successive Prime Ministers of India, P.V. Narasinghma Rao, H. D. Deve Gowda and Inder Kumar Gujral, a cease-fire agreement had been drawn up between the NSCN (I-M) and the Indian Government in July 1997. This was to be followed by peace talks between the rebel leadership and the representative of the Indian Prime Minister in a third country to find a durable political solution. Between May 1998 and November 1999, the leaders of NSCN (I-M) and the emissary of the Indian Prime Minister had held seven rounds of peace talks in Bangkok, Zurich, Paris and Amsterdam. The eighth round of Indo-Naga peace talks was scheduled in The Hague from 19-24 January 2000, where Muivah was going via Bangkok after consulting some of his colleagues. The peace process suffered blows with UNHCR declining his application for refugee status, the framing of a case by the Government against his nephew Grinder Muivah, who was acting as a 'go-between' the Indian and NSCN negotiating teams, in a hijacking case in Calcutta within few weeks of his arrest, and the issue of a non-bail able warrant against Muivah by the Government of Nagaland on the alleged ground that he was part of a conspiracy to kill the Chief Minister of Nagaland, Mr. Jamir. Several NSCN political functionaries and activists were arrested at this time on old and new charges in what amounted to a violation of the ceasefire agreement. The states of Nagaland and Manipur continued to remain under the Armed Forces Special Powers Act, in operation for the last four decades. While the UNHCR refused to explain reasons of rejection, the Indian government declined to intervene and officially communicate to the Thai government that Mr. Muivah was a political figure required for the peace talks, though three former Indian Prime Ministers, Mr. V. P. Singh, Mr. Chandra Sekhar and Mr. Deve Gowda, had come forward to publicly recognize the imperatives for which Muivah, the chief negotiator with the Indian government in a peace dialogue to end postcolonial India's longest running conflict, had to travel on a fake passport. The question therefore is not only whether Muivah should have been accorded the refugee status, but also whether the 1951 Convention itself is of much relevance to a world where the ethic of dialogue and reconciliation is increasingly capturing the imagination of a politics of justice.[4]

Location of Power to Define, Determine and Exclude Persons Relating to Refuge

This brings us to the second problem, that is, the location of power to define, determine and exclude persons relating to refuge. As we shall be showing in course of this chapter, this power meant all along power of an international system (that meant in reality the UNHCR and the states), over the refugees, which indeed was the case here. The office of the UNHCR and the states determined everything from the beginning, and the persons who needed refuge had no voice in it, though the principle of justice demanded that while under the 1951 Convention, responsibility for excluding lay with the states or in case of an application to the UNHCR with the UNHCR, there had to be substantially demonstrable grounds, in other words the responsibility was not to just exclude, but establish grounds of exclusion. In short, the claimant was to be given fair hearing. Again, given the significance accorded to the question of 'well grounded fear of persecution' in determining the status of refugee, exclusion clause was meant to be an extreme measure, and even then the person could be protected against forcible return (*refoulement*), for among other things Article 3 (1) of the Convention against Torture (1984) covered excluded persons and aliens. Article 1F (b) of the 1951 Convention also specifically mentions that crimes to be considered under the exclusion provision have to be 'non-political'. Yet, we must remember that the system grew in the wake of cold war and great power competition, and possessed no capacity to respond to the imperatives of justice and reconciliation. Therefore, the UNHCR, as it has happened in this case, remained deaf to the wider political winds blowing outside the glasshouse at Geneva. In this case, the significant point to note is that Thuingaleng Muivah took a completely political ground in lieu of humanitarian ground in his appeal letter to the UNHCR against latter's rejection of his application for protection under the UNHCR mandate. After all, he said, the UNHCR in their letter of rejection delivered to him had acknowledged that he possessed 'the characteristics of refugees' as defined in the Convention Relating to the Status of Refugees.[5] He cited Article 1 of the Convention which describes a refugee as someone who:

> Owing to well founded fear of being persecuted for reasons of race, religion, nationality, membership of a particular social group or political opinion, is outside the country of his nationality and is unable or, owing to such fear, is unwilling to avail himself of the protection of that country; or who, not having a nationality and being outside the country of his former habitual residence as a result of such events, is unable or, owing to such fear, is unwilling to return to it.

He then went on to argue that if he possessed 'the characteristics of a refugee', then why was the UNHCR denying him the status of a refugee on the ground that he had waged a war against a state? For, he expected that the UNHCR would be surely 'well aware that the international community has criticized the Indian State for its reservation to Article 1 of both human rights covenants which enshrined the right to self-determination as a basic right of all people', (and) it was 'extraordinary that India should have given shelter to the Tibetans, supported the Bangladesh liberation with arms and annexed Sikkim, but it (did) not recognize the rights of the people of the North-east such as the Bodos, Assamese, Khasis etc. to their right to self-determination'.[6] Muivah also referred to Judge Rosalyn Higgins (a judge of International Court of Justice) who had taken up the issue of the Armed Forces Special Powers Act and the Assam Disturbed Areas Act with the Indian Attorney General. He said that despite all appeals and protests from the peoples of the Northeast, the Indian Supreme Court had upheld the Armed Forces Special Powers Act, and condemned the people thereby to live under martial law for an indefinite period of time. He said that he was referring to the background of his flight from India, as there were 'no democratic channels through which the Naga people (could) redress their grievances'. He referred to the Preamble to the Universal Declaration of Human Rights that recognized that 'if man is not to be compelled to have recourse, as a last resort, to rebellion against tyranny and oppression, then human rights should be protected by the rule of law'. He said that he had taken to 'rebellion as a last resort because of the tyranny and oppression of the Indian Government against his people', and denied that he or his organization had ever committed 'excludable offences which may be characterized as war crimes and serious non-political crimes'.

He went on to refute the three possible grounds of exclusion. First, he said that he had committed no crime against peace and security, as he had done no 'planning of or participation in an unlawful war', for his war, the 'Naga national liberation struggle, was a *just war* (my emphasis, *just in method*, not only in aim, in other words lawfully constituted method of war) being fought in self-defence'. Second, he had never participated in 'war crimes such as mistreatment of prisoners or civilians', for 'Nagas which include myself personally and the NSCN have always respected civilians and even though we did not know the existence of the Geneva Conventions I can categorically state with a certain pride that we have not violated either the spirit or letter of humanitarian laws. Naga code of conduct itself would never allow us to attack civilians or treat prisoners with anything less than dignity. If there has been a breach of the code of conduct we have taken appropriate action against any of our members who have indulged in any such crime. Throughout the history of Naga people's struggle we have never hurt Indian civilians or attacked civilian targets. Books by Indian officers speak with admiration for the Naga spirit of honesty and fair play in war and peace.' Third, he or his organization had never indulged in 'crimes against humanity'; on the contrary, 'Nagas are the victims of systematic human rights violations and the Indian Security Forces have been guilty of crime against humanity when they have systematically tortured our men, raped women with impunity. The Amnesty International has recorded the violation of human rights committed by the Indian Security Forces during counter insurgency operation'. He ended his letter by saying that the UNHCR knew that 'if it maintain(ed) its stand, and he (were to be) deported to India for lack of possession of any passport or other travel papers, this act (would) in all likelihood lead to my re-arrest, torture and, as stated above, worse' that 'my people, the Naga people, will suffer tremendously because (1) the Indian authorities will launch a new offensive against the Naga people; and (2) the peace process that I have been engaged in for the past three years will, in all likelihood, break down as a result. The peace, which we have achieved on the ground in Nagalim as a result of the talks I have held with the Prime Minister of India, is now fragile. I am worried for my people. I am worried that even the little opportunity for peace which has been created will disappear once I am in the hands of the Indian authorities and no longer a free spokesman of my people'. He said that he was aware that the Naga war of independence was labelled as war of secession by the Indian government. 'We have always maintained that since Nagaland declared independence before India did, and since we were never part of India and never accepted the constitution of India, there can be no question of secession. This is an issue, which needs to be resolved by an independent judicial body.' He said that he had laid before the UNHCR the grounds the international human rights law had provided for freedom-fighters like him in defence of his politics, namely Article 6 (Everyone has the right to recognition everywhere as a person before the law), Article 7 (equal access of a person before the law), Article 8 (the right to effective remedy), Article 9 (right against arrest, detention and exile) and Article 10 (right to public hearing) of the Universal Declaration of Human Rights. And he concluded his appeal by saying,

> I hope that I have been able to put forward my case for protection by the UNHCR. I have been forced to live in exile, away from my homeland for fear of persecution

> on the ground of my nationality (Naga), social origins (indigenous community), and political beliefs (being the General Secretary of a banned organization). I categorically state that I have never personally committed or been responsible for any act, which would bar me from the protection of the UNHCR as a political refugee.[7]

The question is, in such case, who defines justice? What constitutes just war? If state's deployment of a law that permits summary killing and its assurance of immunity to its armed officers for meting out summary death is just, for law considers it as correct and permissible, who will say that a war of independence is unjust, and therefore a person is engaged in crimes and hence he is to be excluded from the benefit of refuge? The more insidious implication is that who will reckon with the fact that an official international humanitarian organization such as the UNHCR is siding with a (warring) state, as in this case? Finally, who will judge whether the actions of a UN body have been in consonance with the standards of justice set by the United Nations itself as in this case? All these devolve on the significant issue of location of power. [8] James Hathaway has brought out the inherent irony of the situation when a refugee asks for recognition,

> The hallmark of a Convention refugee is the inability or unwillingness to return home due to a 'well-founded fear of being persecuted'...It is generally asserted that 'well founded fear' entails two requirements. The first criterion is that the refugee claimant perceive herself to stand in 'terror of persecution', her very personal response to the prospect of return to her home country must be an extreme form of anxiety that is neither feigned nor overstated, but is rather sincere and reasonable, Second, the subjective perception of risk must be consistent with available information on conditions in the state of origin, as only those persons whose fear is reasonable can be said to stand in need of international protection...(This) two-pronged approach to the definition is neither historically defensible nor practically meaningful. Well-founded fear has nothing to do with a state of mind of the applicant...The concept of well-founded fear is rather inherently objective...[9]

Hathaway has cited Judge Goff who had commented in a judgment involving a number of Tamil refugees fleeing Sri Lanka in 1987 and seeking asylum in Britain after an offensive by the majority Sinhalese government and the Indian army against the Tamil forces in the north of the country (R. v. Secretary of State for the Home Department, ex parte Sivakumaran [1988], I All E.R. 193 [H.L.]), '(The) true object of the convention is not just to assuage fear, however reasonably and plausibly entertained, but to provide a safe haven for those unfortunate people whose fear of persecution is in reality well-founded'.[10]

Humanitarianism and the Political Ethic of Care and Reconciliation

This brings us to the third problem posed by the ethic of care and reconciliation. If Muivah had, as the UNHCR admitted, 'the characteristics of a refugee', it clearly denotes an ambiguous situation. He is an escapee, he needs refugee, but he cannot be given refuge for he has conducted war against a state and has fought the state with his politics of independence. Clearly the Convention of 1951 with its exclusion clause is inadequate to handle the dilemma, because one, the Convention belongs to what already appears as antiquity and times have changed bringing in new imperatives (justice and reconciliation); two, and this follows from the preceding, the UNHCR does not have the standards with which it can resolve the dilemma. Therefore it cannot concern itself with the fact that denial of refuge can impair peace process, or loss of freedom impairs the capacity of the aggrieved engaged in reconciliation to reach just peace. The inability to understand the new imperatives and dialogic implications reflects the crisis of the liberal principles that underlie the moral economy of refuge today – an economy characterized by a unilateral language – the language of protection and the ground reality of rejection.[11] In order to understand this moral economy, we must think deeply how law, that hallmark of a modern regime, has made this combination (of protection and rejection) possible, how it has made fear operate in unison with an arrangement of care.

Let us go back in this context just for a while to Thuingaleng Muivah's case. Let us ask for ourselves for a moment, if the UNHCR had provided 'objective' tests proving that his fear was not 'well-founded', would it have helped the institutional process of care? Unfortunately, James Hathaway's stress on objectivity and the need to avoid judicial (that is of the justices') prerogative in determining the subjective feelings of fear of a person seeking shelter is not of much help here. The fundamental irony of a situation where the victim has to prove his victimization is not lost or compensated by objective tests. Kafka spoke *In the Penal Colony* that the punished experienced pain and punishment on his own body without the formal knowledge of the sentence upon him. The victim asks for shelter, the immigration officer/judge demands, 'Justify your fear, give reasons for it'. He answers: 'My father has been killed by the policy of my country...' 'Ah', says the judge, 'you must accept that fear is valid when it is based on facts...Let us have a look at the facts. People are being killed in your part of the world, some Tamils in particular. But then people are always killed in your part of the world. On the basis of the true facts as I know them I can find no systematic persecution of Tamils or any group amongst you. There is no objective basis for your fear as you are under no "real and substantial risk"...I cannot admit you for a moment.' The role of judge thus changes gradually from a recipient of request to an interrogator of grounds, who arrogates to him the sovereign authority to interpret, assess, and declare the past pain of the refugee and his fear of future torture, and makes his evaluation shareable with the shelter-seeker. Pain that was so because it was un-shareable, resistant to language, hence had found language in escape now arrives in the legal form of understanding to resolve (hopefully) in shelter. If fear is well founded, it must match up to the language of law, justice and the judge; if not, the refugee is lying. If the refugee is inarticulate, he is not in fear. The outcome of this translation of fear into knowledge and then reason, is the extreme injustice to one who in fear had sought shelter and who irrespective of the outcome now symbolizes what Jean Francois Lyotard calls the 'violent double bind', an 'ethical tort' (the *differend*),[12] the extreme form of injustice, where on one hand fear may not lead to care for it has not been legally measured as adequate to be deserving of care, on the other hand the lending of fear to legal measurement has permanently tied care to measuring fear, and therefore to a substantial measure has impaired care.[13]

The geography and politics of fear is thus perched on power that influences in a substantive manner the ways in which discourses of fear are manufactured through institutionalized politics with the effect that fear reinforces the exercise of power. It is significant therefore to look into how fear is defined and utilized within hegemonic discourses, the ways in which fear is constructed around certain social identities, how it is reflected in experiences of public spaces, territories and environments, how fear surfaces in relation to racism, homophobia, sexism, ethno-sectarianism and contemporary patterns of political, legal, and social exclusion. All these indicate the need to re-conceptualize the theme of refuge, one of the most important aspects of humanitarian politics in today's world, indeed the concept of humanitarianism. It means above all the imperative to investigate the politics of humanitarian actions, such as giving refuge, practices which are meant to be guided by the ethic of care and reconciliation, the way in which this ethic has appeared, the way in which the ethic turns into unilateral affair of governing relief and shelter – indeed the method on which humanitarian studies base itself.

An Issue of Method

Dialogue, as we have been trying to suggest in all these pages, is above all a matter of method pertaining to the some of the ways in which contentious politics is played out. The reason why we should pay attention to the issue of method by which 'humanitarian affairs' (a name under which goes on a significant chunk of the international politics of war and peace) is conducted is because, without an examination of the method we cannot probe into the ways in which humanitarianism is linked to the politics of war and peace, and if our aim is to lodge justice at the heart of humanitarianism, we must repeatedly visit the question of method in studies of humanitarianism, and see for ourselves as to how much dialogic spirit pervades the world of the humanitarian. My instance being refugee management and refugee studies, the issue of method in this case has to grapple with the received wisdom of the inside/outside; as I argued in chapter three, in this received wisdom democracy appears as an *internal* problem and geopolitics as a problem related with the *outside*. I try to show how on one hand the issue of refugee is seen as an issue that comes to us from across the borders (the appearance of geopolitics), on the other hand, care of the refugee is seen as an internal task (invocation of democracy) – with little dialogue between the two. The denial of refuge to Thuingaleng Muivah (it also faintly reminds us of the case of Abdullah Ocalan, the Kurd leader who was abducted from his place of refuge by the Turkish state and brought back to Turkey without any international outcry) is only indicative of a larger problem of political ethic with which humanitarian politics has to concern itself today. Concretely speaking we have to refer to the problem of the incapacity of global institutional politics today to connect the inside and the outside – the questions of democracy and geopolitics – and thereby respond in a humanitarian way (that by the definition of the term has to 'exceed' politics) to issues thrown up by population movement – issues brought along with their persons by the forced and unwanted immigrants – because this global institutional politics is anything but dialogic, and therefore unable to cross the limits of current politics.

The issue of method is thus significant. And, I propose to examine it in this context in some details.

The question of method in refugee management raises its head now because of the massive flow of illegal migration, which has brought out the inadequacy of the received notions of refuge, and before which humanitarian arrangements are failing disastrously. The flow of illegal immigration has not only overwhelmed in some cases the flow of refugees, but has got mixed with it to such an extent that we can say that aliens have appeared as a subject in international politics, with which neither international politics has any dialogue, nor any institutional conversation with them is possible. Definition is of course available in municipal laws of who is an alien, and this is not surprising, because an alien is an alien to a State (but can there be an alien to all the states on earth?). But illegal immigrants who are aliens to a state have been in a state of double jeopardy – they do not have the good luck to get protection when they arrive, and they will not benefit from any moral responsibility owned by a state wherefrom they decided to exit, a state that evidently does not care much for the fleeing population. Allowing population to leave is part of its pursuit of a 'nice exit' policy (except in case of migrant workers when foreign remittances to the economy would be going down). We have to remember that unlike the Civil and Political Covenant, the International Covenant on Economic, Social and Cultural Rights does not create obligations on the states to fulfil immediately on signing. Therefore, a state can get away by arguing that its allocation of resources is insufficient, but is non-discriminatory. An illegal migrant who is forced to move out from his country is seen as showing desire for 'good life', and thus not eligible for his/her right to protection of the social, economic and cultural attributes necessary for his/her dignity.[14] The mix of the two flows, of the refugees and illegal immigrants, now accentuates all the issues that face humanitarian politics today. These are at their bottom, as I have indicated earlier, issues of power, fear, and ethics – issues that indicate the existence of a sphere of politics, which is marked by an almost complete absence of dialogue, because aliens are the denizens of this sphere, and with aliens you do not speak.

Here is an instance of what is happening therefore round the world, marking in a very condensed form the non-dialogic segment inhabited by the aliens who suffocate, perish, and die in the most silent way, without any chance to talk to the world of the international that they wanted to enter:

> Last month fourteen men and women left their coffee farms in Veracruz, and began the journey north. Within days, their bodies were found on the hardpan of the Sonora desert. On first look, they died of agonizing dehydration, like hundreds more over the last few years, trying to cross the same forbidding border.

But their deaths were caused by more than lack of water. These farmers left their beautiful Veracruz Mountains because free-market reforms – no rural credit, no crop subsidies and others – drove them off their lands. And having made the hard decision to look for jobs and a better life in the north, U.S. immigration policy made their deaths practically inevitable.

No visas were available for them – the waiting line for green cards at the embassy in Mexico City goes back to 1976. A draconian border policy has closed the safer routes across, pushing migrants further and further into the desert and mountains, making the great migrant stream less visible, along with its human cost.

And if they had arrived safely, what life would these farmers have found?

They would have become part of a migrant workforce with conditions and wages at the bottom, denied the most basic rights – no unemployment insurance, no medical care, no social benefits of any kind. Because of employer sanctions, the very act of working would have been a crime. Ironically they might easily have been employed by the same corporations relocating jobs to Mexico, attracted by the very free-market conditions, which force migrants to leave.

But perhaps the worst thing about their deaths is the way they'll be used, not to advocate for humane changes in U.S. immigration policy, but to justify a new *bracero* program making border-crossers like them a permanent, second-class workforce for the profit of U.S. business.

President George Bush and his fellow free-market advocate, Mexican President Vicente Fox, are both under pressure to reduce border deaths. Vastly expanding guest worker programs, they argue, would open the doors of legal immigration to those now forced to cross in secret.

While guaranteed labor rights on paper, however, guest workers depend on the continuation of a job to remain in the country. Employers therefore not only have the power to fire workers who organize or protest bad conditions, but in effect to deport them as well.

Beneath a humanitarian cover, business gets what it wants – workers at lower wages with fewer rights...

Twenty years ago, most unions wrote off immigrant workers. In 1986, the AFL-CIO supported employer sanctions. But today unions are rethinking that attitude and as a result, the political alliances that limited the possibility for immigration reform have changed. Amnesty for the country's 9-11 million undocumented immigrants, which was off the radar screen in Washington just a few years ago, is now a realistic goal.

'Most unions today are at least trying to organize,' explains Hotel Employees Union President John Wilhelm. 'And no matter the industry, they run into immigrant workers. That's what brought home the failure of the AFL-CIO's old immigration policy.'

Last year, the percentage of U.S. workers belonging to unions dropped from 13.5 percent to 13.3 percent, and fell to 9 percent in the private sector. For the overall percentage to stay constant unions have to organize 400,000 workers a year; to increase by 1 percent, they have to organize twice that number, a rate not achieved since the 1940s.

Over the last decade, immigrant workers have proven key to labor's resurgence. 'Every period of significant growth in the labor movement was fueled by organizing activity among immigrant workers,' Wilhelm says. 'We're a labor movement of immigrants and we always have been.'

Reflecting this new attitude, unions are proposing an alternative to a new *bracero* program. 'We're putting forward a comprehensive agenda, including legalization,

> repeal of employer sanctions, and workplace protections regardless of legal status,' says Service Employees Union Vice-president Eliseo Medina. The new president of the Laborers Union, Terence O'Sullivan calls for opposition to contract labor, and for increasing the ability of immigrants to reunite their families in the U.S.
>
> Illinois Congressman Luis Gutierrez has introduced a bill taking the first step – expanding legalization opportunities for immigrants who arrived before this year.
>
> From the opposite end of Congress, Senator Phil Gramm, a recent convert (like Jesse Helms) to the *bracero* cause, is introducing a bill to permit recruiting guest workers for a year's labor, so long as they have no right to stay. At the same time, he proposes increased enforcement of employer sanctions to force workers into the program, making the undocumented even more vulnerable, their labor cheaper and their conditions worse...
>
> The choice is not over what will or will not stop people from coming across the border, but over their status in the U.S. It's the age-old American dilemma: bondage whether as slaves, indentured servants or *braceros*) or freedom (even if that still leaves workers with the need to organize and fight to improve conditions).
>
> Behind the debate lies a fundamental question: Is the purpose of immigration law to supply labor to industry on terms it finds acceptable, or is its purpose to protect the rights and welfare of immigrants themselves?
>
> There is another framework for dealing with migration, other than contract labor and death on the border. The UN's International Convention on the Protection of the Rights of All Migrant Workers and Members of Their Families supports the right of family reunification, establishes equality of treatment with citizens of the host country, and prohibits collective deportation. Both sending and receiving countries are responsible for protecting migrants, and retain the right to determine who is admitted to their territories, and who has the right to work.
>
> The Convention recognizes the global scale and permanence of migration, and starts by protecting the rights of migrants themselves. That's where an immigration policy based on human rights begins.[15]

This is more than a refugee situation that is legally defined and assumed. The non-dialogic condition has prevented studies of humanitarian law and actions from opening up to currents of critical social thinking, making it almost an appendage of international donor agencies, multilateral protection organizations, and the discourse of humanitarian crisis. The door of the humanitarian, which had been open at all times, has been suddenly locked from inside. The death of tolerance, the absence of dialogue between municipal and international legal institutions and the aliens, the hyper-active movements of the global managers along diplomatic and sometimes military corridors to save the world from humanitarian crises, and the denial of any subject-hood to the aliens form the condition which along with the lock and the closed door have formed an enigmatic triangle where humanitarian studies, in particular refugee studies, are being both offered to and withdrawn from us.

The new mix of forced and unwanted population flows and the inadequate appreciation of the new phenomenon in refugee studies raises the problem of method. While forced population movements have been hitherto studied from economic and demographic angles, its link with the politics of citizenship is still inadequately appreciated. Similarly, the notion of *forced* is so narrowly defined, that the structural violence continuously producing aliens escapes our attention, though violence and coercion are considered as benchmarks in determination of refugees. Then again, the politics of humanitarian regime has subsumed within it the ethic and sentiment of care, protection and hospitality so much that we are still not attentive enough to find out how the politics of the humanitarian has successfully created vulnerable sections of population as critical categories for political management, for which the regime has to care, but with whom it cannot talk, because the vulnerable has no voice, and this apart from the fact that more than half of the vulnerable is not even legally defined subject – they are simply the aliens. The closure in humanitarian affairs to the subjectivity of the aliens is largely due to a neglect of the *subject* in forced migration studies, the subject who moves, who makes the movement, particularly under conditions of globalization. There is now a great urgency to address these questions because, as indicated above, international relations theory in face of the magnitude of the mixed flows appears to have made up its mind to become irremediably a managerial exercise, and we seem to be reconciled to the presence of a large chunk of population who are a burden, who are to be cared for, but who can have no role in politics, with whom there is no role for dialogue even for their own lives.

How did the humanitarian become so non-dialogic? For that we have to briefly go back to a history that has made the non-dialogic conduct of humanitarianism a permanent feature of the current international life.

A History that Raises the Issue of Method

The Convention of 1951 relating to the status of refugees recognized the obligations of the contracting states to grant protection and set standards for treatment of refugees in the asylum- states. It also established the principle of *non-refoulement* according to which persons may not be returned against their will to a territory where they may be persecuted.[16] The institutional governance of the convention was assigned to the office of the United Nations High Commissioner for Refugees (UNHCR) that was to encourage governments to subscribe to the convention and make regional conventions also. It was to promote the grant of asylum to refugees. It was to secure due process of law in examining refugee applications and enforce international standards in the enforcement of legal status. The UNHCR was to secure durable solutions also through voluntary repatriation or eventual grant of nationality of the country of residence. The protocol of 1967 enlarged the scope of the convention. The functions of the UNHCR increased. The need to have a separate set of rules and recommendations for the category of internally displaced persons was gradually recognized.[17] The UN human rights machinery seemed to be paying greater attention to the condition of the refugees.

The convention also led to regional initiatives. The Council of Europe recognized in 1994 that the refugee question called for a regional approach. Methods such as information-sharing, parliamentary attention, defining humanitarian standards for acceptance of refugees emerged as parts of a regional framework of refugee protection in Europe. The OAU convention of 1969 broadened the scope of definition of a refugee. The 1984 Cartagena Declaration of the Organization of American States took note of 'generalized violence', which could produce refugees. The interaction between the 1951 Convention and the OAU Convention and Cartagena Declaration and also discussions in countries (refugee receiving and refugee producing countries), which are not signatories to any Convention, in this regard seem to have taken forward the humanitarian work in protecting the refugees. The Convention, essentially a European cultural product, seems to have reached out to the world outside. Such interaction has also brought into notice the relevance of harmonization of laws and practices both at governmental level and within judiciaries to an extent where judges will find it difficult to adopt an isolationist approach. The OAU Convention has specifically encouraged spread of information for arriving at rational judgment. Today, the benefits of collation of material, sharing information, common package of instruction and advice, and various memoranda of understanding with the UNHCR have resulted in initiatives under municipal regimes also. For example, the Model Law on Refugees drawn in a Roundtable Workshop on Refugees in the SAARC region has been able to link the issue of group determination with individual status recognition. It has recognized that the issue of ethnic identity in regions such as South Asia is often confused with nationality and passport. It has been able to propose sensible and practical proposals for procedures to obtain recognition as a refugee. Finally it has demonstrated understanding of the complex issue of 'sudden and mass influx'. Lawyers and public groups have proposed even further improvements on the model law. The effect of all these has been that it is increasingly difficult to throw out persons whose claims have been rejected, and demands are in the air to grant even after rejection 'exceptional leave to remain' (ELR) that can save thousands from ensuing distress. All these appear as providing the humanitarian world with dialogic procedures and spirit.

But at the same time these developments in fact indicate that, the system is virtually stretched to its limit in face of the unprecedented population flows in the nineties of the last century. Massive refugee flows, forced repatriation, the decline of liberal asylum practices, the creation of *non-entrée regimes*, and the security mentality overwhelming the refugee care arrangement have made the vision of 1951 inadequate for negotiating the population flow in the world of today. In any case the convention of 1951 was intended for Europe, and did not consider the millions fleeing homes in the wake of de-colonization, partition, and the emergence of new states. No UNHCR took care of the refugees in South Asia that was being de-colonized. No UN administration took exception to the walls of Europe being raised higher to stop the refugees. A report in the *Human Rights Tribune* describes in details the charade of the Dublin Convention, the Amsterdam Treaty and the new Charter of the Fundamental Rights of the European Union that do not take into account the rights of the non-European citizens – and this despite the passage of immigration and asylum matters to the first pillar under the Amsterdam Treaty. Each state is free to devise guidelines for the conduct of its security forces. The same report then says,

> In 1998 Semira Adamu was being deported from Belgium to her native Nigeria. Although only 20 years old, 11 gendarmes and two airline company employees accompanied her with two of the gendarmes staying with her on the plane. To stifle her protests, one of the men held her head down into a cushion for about 20 minutes while, as surveillance tapes later revealed, he and his colleague joked and laughed. Ms Adamu lost consciousness and was taken to hospital where doctors were unable to revive her.

Strangulation has not been enough to keep the tide away. Men and women have to be tied and chained. The same report then says,

> In 1999 Marcus Omofuma, 25, was being deported from Austria to his native Nigeria via Bulgaria. Police had bound his arms and legs on the way to the airport. As he continued to protest on the airplane, they covered his mouth with tape. When the officers removed the tape after landing in Sofia, they realized he was unconsciousness. By the time a doctor arrived. Mr. Omofuma had died. The police was apparently unaware that Mr. Omofuma suffered from chronic bronchitis.[18]

Besides not being exercised with the abuse around deportation, the UNHCR is still unable to administer quick and proper care of the millions fleeing from ethnic violence and state terror in a situation when some states by way of prioritizing refugee issues decide to ignore some cases while taking up the godfather's role in others in a completely disproportionate manner. In face of the structural adjustment policies that produce what the historian of indentured labour Hugh Tinker called 'new slavery',[19] and with an entire affluent continent being populated today by aliens leading to a submerged existence – whom following Eric Wolf we may call 'people without history',[20] it is meaningless to refer to 1951 as the defining moment in humanitarian affairs. The phenomenon of forced migrants, unwanted migrants and migrants of the underworld within the country, across the country, not only makes the relevance of the year of genesis suspect, its presence indicates the existence of a huge chunk of people with whom international liberal democracy cannot converse, it has nothing to dialogue with the aliens on the possible forms of a political society that will be able to expand the horizon of citizenship, meet the aliens on the ground of participation and political subject-hood. And though, international jurisdiction and growing jurisprudence in issues relating to the consideration of claims and recognition of refugees have contributed to the widening of definition of refuge-seekers and Convention refugee claimants, as Judge Geoffrey Care of the International Association of Refugee Law Judges has demonstrated, the peculiar nature of the Convention leaves a tremendous burden on jurisprudence and enormous freedom to the state. Referring to one particular case (R v. SSSS ex p JCWI: R v. SOE ex p B [1996] 4 All ER 385), he commented,

> Increasingly the courts and tribunals have given interpretation of the states' obligations under the Convention, which are not always in conformity with the views of the executive of the state itself. The manner in which the Convention is applied has also frequently been with the reality for the individual himself or

> herself very much in mind…To put it another way, the courts have distanced themselves from the litigant parties, as courts do...I have in mind...two decisions of the Court of Appeal (in UK). The Courts were faced with refugees who, under the interpretation of the regulations, which the state would have followed, would put the refugees on the street. The courts interpreted the regulations to prevent this saying that 'parliament could not have intended such an inhumane result'. Parliament did however, because the regulations were promptly amended.[21]

Refugee determination procedure on individual basis and the unequal sharing of burden of care have now produced confused, traumatized, and nervous shelter-seekers who travel rarely with supportive documents, false or no papers, and land in alien systems which are in Judge Geoffrey Care's words 'frequently hostile or incredulous' hosts.[22] The more fundamental question is, how much can decisions of determination of refugee status on individual basis but taken on broad collective political, economic and cultural considerations be able to set up an international ethic and framework of care, hospitality, and kindness? Also, how would law in such cases settle on proofs? Judge Care has reminded us again while referring to another case of similar nature involving a Sri Lanka Tamil who had fled persecution allegedly at the hands of the LTTE (R. SSHD ex parte Karunakaran 25 January 2000, unreported) that,

> The civil standard of proof, which treats anything, which probably happened, is part of a pragmatic legal fiction. It has no logical bearing on the assessment of the likelihood of future events or (by parity of reasoning) the quality of past ones...The method of evaluation is itself not one of hard facts. But it requires knowledge not only of applicant's own tale, and what is accepted of it, but a whole range of other factual matters.[23]

Therefore, it is not merely the issue of narrowness of determination-principle as indicated in the Convention that is at the root of the authoritarian nature of the humanitarian arrangement; a more fundamental structural issue is here at stake. The refugee flees from the induced violence of the state, private armies, systemic discriminatory institutions, environment disaster, and developmental catastrophe – but s/he flees from a *state* where s/he is a citizen, and lands in another *state* no matter where it is, and here is an international system that tries to frame rules of the game – escape and refuge, how you are to escape and how you are to get refuge. How can a dialogic ethos emerge from that? In other words, how can an ethic of care and kindness and a framework of responsibility grow from that? Daniel Warner has asked in the context of state and international responsibility,

> The problem we are now confronting is the nature of the exceptionalism of the refugee situation. Is the refugee situation exceptional because the refugee is merely outside some state responsibility? Or, and this is what we are implying, is the refugee situation exceptional because of the de-territorialization of the trinity, the inherent violence of the state, and the incapacity of all states to fulfil their human rights obligations consistently?[24]

Consider the evolution of the rules of the game little more. It is not enough to note that these rules begin by unilaterally defining who a refugee, that he is not a fugitive fleeing criminal persecution and breach of law, that he is not an economic migrant, not an escapee from developmental disaster or natural or man-made catastrophe, but is one fleeing from violence, persecution or threats to persecution to freedom and safety.[25] The rules began not so much with humanitarian intent, but it began under the League in 1926 to shelter persons of Russian origin fleeing Soviet Union. A similar approach was employed in 1936 for those fleeing Germany. It developed later into Article 1 of the 1938 Convention. But from the beginning the Convention while framing rules of shelter began framing rules of exclusion also. Thus Article 1 (2) excluded from definition persons who left Germany for purely personal convenience. The European scenario continued to dominate the development of refugee law (Bermuda Conference of April 1943), the formation of the International Refugee Organization (IRO) and subsequently the UNHCR. Thus the massive population flows in the period of post war de-colonization passed un-addressed. Only in 1957 the General Assembly for the first time instructed the High Commissioner to assist refugees who did not come fully within statutory definition, but whose situation was of 'concern to the international community'. That is how also developed the idea of High Commissioner's good offices. Regional initiatives came not because of the UNHCR, but besides it, and regions reflected their particular concerns. Thus while Africa concerned itself with the ruins after de-colonization, Europe concerned itself with erecting higher walls to stem the tide of inflow of the escapees. 'Humanitarian considerations' or 'displaced persons' were terms that started to be slowly used by the General Assembly, which wanted to address situations of civil war as in Sudan, Laos or Vietnam. The 1967 protocol situation improved the situation only marginally, for the expanded definition was still woefully inadequate in addressing either the political issue of negotiating the population flows from *all* kinds of duress or the ethical issue of erecting a framework of care, hospitality, and acceptance of burden. In this double jeopardy, the international legal humanitarian framework could be at best a hopeful presence, at worst a system that was open for cynical use of great powers.

In any case, just as states have never borne the brunt of caring and sheltering the evicted which has been always greatly carried out privately, that is publicly by communities, families, villages, towns, and other solidarities, in this case also the international public authority could carry only a minimal burden. And as common sense rules the political world most of the time, in this case also an amazing array of practices – regional, national, international, common, customary, judicial, constitutional – inhabit the world of the humanitarian, turning the field of the humanitarianism into one of governmentality, a field that only produces governing and governable relations, a regime of controlling guidelines. Nor does it prevent certain practices from being the master ones. The scenario is one of a hierarchical order, where at the top major decisions, policies, laws, formulations, and practices can be invoked at will by the international regime that may not have a necessary organic justification for such invocation to negotiate situations as warranted by the major powers. Below the top we have the second tier represented by regional decisions, formulations, policies, laws, and practices. Finally, countries faced with hosting most of the flow do as common sense demands. The maintenance of this hierarchy above all demands the ideological discipline of definition and arrangement, and the expulsion of any dialogic principle that would have impelled dialogue with the victims, burden sharing communities, human rights organizations, and the humanitarian world.

It should not cause surprise therefore that, in spite of the attempts to widen the definition of a refugee, and widen the institutional base of refugee protection, make it participatory, accountable to the victims, and make accountability, responsibility, and dialogue the three pillars of humanitarian conduct, the vice-like grip of governmental practices over matters of care and hospitality remains unshakable.

The Paradox of Ethics

After all, care, kindness, and hospitality are issues of ethics that remain always beyond the assured horizon of law. Yet the paradox is that while the humanitarian law has drawn its legitimacy from the task of protecting humanitarian virtues of care, kindness, hospitality, and justice for the wronged and the victim, the institutional practices of protection build up on the legal compulsion of determining the need and the quantum to care, the quantum of how much one can be kind, and hospitable. In this institutional compulsion that is invariably non-dialogic in nature, we have the reappearance of the problematic of how to make law dialogic, that is to say, responsive to ethical considerations that flow from considering the recipient of care and hospitality as the subject, and not merely the object of charity.

In order to understand the problematic we must distinguish between what are *legal doctrines*, *systems*, and *sentiments*. Though in the world of refugee studies, sentiments come last, in the world of humanitarian ethic sentiments come first.[26] Care, kindness and hospitality – the most acceptable route through which people desire minimal justice for the wronged, deprived, and the victim (I am leaving aside for the moment the element of punishment, also we all know that retribution un-tempered by kindness never succeeds in restoring human virtues), are first of all emotions. They belong to what can be called collectively a highly developed emotion, 'a thick feeling' – *a feeling without an object*. Such sentiment carries the message of a good life, sometimes a unitary concept of such good life, which also indicates nostalgia for what 'was good in the past but existent no more', an 'informed emotion' within the sphere of lived life, and to be more precise, an emotion invested in organization of social life', and which 'is a state arising out of present conditions as much as out of the past itself'.[27] Thus it may be that irrespective of the identity of the shelter-seeker, by and large some people may be inclined to offer shelter. Again such sentiment does not carry a doctrine, but a notion of happiness and a moral satisfaction in being altruistic. The way common people of West Bengal and Tripura offered shelter to the refugees of East Pakistan was an indication that such sentiment resides in one's morality as exercised in judgment. It indicates a feeling that is not negative, meaning that the host not only does not want to do ill to the neighbour, but shelters the neighbour in distress also. In this evocation of kinship, there is respect for the limits of the self that in fact is trying to transcend them. Such sentiment, as we can see, carries a diffused sense, implicitly an acknowledgement of variety, it belongs to the genre of what we call 'mood'. It indicates sensibility, hence sensitivity, a good sentiment. Such sentiments can be at times remarkably a political emotion, but mostly a moral emotion, sometimes without being noticed a mix of both.[28] Thus care for the refugees from Nazi Germany, from East Pakistan, from South Africa under apartheid, have been a mix of both; care for refugees from Bolshevik Russia or Communist Cuba a case of political emotion (I am leaving aside again the issue of political choice); care for the people fleeing from hunger a moral emotion. One need not look for three distinct categories; they are mixed in real world, but suffice it to notice that there are different types of emotions involving care, kindness, and hospitality. One need not even look for concrete evidences of these sentiments, their body-forms, for organic metaphors have always been dangerous. But this much we can say that many of the emotions are dialogic in their impetus, they impel individuals to reach out to the world outside, and they lead individuals to converse with the victims and act.

But sentiments, powerful as they may be, are not enough as governing tools. The same ethic that forms the emotions becomes the material (this is what is called by the strange term *applied ethic*) for government, the political technology of administering, in this case, population flow. Therefore, we have the next set of issues around who deserves entry into neighbourhood, who is eligible for kindness, and who will be kind and share the burden of hospitality. Diffused sentiments suffer deficit in order to become elements of a humanitarian system. We have here the classic problem of Dostoyevsky, *we love humanity, but hate human beings*. The system that builds up on convention, protocol, office, role of special rapporteur, budget, grants, relief-rehabilitation-resettlement allocation, inspection, determination, and repatriation, is one of love for humanity and often of hatred of human beings. Fifteen years ago, David Kennedy like a sleuth had brought out the foundational history of the international system of political care, and the historically contingent nature of the idea of asylum as a solution contrasting the current notion with notions of asylum before 1700. He showed, how while hospitality and protection accorded by princes and other authorities were quite prevalent, and how this was without a coherent doctrinal notion of asylum. They were neither uniform nor tied to a system. 'One found merchants fleeing debts, peasants seeking to escape feudal bonds, or replace collapsing allegiances, children of mixed-up parentage seeking license to trade, members of religious orders and their flocks seeking princely support or freedom to practice their faith, and traders seeking military assistance against pirates or in recovery of property at sea. Many of these individuals received the protection they sought merely by being present elsewhere, for example in a different religious area, while others received special assistance...These people were referred to as exiles (voluntary and involuntary) and suppliants, but these were not status categories...Princes welcome exiles, not because their status triggered a duty, but because it was just to do so unless there was some reason rendering it unjust to do so. Since the welcome granted was not triggered by the doctrinal structure of the exile, it was never doctrinally elaborated.'[29] One can see here traces of the gradual denudation of the interacting element in humanitarian affairs through the metamorphosis of humanitarian feelings into a system, and then into a doctrine.

The international notion of refugee protection came hand in hand with the notion of status. 'To be beyond legal status is to be nothing...the need for a legal status simply did not arise in a world which neither placed the sovereign at its centre nor distinguished law from politics and morality.'[30] The regularized system of international protection came in the wake of the development of a unified notion of unilateral sovereign capacity (of states and an international authority) to grant asylum. The positivist development of law resulted in making care and protection dependent on a set of jurisdictional boundaries instead of a set of notions about justice. The further consequence has been in form of a colossal loss of history and memory of a vast variety of experiences of care, protection and responsibility that included those of public care, disorganized care, community care, and sheltering by various political societies through decades of persecuted populations of many countries, simply because all these went on outside the system. Faith in an infinite legal process rather than closely investigating the unilateral nature of institutional actions for care of the victims has resulted in the loss of the dialogic element in the conduct of humanitarian arrangements for the victims, also a loss of that vision which sees the penumbra of human rights law developing into valid rights.

How are we to explain this paradox, where law, name of that master non-dialogic process that runs by command, results in loss of 'rights', instead of their validation? To study that deficit, it is clear by now that analysing the process whereby a system supersedes or overwhelms emotions is not enough. Or, probably, such an analysis will show how systems bring in their wake doctrines that subvert emotions. The doctrine of international care and relief draws its strength from other sources as well besides the emotions we have referred to, such as notions of sovereignty, test, harmonization, and law. Obviously, vast amount of care and hospitality, as we mentioned just now, is borne by the public, *privately*; as studies of refugee care and protection show that in the profile of protection, non-state protection is more than the amount of state protection; and refugees and immigrants get shelter more from private (clan, other affinitive ties, extended families, and what Charles Tilly calls *the transplanted networks*) sources.[31] Yet in order to bear a less share of burden, why does a system need a doctrine? Not only knowledge produces power, its availability disciplines emotions and thick feelings, on which humanity thrives. Thus studies of ethnicity, violence, law, humanitarian intervention, mass policing, safe zoning, asylum, repatriation, resettlement, exile status, strategies of conflict management, form the core of what we know as humanitarian management of the victims of forced displacement. Doctrine, system, emotion – this is the order of elements that helps the political technology of rule by turning the humanitarian into a non-dialogic affair. All the more hence, this order needs to be inspected.

People displaced by agencies/other people are, *prima facie*, wronged. If to be dialogic in humanitarian affairs is to acknowledge the existence of the victim as a (political) subject, to be responsible to the other, the question is: How shall we judge the moral obligation (of others, states, institutions, etc.) towards the wronged, the actual and potential victims? We have here not only the relevance of the ethic of care, kindness, and hospitality, but an added ethic, that of *responsibility*. Responsibility has two implications and conveys two senses – responsibility of the international institutions; also responsibility of the host and that of the expelling state.[32] Ethically then the issues are – what is morally owed to the victims of displacement? How do we place and apportion responsibility for man made disasters, say from a developmental process (a dam, a highway, a famine, a flood)? What is the politics of reparation in these cases? And what will be the norm of responsibility in the more complex cases of indirect displacement and repeated displacements? What will be the method? How shall we settle for what is known as *minimal justice*? These questions of ethics staring before us can be addressed by only a dialogic procedure and order, and certainly not through the hierarchical arrangement of doctrine, system, and emotion. While there is substantial agreement about preferred values, systems cause deficits in agreements, and doctrines render them absolute elements of a strategy.

We can see the ethical dilemmas inherent in current unilateral order of humanitarianism from another angle. What can be the principles guiding our choice if we are to move from the angle of marginality and are not to say that our choices are guided by considerations of marginal gain in improving the 'humanitarian situation'? Should we have an equal-sharing perspective, that is to say, a perspective of egalitarianism in distribution of costs (and benefits)? There are considerations of horizontal equity, compensation, and finally maximization of steps for the conditions of the worst off. In considerations of minimal justice and in judging the effectiveness of an institutional system for care, kindness, and hospitality, we cannot ignore these questions of ethics, for the notion of minimal justice hinges on precisely these questions. And these are issues that form the core of care, far removed from the world of international affairs. Sovereignty is a significant problematic in this, but only one. What can be the principle in the choice between national sovereignty and international intervention when we know that it is a closed game, and yet we are forced to opt for one of the two? What is preferable – expulsion of one community from Kosovo by a majority-centric state or mass bombing to stop that? If neither of the two is preferred, what can be proffered in stead? All these question suggest that the most *humanitarian* task has to become most *political* today in its judgement, namely, how can the humanitarian task re-validate its mandate by rejecting the non-dialogic mode of international actors, and the bureaucratic use of the humanitarian agenda by states and immigration officials all over the world? Therefore, humanitarianism must grapple with the paradox of law – law as morality and law as self-organization, in this case the morality of care that appears as the law of protection and care, and the law of organization, which gives form to this morality by transforming this morality into political reality. If humanitarianism, as seems inevitable, has to subject itself to law and some form of sovereignty, the issue is how can this subjection be minimized? Beyond sovereignty and law, the more urgent issue confronting us, therefore, is the ethical issue of responsibility and dialogic procedure. Without these, as we can see, hospitality is now subject to rules, procedures, laws, regimes, systems and doctrines. We are now witnessing the last act in an opera on the death of emotions.[33]

The Outcasts with Whom There Cannot Be Any Dialogue

Yet to speak of victims being denied of subject-hood through a complex arrangement of doctrines, systems, and feelings is not enough. The irony of the situation should be apparent. For, after all humanitarianism in modern world could arise only with the emergence of the category of the 'victim', consequently the appearance of the virtue of humanitarian feelings. Not only many virtues such as this are historically formed and subverted but that this history provides us with abundant tales of how the tune of empathy for the victim has been for long a convenient strain in diplomacy. After all, the single most important contribution of the 1951 convention on the refugees was that it made the refugee a victim. Knowingly or unknowingly it celebrated victim-hood, thereby celebrating the authority of the saviour. Before that we knew that, victims of state repression fled, and by heart the lives of Rousseau, Marx, Lenin, Rosa Luxembourg, Leon Trotsky, or at home M.N. Roy, Rashbehari, Subhas Bose, communities such as the *Gadar* – they were the refugees, they were exiles. Or, we can go back to Babur, the greatest refugee of his time, yet we thought of them rarely in terms of victims or only victims. Why? Or, yet again, we knew of areas like the region around Black Sea chronicled in the fascinating account by Neal Ascherson,[34] where fugitives arrived finding their way through astonishing times, where from Pushkin to various mutineers jostled with each other, where Greeks from the age of Herodotus, Romans, Russians, Poles, Scots, Hungarians, Tatars, Cossacks and Turks assembled to make what is today possibly the most variegated humanity concentrated in one area. Yet, we have never thought of Black Sea as the refugee-land on earth. Jostle over space was never an exceptional thing. It went on with a deep sense of cosmopolitanism. Space did not make a tribe, time made it. Today, space has taken over time, or at least has combined with time on an equal footing, so much so that population movements raise the spectre of squeezed space, conflictual space, rattled space, damned space, disturbed space, negotiable space, defeated space, and conquered space. The refugee is seen as a victim of territorialization of identity, who in turn contributes to further territorialization of identity.[35] The humanitarian logic is a closed circle, and it may appear that 'from victim-hood to agency' is the solution that a dialogic vision offers to us. Yet we must remain aware of the complexity of the problem. One may ask for instance, whose agency? If we consider the highly differential situation relating to refuge, the escape-formula will not be found as simple. In some cases as in Afghanistan, the UNHCR admits, refugees are neglected, while Kosovars are with lighting speed returned to their homeland;[36] in some other cases gangs of drug running refugees are shown as victims (one only has to read the history of Kuomintang refugees from modern China to have some idea of the duplicitous nature of the proclamation, *from victim-hood to agency*). Still, probably the most productive way will be to look at the refugees as active agents, as subjects with their plans, desires, dreams, conspiracies, piety, folly, intrigues, revenge, and spirit of reconciliation. Revolutionists, restorers, fatalists, – refugees are of all types. They are victims no doubt – victims of statecraft and globalization, but also victims of the discursive power of nationalism, human rights, and a most partisan kind of humanitarianism. But they are agents also, playing their bit.

One way to get out of the bind may be, therefore, to firmly anchor the study of the humanitarianism to the study of the conditions of political modernity, which means, first, to anchor it to the study of modern citizenship, second, the study of modern governmental practices of administering care, and third, looking at the ways modern political conditions offer us towards the restoration of dialogic procedures, for only dialogic procedures can step beyond the limits of modern politics towards making humanitarianism what it aims to be – to be beyond politics, beyond the unilateral acts of authority and guardianship. Political modernity has placed humanitarianism in a field that is marked by a focus of loyalty, requirement of duties, civil equality, institutionalization of political rights, legal rationalization of the distinction between foreigners and aliens, and direct relation between the state and the citizen as contrasted with the indirect relation existing in the ancient time. Dialogic politics implies in this context the interrogation of the distinction between citizen and the alien – a distinction that has been made so fundamental today,

Citizenship is like a *caste* – an inscriptive status, an ascribed quality following a line of descent, a membership of not only a state, but 'an imagined political community' much in the way Benedict Anderson has spoken of nations as imagined communities.[37] Christopher Pierson in a related discussion cites the examples of legal disputes on rights of inhabitants of erstwhile British colonies to show how citizenship has transformed from membership to status,[38] and only historical examinations will tell us, while earlier population movement did not create 'refugee-crisis', today why and how such movement produces crisis, with states having become almost without exception majoritarian polities. If citizenship were to be a sliding set of rights, beginning from right to vote to right of access to social welfare, the inhabitant can be a citizen without becoming a *maximum citizen*. But status makes the ownership of it a maximal concept, and the practices of citizenship today fly in the face of Article 13 of the Universal Declaration of Human Rights, that says that citizens should have the right to leave a particular jurisdiction, though the provision does not guarantee the corresponding and general right to take up residence elsewhere. The threat of becoming an alien or stateless is real and reinforces the position of membership of a state as lesser evil than becoming an alien. We need hardly any further comment on this diabolical situation that produces refugees and then denies their parentage. Again, we hardly require any comment on why the aliens, products of such situation, are left with no dialogic power to influence the care structure that political modernity has built for them. Indeed, the existence of a huge segment of world population denied of any dialogic role in politics, therefore of any presence, like a huge continental shelf sliding under the water, can be evinced only from the elaborate arrangement of institutional practices aiming to contain the immigration flow. In this respect we need to read also into the evolution of the customary international law.[39] The point to note here is that the 1951 Convention on refugees while being a landmark in humanitarian protection of victims of violence, made the virtue of care a part of governmental and authoritarian politics, in which the victims would have no role to say, precisely because the Convention did not any way question the prevailing politics of citizenship, statehood, and authority; it was not in any way marked by international democratic obligation, or any imperial shame.

The link between politics of citizenship and the humanitarian protection of the alien is clear from recent European experiences, which are well documented to lend us an idea of the double reading. The pressure of immigrants on the closed door of the affluent home was not a sudden development. The French political philosopher, Etienne Balibar, in a remarkable essay, 'France Algeria – One Nation or Two?' has drawn attention to the intertwined fate of Europe with erstwhile colonies even in terms of spiritual destiny.[40] He has also drawn our attention to the paradox in the emerging situation where to be a citizen of Europe and gain access to the common homeland, one does not need a nationality, but then what does one need? White skin, or residency in a constituent state of Europe, or being just a resident of Europe? Both within the common borders raised high and outside, where thousands stand before the fort gates to open for them after a long wait, the irony is unfolding. From Maastricht to Amsterdam is a journey that witnesses on one hand the sovereignty transfer from intergovernmental decision-making process to the community, on the other hand a growing realization that setting the house within is difficult, for the without is there within not as separate category, but very much inside, inside the frame. Thus, because the non-European is there as European for a long time, the new borderline between the European and the non-European is unstable.[41] The instability plays havoc with policies of common risks, border controls, coherence of immigration policies, and synchronization of asylum practices. The un-harmonized nationality laws, Catherine Withol de Wenden tells us, remain as an evidence of a complicated scenario marked by the erasure of boundaries between the inside/outside. She also tells us that the phenomena of trans-national networks, minorities, Diaspora, globalization of mobility, weakening of legal barriers in face of flows, the Europeanization of decisions, and the sensitivity of 'national public opinions' to immigration, show how difficult it is to humanize the issue. In fact, the political and the humanitarian are defining each other today.[42] The classical attribute in Jean Bodin's definition of national citizenship as a mutual contract involving obligation gives way to a world where the historical realities of the over-determination of inside/outside assert themselves to the extent of pushing aside the canonical idea. In any case, citizens embodying a republican spirit, forming a nation whose self lies in the citizenry, whereby nationality and citizenship become closely related notions, was even then a partial reality. And now, with renewed interrogation of nationality and citizenship laws, the continuous flow of population across continents thereby influencing the evolution of the continental shelf is being acknowledged, albeit grudgingly, and often in an indirect way. The presence of the alien speaks of double deficit – a *democratic deficit*, as we cannot conceive of democracy without citizenship, and of citizenship without the aliens who have no democratic voice and can only expect charity, and a *humanitarian deficit*, as we cannot have a system of humanitarian care without a non-humanitarian structure of rule, procedure, law, and arrangement to designate, determine, fix standards for, assign, and deny care and refuge.

Who is then the alien with whom you cannot talk, who cannot be your dialogic partner in your society? Is he/she the disfranchised, or the refugee, or the illegal immigrant, or the *gasterbeiter* (the guest-worker), or the frontier-worker, or the non-state person?[43] Or, the ones who have been there with us through the times, who were all along with us as the shadow so much familiar that we did not notice, yet on whom we cannot count for loyalty in time of crisis, who are the unfamiliar faces, and therefore at those times we realize that *they* and *we* belong to two different races? Who are the aliens – these minority groups stranded in time? All these are to be sure; and as aliens they are the soul mates of political modernity, they are the alien soul mates. The logic of political familiarity on which democracy as a concept and as a system thrives produces alien-hood. And then democracy has to wrestle with that. Indeed, it cannot do without. One can go through the recent official texts of migration management in Europe, even a glance will tell how immigration flow is made governable by a widening of the concept and practices, by acknowledging the inadequacy of hitherto prevalent ideas. In 1959 the Council of Europe had an agreement on the abolition of visas for refugees within the territory of the contracting states. In 1961 the provision in the European Convention on Human Rights relating to the right to asylum was strengthened. In 1967, the Committee of Ministers in the Council took the decision to act in a 'liberal and humanitarian spirit' (Article 1) in treating persons seeking asylum. The Council further declared after ten years in 1977 that, 'grant of territorial asylum is a peaceful and humanitarian act' (Article 3). The European Conventions on Social Security were to be applicable to refugees and stateless persons also. The same declaration (1972) emphasized the protection of 'frontier workers', whom it defined as 'an employed person who is employed in the territory of one Contracting Party and resides in the territory of another Contracting Party where he returns in principle every day or at least once a week' (Article 1, n). Meanwhile the convention on extradition (1957) synchronized the rules of extradition existing in the contracting states. The legal status of migrant workers was clarified in a resolution of 1977. They were not to be discriminated against in conditions of work, forms of recruitment, health and medical services, travel and other related issues. Noticeably, it could not go into the structure of job market that would have shown the limits of this principle of non-discrimination. In two major decisions, the European Agreement on Transfer of Responsibility for Refugees adopted in 1980, that decided to stick to the definition of 1951 and 1967 (Article 1), and the Convention on the Suppression of Terrorism (1977) laid down the basic contours of the shelter policy.

Realizing that all these were limited in the face of the massive presence of discriminatory practices, an Additional Protocol to the European Social Charter of 1961 was passed in 1988. Meanwhile, the Parliamentary Assembly of the Council of Europe, concerned with fresh arrivals of entry-seekers in the continent, went on demanding more coherence and synchronization of asylum practices (Resolution 1088/1988). As the Arab-Israel conflict flared up in the second half of the nineteen eighties, the assembly was again concerned, and instructed the member countries to 'continue to coordinate their Middle East policies to encourage a comprehensive, just and lasting solution to the conflict, and a just resolution of the refugee problem; and to cooperate on a programme of international aid for economies of the affected areas, on the basis that economic stability is a foundation of peaceful coexistence between different people'.

Here we have the basic outlines a liberal asylum policy (after having been instrumental in creating the Palestinian refugees) – give money so that they do not fight and give some more to the vulnerable groups so that they do not have to leave because of the fight. If they arrive, give them shelter, and meanwhile keep on seeking solutions. Here we also find the strong influence of economic thinking, which tells us that it is primarily economic considerations, at a cruder level, financial consideration that leads a person or a group of persons to leave or to stay back. Then came the Schengen, Amsterdam and Dublin agreements that freed the inside from a host of controls while raised the wall higher against the outside. The CSCE policies anchored the aspects of security, human rights and asylum in one comprehensive approach so effectively that the General Assembly noted it in its resolution (47/147) of 1992. These resolutions coupled with ideas and measures on internal displacements, as one can see, present before us a picture of progression of ideas and steps, the mark of a passage from the antiquity of 1951 to the time of today, when refugees, immigrant workers, guest residents, racially abused minorities, non-state persons and the illegal immigrants present a combined picture of the alien.[44] The appearance of the alien is a testimony to continuity and break both, the passage of a category essentially symbolizing the outside to one that is now there, within.

Or, we can take the experiences of Bengal. The ways refugees were hosted here, cared for, and managed as a problem, throw light on notions of citizenship in a post-colonial polity. They show how citizenship is mixed up with ethnic assertions. They show how the idea of 'insecurity', at once definable and amorphous, continues to haunt the political universe. In the emerging public sphere of post-colonial Bengal, refugees played a role that defined other roles. And, now with the issue of partition-refugees being on the whole solved (as a political problem), the question is how did the refugee issue in West Bengal vanish, what was the reason – the ground of the erasure? Can we ascribe it to the best practices, or to the tolerance of the host society, or the de-centring of the refugee self, or, the conquest of the civil sphere by the refugees, whereupon there remained no more the special need for the persistence of the refugee issue, or what we can describe for want of better words, to the dialogic politics that indicates all these? As the bunch of extra-ordinary narratives in *Refugees in West Bengal – Institutional Practices and Contested Identities* suggests,[45] the issue survived by submerging itself in the host society, in the process changing the latter politically and culturally, permanently and quite substantively, so that Bengal has now a grey political society, something akin to the profile that emerges in *The Marginal Nation*. On one hand we can say that, population was being regulated top down, population became a subject of information, and it produced changes in administrative policies of managing the population, in turn changing the nature of the population – both host and immigrant – sometimes durably, on the other hand, as these papers show, the negotiation of the new world, to which the refugees had arrived, went ahead with dialogue between the immigrants and the host, refugees and the government, between notions of charity and rights, all in all a paradoxical combination. On one hand, agony and victim-hood were the sign-words of the negotiation; on the other hand there was tremendous activism that rested on the discourse of the victim.[46] This was no doubt, we must remember, a contentious politics, but a politics that allowed conversations and negotiations. The lesson is that once we start looking at the refugee as a political actor, and not just a victim, we shall have to admit that there remains the pressing need for dialogue within the humanitarian world on issues of his/her right to go back, right of the stateless to claim a state, or the right of the expelled to reconcile on an equal footing. In short we shall have to take note of what Edward Said calls in the Palestinian context *the politics of dispossession.*

The Palestinian experience is relevant to the South Asian case for two reasons. First, the administration of Palestinian refugee care and rehabilitation was never brought under general supervision similar to the situation in South Asia where there was no general supervision by and no general responsibility of the international community on the care and resettlement of hundreds and thousands of refugees in the time of partition and thereafter.[47] Second, as Edward Said notes, the two politics, of the return of the expelled and that of the occupied areas in Palestine-Israel, were always kept disconnected.[48] (The reason might have been to avoid scrutiny and keep the matter a part of the politics of settlement on Palestine). Said notes in this context that the occupation policy of successive Israeli governments in colonizing the occupying areas was not prevented by the international community, and was made *a fait accompli*. It ensured the fact that in the event of a future settlement, as is proved today, the refugee would ever remain the permanent subject, the vanquished, the refugee never to reclaim 'the lost sky'. This situation is exactly similar to the phenomenon of dispossession in South Asia. Reports on colonization of land of the South Bhutanese émigrés, the returnees in the Chittagong Hill Tracts, and the escapees of violence in North and east of Sri Lanka are evidences of the same phenomenon.[49] A permanent victim needs a permanent care administration, and it is better that way when the strategy is to deny the political autonomy of the dispossessed. Of course, as Said shows, victimization was of enormous magnitude. Over 55 per cent of the occupied area in Israel were earmarked for expropriation, 25 per cent of Greater Jerusalem was annexed, 30 per cent of Gaza strip was settled, and 80 per cent of water of this area was for the colonizing settlements and Israel proper. Exodus continued, so continued resistance from the people waiting to be driven out, in fact a huge number – about fifty percent of total Palestinian population – and from people driven out, about 4,00,000 Palestinians in Lebanon alone. To create a firmer security ring, Israel drove the refugees out of a greater part of Lebanon also, mass murdered refugees in camps, and created the black year in the political history of the dispossessed, the year of 1982. The Arab resistance collapsed in face of concerted western military and political might, under the deadweight of corruption and oligarchy, and the let down by an inappropriate leadership of a stateless people, who could float, organize raids, mount heroic resistance, but was inadequate for reclaiming political history for themselves on their own. The archetypal victim would be thus considered fit for few dollars to stay in the squalid camps, and the expelling state would get an annual help of 5 billion dollars to keep the refugees from returning. Resistance continued, so went along with that power play with the *victim*.[50]

Women and the Humanitarian Deficit

Nowhere the humanitarian deficit is more apparent than in the way women feature as refugee-subjects in refugee management. The critique of citizenship as a status marked by universality and equality flows above all from the fact the structure of the liberal-humanitarian is characteristically patriarchal. This implies that the humanitarian system by itself dominates women, or is constituted by forms of rule in which men will systemically dominate women. The humanitarian law is almost a male agency. The way a refugee is defined, or the way in which the Geneva Conventions define humanitarian protection in war, shows that such laws see and treat women the way condescending men see and treat women. It is the male gaze that frames the humanitarian. The non-gender specific nature of fear, persecution and threat as outlined in *The UNHCR Handbook on Procedures and Criteria on Determining Refugee Status* does not therefore think deeply on the *fear*, the fear of violence, fear of state, of law, assault, and breakdown of security and life around, faced by more than half of the refugees on earth.[51] The gender-blind concept and application of humanitarian law can go to harsh extent. The official guideline of a country for the asylum officers adjudicating asylum claims by women says,

> Women make up a large percentage of the world's refugees. In order to qualify as a refugee under our laws, female applicants must – like any applicant – show that they cannot return home and cannot avail themselves of the protection of their country because of 'persecution or a well-founded fear of persecution on account of race, religion, nationality, membership in a particular social group, or political opinion'...Often of course, the asylum claim of a female applicant will have nothing to do with her gender. In other cases, though, the applicant's gender may bear on the claim in significant ways to which the adjudicator should be attentive...Such claims must be analyzed within the terms of United States law, but gender related claims can raise issues of particular complexity, and it is important that United States asylum adjudicators understand those complexities and give proper consideration to gender-related claims...As in all asylum cases, the asylum officer must assess whether the harm that the applicant fears or has suffered is serious enough to be regarded as 'persecution' as that term is understood under the relevant international and domestic law...The forms of harm that women suffer around the world, and that therefore will arise in asylum claims, are varied...have included sexual abuse, rape, infanticide, genital mutilation, forced marriage, slavery, domestic violence, and forced abortion. The form of harm or punishment may be selected because of the gender of the victim, but the analysis of claim should not vary based on the gender of the victim. Asylum

> adjudicators should assess whether an instance of harm amounts to persecution on the basis of the general principles...[52]

This is a classic case of procedural neutrality. It recognizes that there may be a specific nature of abuse suffered by a 'social group' (it is primarily this route by which vast number of abuses are identified), but upholds general principles as the basis of *adjudication of claims*. Hence the humanitarian law may be procedurally neutral, as law under patriarchy is often so; but this procedural neutrality on which the humanitarian law is based bases itself on a prior construction of a liberal-humanitarian regime where male power is entrenched. Thus rape, violence, mass murder, eviction, girl-sale, illegal underpaid female labour, sex-slavery, pity – all these flourish because the humanitarian only aims at protecting the population facing abuse, turning them into victims whose appeals (in law *claims*) have to be adjudicated, while the very gender-structure of abuse continues.

Paula Banerjee in a detailed study on the patterns of institutional response to the conditions of refugee women in South Asia (response of state, relief agencies, UNHCR, and other international humanitarian missions) shows how the policy of conferring victim-hood on women by the humanitarian regime is a male game in the international politics of today.[53] This game, she reminds us, reflect the 'fissures', which Etienne Balibar has argued, are in the 'modern political community' – fissures that emerge from 'practical and ideological sexism as a structure of interior exclusion of women generalized to the whole society...(and leading) to the 'universalisation of sexual difference.'[54] Through case studies of population flows she shows how state formation in the de-colonized era, population movements, and gender-subjection have been closely connected; and how the humanitarian regime built around such states and gender differences develop a precarious relation with women. The nation building projects in South Asia have led to the creation of a homogenized identity of citizenship, denial of spaces of difference, and the segregation of minorities on the basis of caste, religion and gender from the collective. In such denials, she shows, women lose most their individualities and abilities to make choice. As political non-subjects refugee women emerge as the symbol of difference between us/citizens and its other/refugees/non-citizens. And what the humanitarian regime does not recognize, whom it has turned into victims, begin as dislocated subjects negotiating spaces, retrieving agency in the face of institutional apathy. Her studies of women as dislocated subjects deal with institutional policies, responses and control with regard to abducted women in the Indian partition, displaced women from Sri Lanka in India and in Sri Lanka, statelessness of displaced women from Burma, and Afghan women refugees in Pakistan. In all these cases, she shows, how the institutional structure of humanitarian regime is built on denials of experiences of women as dislocated subjects and their political agency. She cites the example of Angela King's mission to Peshawar and Islamabad. When Afghan women requested the UN through Ms. King that they would like to mobilize educated Afghan women in peace-making, Ms. King reportedly asked them to apply for UN jobs instead. After the meeting the women felt 'confused, insulted, hurt, angry and substantially ignored.' But they noted bitterly 'this is not an unusual situation – neither within our societies, nor within the UN agencies'.[55] This is where we can draw on feminist studies of management of care lessons for our consideration towards a dialogic politics.

First, taken collectively these cases reflect on specific characteristics of loss of women's political subject-hood due to dislocation; the overwhelming presence of women among the refugee populations is not an accident of history, it shows the way by which states have made women political non-subjects with whom they have nothing to dialogue, nothing to reckon with.

Second, it should be recognized that even in their marginality refugee women are never merely victims, and they exert agency even in situations of marginality. Dislocation is a debilitating experience no doubt, but women have many times transformed it into an empowering one.

Third, and this seems to me significant in the context of our discussion of method, a system-centric narrative on the situation of victim women trivializes the situation, and shifts our attention from the fact that these unilateral and paternalist practices have been by themselves gender-specific to a strange extent where these 'humanitarian' policies and practices consider death or threat to liberty a reason for asylum, but not rape or vulnerability to human trafficking a sufficient reason for care and protection. The only way to turn our back on such system-centric discussion is to retrieve and firmly focus on women's experiences from such marginalities, recognize individual voices of refugee women in any narrative of displacement, for in Paula Banerjee's words, 'only by retrieving refugee women's own voices and not dismissing their individual experiences as anecdotal can we centre the marginal'.[56]

Experiences have been central to a dialogic situation. The implications of an experiential argument for restoring dialogue at the heart of humanitarian politics, which can become humanitarian only by becoming gender-sensitive, are to say the least immense.

'Truth Games'

Given then all that have been said above – the way humanitarian affairs have been turned into pure governmental politics, the way in which power, fear, and ethics have combined as elements of a humanitarian structure, the history of an international humanitarian agency such as the UNHCR, the humanitarian convention such as the 1951 convention, and the humanitarian practice of giving asylum, the appearance of the aliens as the political non-subjects, and the anti-women nature of the humanitarianism that the victims face daily – given all these facts, the non-dialogic nature of the humanitarian world speaks not only of the deficit in humanitarianism, but the nature of the 'truth games' being played out in the form of care, protection, and humanitarian conduct in this crooked world.

That is where our dialogic critique must concentrate, the games that humanitarian politics builds around the duality of citizenship and alien-hood. An alien is never permanently alien. In *Black Sea* we have the picture of a strange world where aliens are becoming familiar animals, and then they become indigenous, branding others as aliens, whereas sometimes these others who were indigenous become unfamiliar eventually to become aliens. In the experiences of Bengal that are before us, the refugees arrived as aliens, aliens in a land, aliens in families which were often their own, aliens in language (the eastern diction was always a matter of derision on the western side of the river Padma-Ganga), aliens as citizens. Victim-hood was the signature of the alienation. And yet, we have to only appreciate the irrevocable reality of partition (that is to say as it appeared in course of time to the host and the alien as irremediable, irrevocable), in which victims became active agents. As victims they became the agents. This was not certainly a straight process; it was another instance of contentious history. From forcible occupation, eviction, export of immigrant labour to faraway lands, forests and seas, organization, street politics, self-help amidst odds, community settlement, community help, and the evolution of governing policies and the governing *gaze*, this was a contentious process. It made West Bengal governable, but it also rescued the refugees from the destiny of a permanent victim. Joya Chatterjee in two consecutive fascinating essays shows the pre-history of this passage. In one, she shows how the Radcliffe line was carved out, *thana* (area of jurisdiction of a police station) by *thana*, producing the victims, she also shows how this process had the early signs of the passage.[57] In the second essay, she outlines the development of state policies of controlling the immigrant population that made the population governable, but also occasioned the development of the proud identity of the immigrant on the basis of the ideas of justice and rights, that helped him/her to survive, shake off alien-hood, and leave the stamp of the subject on the politics of West Bengal.[58] We can note here the passage from a unilateral order of humanitarian conduct by the government to a dialogic situation. Responsibility and reconciliation heralded the passage. If piety, virtue, trust, care, and compassion played a part in the process of reconciliation, hard politics was also significant. The 'truth games' were being played out with uncanny felicity. Morals had their place, but were employed strategically. Trust, which effects reconciliation and which reconciling reinforces, came as a product of morals and power(s) lodged in the conflict patterns of a society. Reconciliation was a political act that evolved in a contentious process lasting for almost half century.[59]

Therefore as against a seamless picture of humanitarianism suggested to us by international conventions and declarations, we have as the Bengal experience suggests, a scenario of a contentious process of dialogue against a background of redrawing of geo-political and ethno-political boundaries, and a scenario of new strategies of reconciliation. Those studying the politics of dialogue would do well in learning from critical geographers working on reappearance and re-negotiation of boundaries, in understanding how dialogic situation emerges in one context and evaporates in another.[60]

Studies of humanitarianism reconnect us with long-lost experiences whose few traces only remain to evoke them. Strangely, time passes sometimes quickly, for it is just half a century of de-colonized South Asia, but things have changed so much at least at certain levels. The traces that remain have to be followed back to reach the experiences re-animated before us in the available accounts of forced displacements and humanitarian protection of the victims. Despite all cries of agony and victim-hood, the recovered experiences tell of challenge, contest, and a delightfully ambiguous activism which narrate how refugees become subjects that designate relations, negotiation, re-negotiation, also a subject that is involved in the simultaneous pursuit of trust and security – trust in security and security in trust. Through the lens of these experiences, the history of humanitarianism actually conveys two stories – one, how aliens become governable, and two, how there is a transition from victim-hood to the subject. These two sets of stories can be found anywhere in the world, certainly in many post-colonial regions; they are actually overlapping, making up one body of chronicle. If the subject of humanitarianism who is a non-subject in politics is a sign of anything, it is the sign of a profile of superimpositions of many experiences in which the political non-subject breathes, lives and loses his/her (non) identity.

All these experiences play with the division of victims and non-victims, the inside and outside, and proffer new content signified by the word alien. The word is open for exploitation, for it cannot empty itself so easily. Giving itself to eternal mediation it always recomposes. This is important, for it lends to contestation and puts the seal of sovereignty on experiences that we have been talking about in this chapter. The Indian Foreigners Act, 1946 controls the movements of aliens, though expectedly it does not define an alien. But more important, basic rights had be given to non-citizens also, such as Articles 14, 21, 25 – rights to equality, life and religion. Probably other basic provisions, such as Articles 3, 4, 7, 16, 17, 27, 28 also apply to aliens. The courts, free from the obligation of defining a refugee, extended protection on various occasions and in several ways in matters of physical security, *non-refoulement*, and certain basic amenities (food, medical care, other humanitarian provisions). These case laws[61] show the strategic presence of the category of *alien* that eludes definition but invokes humanitarianism. In the mirror of the image of alien, we can see the limits of the tools with which we had been carrying on practices of humanitarianism. The need to define who requires assistance begets the definition of the victim. The need for definition arises, with that the need for specifications. In the present instance of humanitarianism we are studying, that is protection of the victims of forced displacement, the scandal over determination procedures, particularly with having to satisfy the criterion of innocence as far as criminal acts are concerned, the de facto onus on the victim of torture, seeking shelter, to prove torture, and the various ingenuous methods of quarantine at the airports, seaports, railway stations – all these show what happens when you trust a definition, a beginning, a source, an origin. If we had the time to do a kind of counter-ethnographic exercise, we could have shown in letters the un-narrated experiences of the quarantined of continents. Therefore, the problem is not so much of determining whether the humanitarian is 'truly humanitarian', but how certain feelings can become 'humanitarian' only by becoming resorting to certain methods. As the philosophers say, the problem is not about what is true, but truth itself.[62] Basically this means reflecting on a crucial phenomenon of our time, namely, the strange dialogue between humanitarianism, geopolitics and law.

The marks of the present day humanitarian order, which we have discussed in these preceding pages, are actually signs of the compulsive legacy of geopolitics to which we shall be returning one more time by way of closing this book. But what is more significant here is that these are also signs that create a loss of meanings, in this case the meaning of humanitarianism. The tumultuous events of de-colonization, new state formation, revolt against racism, mass of humanity seeking better life, neglected populations leaving country after country as mark of protest, 'voting with their feet' and rejecting the 'calculated kindness'[63] of states, donors, great powers, and international organizations – all these were signed into the corpus by the term *victim*. The term *victim* has been successful in emptying meanings related to democracy, popular rights, and federality of politics, which had gone in the beginning into the making of the sign. Now we can see, how signs do not operate singly, they function in groups. Refugee, victim-hood, ethnicity, third world, and similar other signs make a group in which the humanitarian policies operate. These are the signs that rule the 'humanitarian world' today having emptying themselves of their meanings. Courageous human rights activists and some scholars, sometimes make effort to dislodge these signs. But these efforts marked by indecision about the status of humanitarianism in politics are like, as Roland Barthes said in *Empire of Signs*, 'trying to destroy the wolf by lodging comfortably in its gullet'.[64] In this aberrant exercise, the mystique of the humanitarian is being slowly displaced, but very slowly. We frankly need big logs from outside to displace the mystique that remains as so only by being at the centre of the corpus – and that it does by becoming a sign, which has emptied all meanings and has become an empire unto itself.

Notes

1 Letter of T. Muivah to the UNHCR against the latter's decision (UNHCR's letter, 29 August 2000) to reject the appeal for grant of refugee status to him, 27 September 2000.

2 The application of the 1951 Convention is subject to sections D and F of Article 1. They list categories of persons undeserving of protection. Clause F excludes those committing crimes against peace and security, serious common law breaches, and acting contravention of purposes and principles of the United Nations.

3 Appeal letter issued by the South Asia Forum for Human Rights to the UNHCR and others, 4 February 2000, available on www.safhr.org.

4 Facts cited in the dossier, *Naga Resistance and the Peace Process*, published and circulated by Other Media Communications, Delhi, 2001.

5 Letter of Muivah, 27 September 2000.

6 Ibid.

7 Muivah attached thirty-three documents as annexure in support of his legal-political stand of just war, just method, and need for peace. The list of the documents is by itself significant because it throws light on the ground that Muivah occupied in his engagement with the UNHCR. Notable among these (titles) were – Statement of prominent intellectuals and bureaucrats in 1984 on virtual army rule in Naga inhabited areas, Amnesty International Report on Operation Blue Bird, NPMHR goes to UN (Proceeding of the Human Rights Committee), March 1991, Appeal by professors and journalists asking for my release, Three Ex-Prime Ministers demand Muivah's release, Steering Committee (UNPO) Statement concerning Nagaland, Letter to the Foreign Minister Thailand from the Society for Threatened Peoples, Chronologies on the current peace process between the Government of India and the National Socialist Council of Nagaland (NSCN) vis-à-vis the Indo-Naga issue, Statement on behalf of the Society for Threatened Peoples by Mr. Isak Chishi Swu at the UN Commission on Human Rights Sub-Commission on Prevention of Discrimination and Protection of Minorities, Speech delivered by Isak Chishi Swu Chairman National Socialist Council of Nagaland at the UN Commission on Human Rights, Declaration on Nagaland by Support Group for Indigenous Peoples, Belgium, Statement of the NSCN to the Fifteenth Session of the Working Group on Indigenous Populations, Economic and Social Council, The human rights situation in Nagaland, Statement of Isak Chishi Swu, Chairman, NSCN, given at the Sub-Commission's Forty-seventh Session on behalf of the Society for Threatened peoples, PM-Muivah Davos meet sets stage for Naga talks, Letter to the Thai Prime Minister from International non-Governmental Organizations, Letter to the Thai Prime Minister by the Chairman, Congress of the United States Committee on International Relations, and the Statement by the NSCN Collective Leadership on the current Peace Process.

8 Patricia Tuitt has commented in *False Images – Law's Construction of the Refugee* (London: Pluto Press, 1996),

> The well-founded fear concept has evolved from a relatively simple inquiry within which the refugee's subjective feelings of 'terror' were prominent, to a much more complex and wide-ranging inquiry within which concepts such as the 'safe state' will increasingly be sole determinants of the issue of the well-founded fear.
>
> It is argued here that refugee claimant is becoming increasingly disenfranchised from the decision-making process relating to this legal inquiry...(This chapter) examines how the response to these challenges has portrayed the refugee as being a fundamentally unreliable base point of inquiry such as to justify the shift to a generalized or group-based evaluation of a well-founded fear (p. 80).

Daniel Warner has put the problem in a manner reminding us of Nietzsche, 'What would happen if the refugee interviewed the government official? Would the official be able to comprehend that many of his or her deepest longings were the same as those of the refugees, but without the physical dislocation? Would UNHCR field officers understand that their own doubts were the same as (those of) the refugees? Perhaps the government official and the UNHCR field officer do understand, and that is why they have chosen their professions. – 'We are all Refugees', *International Journal of Refugee Law*, 4 (3), 1992, p. 372.

9 James C. Hathaway, *The Law of the Refugee Status* (Toronto: Butterworths, 1991), p. 65.

10 Ibid. p. 74.

11 The best guide to an understanding of that economy is still the collection of legal texts, *Collection of International Instruments and Other Legal Texts Concerning Refugees and Displaced Persons* (Geneva: UNHCR, 1995) and the annual surveys of the UNHCR; also see www.unhcr.ch. For an able defence of the continuing relevance of the Convention of 1951, see the submission by Ruud Lubbers, 'Asylum for All', *Harvard International Review*, Volume 24, Spring 2002, pp. 20-24.

12 Jean Francois Lyotard, *The Differend – Phrases in Dispute* (Manchester: Manchester University Press, 1978), p. 5.

13 Costas Douzinas and Ronnie Warrington, 'A Well-founded Fear of Justice – Law and Ethics in Postmodernity' in Jerry Leonard (ed.), *A Reader in (Post) Modern Critical Theory* (New York: State University of New York Press, 1995), now included in B.S. Chimni (ed.), *International Refugee Law – A Reader* (Delhi: Sage, 2000), pp. 21-27 (hereafter *ILR – A Reader*). Douzinas and Warrington refers to Lyotard (n. 12) in writing on the un-translatability (impossibility) of justice in the language of law.

14 James C. Hathaway, *The Law of Refugee Status*, p. 117.

15 'Fourteen Dead on the Border – For What?' a report by David Bacon, Oakland, California (6/15/01) on: portsideMod@netscape.net, 20 Jun 2001. I am indebted to Paul Joseph of the Tufts University for bringing the news to my attention.

16 *Refugee Watch* addresses in many entries the mythical nature of this. See for instance, Didier Bertrand, 'The Forced Repatriation of Tamil Asylum Seekers from Europe', *Refugee Watch*, 12, December 2000.

17 The widening of the issue of forced migration is best understood in a study of the discourse on the IDPs (internally displaced persons). Though to the regions de-colonized through partition the issue of internal displacement was always as important as the issue of refugees (as defined by the 1951 Convention), it took forty years for the guardians of international community to realize its significance. Then too in their reports, images of violence causing internal displacement dominated over those of development-induced internal displacement, because the knowledge had emanated primarily from experiences of global management of population flow in Africa and Latin America. For a comprehensive study on this, see the two volume study by Roberta Cohen and Francis M. Deng, *Masses in Flight – The Global Crisis of Internal Displacement* and *The Forsaken People – Case Studies of the Internally Displaced* (Washington, D.C.: Brookings Institution Press, 1998); Walter Kalin, 'Guiding Principles on Internal Displacement – Annotations', Studies in Transnational Legal Policy, 32 (Washington, D.C.: The American Society of International law and the Brookings Institution, 2000); *Report of the Representative of the Secretary General Mr. Francis M. Deng, submitted pursuant to the Commission on Human Rights Resolution 1999/47 – Internally Displaced Persons*, Economic and Social Council, E/CN 4/2000/83, 26 January 2000; *Report of the International Colloquy on the Guiding Principles on Internal Displacement* (Washington, D.C.: The Brookings Institution

Project on Internal Displacement, 2000); 'Handbook for Applying the Guiding Principles on Internal Displacement', Inter-agency Standing Policy Paper Series 1, UN, 1999; and 'Handbook for Applying the Guiding Principles on Internal Displacement', Manual of the Brookings Institution Project on Internal Displacement, 1999.

The attention of the UN system on the inadequacy of protection regime relating to the displaced persons began with the beginning of the last decade of the last century. For example, the UN Commission on Human Rights adopted by Resolution 1992/73, which took note of the analytical report of the Secretary general on internally displaced persons and next year by Resolution 1993/95 suggested several measures. The UN General Assembly adopted a resolution in 1990 on assistance to the internally displaced persons in Africa (45/137) and in 1991 a similar resolution relating to those in Central America (46/107).

18 Robin Clapp and Tihana Majcen, 'Deportation of Rejected Asylum Seekers from the European Union – Avoidable Deaths?', *Human Rights Tribune* 7 (4), December 2000, pp. 30-32.

19 Hugh Tinker, *A New System of Slavery – The Export of Indian Labour Overseas, 1830-1920* (Oxford: Oxford University Press, 1974).

20 Eric Wolf, *Europe and The People without History* (Berkeley, California: UCLA, 1982).

21 Geoffrey Care, 'Transnational Refugee Recognitions – A Judicial Perspective', unpublished paper, presented at the International Conference on Forced Migration, Displacement and Conflict Resolution, University of Jadavpur, Calcutta, 20-22 April 2000.

22 See in this context the reviews of the some of the practices by countries that are ahead of others in respecting refugee rights, published in *Refuge*, special number on 'New Directions for Refugee Policy in an Era of Globalisation, 19 (4), February 2001, particularly the following essays and fact-sheets – Janet Dench, 'Controlling the Borders – C-31 and Interdiction' (pp. 34-40); David Matas, 'Refugee Determination Complexity' (pp. 48-54), and Robert F. Barsky, 'An Essay on the Free Movement of Peoples' (pp. 84-93).

23 Ibid.

24 Daniel Warner, 'The Refugee State and State Protection' in Frances Nicholson and Patrick Twomey (eds.), *Refugee Rights and Realities – Evolving International Concepts and Regimes* (Cambridge: Cambridge University Press, 1999), p. 265.

25 It is significant to note how frequently complaints have arisen that the refugees who are the escapees of violence are engaged in promoting violence or are being used to promote violence, and that the international community is turning a blind eye by being simply discriminate on this. The representative of Afghanistan in 1979 in discussion on the High Commissioner's report in the Executive Committee referred to the assistance of the UNHCR to the fugitive insurgents in Pakistan and observed that assistance to those committing acts of aggression against Afghanistan contravened the UNHCR Statute, the 1951 Convention and the UN Charter. The Committee rejected the contention. Similar allegations have come from Bulgaria, Morocco, Iraq and other countries. See for example, UN Docs A/AC. 96/SR.371, paragraph 107, SR 372, paragraph 60; SR 373, paragraphs 34, 78.

26 *Generosity* is a notion that cannot be defined by law. Thus when the system of international protection was in its very early stage of being built up, during the discussions leading to the final adoption in 1946 of the constitution of the International Refugee Organization and the debates on international obligation for resettlement/repatriation, there was one key word missing – *generosity*. The two sides of

the forthcoming cold war argued as to if it should be repatriation or resettlement, oblivious of the fact that on both counts, the states needed to be generous. This was one of the early examples of systemic hierarchy. See in this connection, John George Stoessinger, *The Refugee and the World Community* (Minneapolis: The University of Minnesota Press, 1963), pp. 61-66, cited in *International Refugee Law – A Reader*, pp. 337-340.

27 Tim Strangleman speaks of these in his study of railway industry, 'The Nostalgia of Organizations and Organization of Nostalgia – Past and Present in the Contemporary Railway Industry', *Sociology*, 33 (4), November 1999, pp. 725-746.

28 Related to this, see Victoria Kahn, 'The Duty to Love – Passion and Objectivity in Early Modern Political Theory', *Representations*, 68, Fall, 1999, pp. 84-107.

29 David Kennedy, 'International Refugee Protection', *Human Rights Quarterly*, 8 (1), pp. 42-43.

30 Ibid.

31 The traces of group care (care of groups and care by groups) as the basic pattern of protection of refugees are still there in the way states in South Asia respond to inflows of refugees and other kinds of forced migrants. Thus as an Indian scholar of refugee law, Veerabhadran Vijayakumar, has noted in 'A Critical Analysis of Refugee Protection in South Asia', *Refuge – Canada's Periodical on Refugees*, 19 (2), December 2000, a difference between what he terms as 'Northern approach' and 'Southern approach'. By northern approach he implies individual determination, emphasis on law and procedure, host of administrative and quasi-judicial mechanisms, and attempts to adhere to 1951 Convention and the 1967 Protocol. By southern approach he indicates the prevalence of heavy reliance on bilateral method to resolve refugee crisis (India-Sri Lanka Accord being one such example), group determination, community care and protection, and an almost deliberate reluctance to legally stabilize a related situation. (pp. 14-15) Also see *The Marginal Nation* for an account of how responsibility is at an essential level a community norm.

32 Chaloka Beyani, 'State Responsibility for the Prevention and Resolution of Forced Population Displacements in International Law', *International Journal of Refugee Law*, special issue, 1995, pp. 131-137, reprinted in *IRL – A Reader*, pp. 296-300. The principle of responsibility leads to that of burden sharing.

33 The *New York Times* on 26 May 2001 carried two news items on the death of 14 illegal Mexican immigrants in the desert of Arizona in the United States ('Rights Groups Urge Change in Border Policy' by James Stranglehold, and 'In Mexico, Grim Resolve after Death' by Tim Weiner, p. A11), and spoke of thousands of deaths on the border through the years. Both news items while typically refraining to comment on the dynamics of immigration control policy of the country the immigrants wanted to enter reflected on the predicaments of a humanitarian policy of hospitality in such situation. Should the humanitarian policy in such situation be to provide more job quotas 'so that they don't have to live in shadows' (Weiner's report), should migrants be provided 'survival kits (that) include re-hydration salts and food' (Weiner), or, should there be humanitarian vigil in the desert (Sterngold's report), less strict border control policy than the 'Operation Gatekeeper' (Sterngold), or petitions to Inter-American Commission on Human Rights or some other international human rights tribunal against US border control policy and practices (Sterngold)? I am again indebted to Paul Joseph for these two reports. (See note 18).

On the new literature on values of trust, care and hospitality in refugee studies, see E. Valentine Daniel and John Ch. Knudsen (eds.), *Mistrusting Refugees* (California: University of California Press, 1995); in a review essay on the book, Pradip Bose writes,

'In fact the whole structure of the humanitarian regime is fraught with competition, suspicion, and mistrust...Probably that is why the aid authorities are so averse to what would appear to be the most efficient method of distributing aid, namely the community itself. It might as well be the case that the western notion of 'trust' is capable of only producing mistrust in the end'. – Pradip Kumar Bose, 'Trust and the Refugee Experience', *Refugee Watch*, 7, June 1999, p. 34. It is also significant to note in this connection how Jacques Derrida connects hospitality, responsibility, and a renovation of the concept of friendship in a discussion on his book *The Politics of Friendship* with students of the University of Sussex, 1 December 1997, www.sussex.ac.uk/Units/frenchthought/derrida.htm.

34 Neal Ascherson, *Black Sea* (London: Jonathan Cape, 1995).

35 The significant point to note here is that, the particularity and the exceptionality of the population flow of the present time is brought out more in migration studies than in refugee studies. I have discussed this at length in *The Marginal Nation* and in 'Conditions of Globalization and Conditions of Peace' in R. Samaddar (ed.), *Space, Territory, and the State* (forthcoming, Orient Longman). For a detailed study see, Stephen Castles and Mark J. Miller, *The Age of Migration – International Population Movements in the Modern World*, second edition (London: Macmillan Press, 1998); Nicholas Van Hear, *New Diasporas – The Mass Exodus, Dispersal and Regrouping of Migrant Communities* (London: UCL Press, 1998); and for a wide ranging review of how political literature has negotiated the issue of immigration, Ishtiaq Ahmed, 'Exit, Voice and Citizenship' in Tomas Hammar, Grete Brochmann, Kristof Tamas and Thomas Faist (eds.), *International Migration, Immobility and Development – Multidisciplinary Perspectives* (New York: Berg, 1997), pp. 159-186.

36 *Refugees*, 3 (116), 1999, p. 1.

37 Benedict Anderson, *Imagined Communities – Reflections on the Origin and Spread of Nationalism* (London: New Left Books, 1991).

38 Christopher Pierson, *The Modern State* (London: Routledge, 1996), chapter 5, pp. 127-154.

39 It is important to trace how democratic politics infused in the states a sense of responsibility to shoulder the presence of refugees, where there was no obligation flowing from an international compact. On the issues under debate, see Kay Hailbronner, 'Non-refoulement and 'Humanitarian' Refugees – Customary International Law or Wishful Legal Thinking?' in David Martin (ed.), *The New Asylum Seekers – Refugee Laws in the 1980s* (Dordrecht: Martinus Nijhoff Publishers, 1988), pp. 128-34.

40 Etienne Balibar, 'France, Algeria – One Nation or Two?' in Joan Copjec and Michael Sorkin (eds.), *Giving Ground – The Politics of Propinquity*, London, 1999; Balibar, 'Is European Citizenship Possible?' in James Holston (ed.), *Cities and Citizenship*, Durham, 1999 and 'The Borders of Europe' in Pheng Chea and Bruce Robbins (eds.), *Cosmopolitics – Thinking and Feeling Beyond the Nation*, Minneapolis, 1998.

41 On the non-European as a European for a long time, Rila Mukherjee, 'Leo The African – An Immigrant in Europe', *Refugee Watch*, 12, December, December 2000.

42 Catherine Withol de Venden, 'Post-Amsterdam Migration Policy and European Citizenship', *European Journal of Migration and Law*, 1, 1999; see also her, 'How Can One be Muslim – The French Debate on Allegiance, Intrusion and Transnationalism', *International Review of Sociology*, 8 (2), 1998.

43 Researches in race relations have brought to light the impact of foreign workers in the class structure of a society and the works of Stephen Castles have been compelling in this regard. See Stephen Castles and Godula Kossack, *Immigrant Workers and Class Structure in Western Europe*, published on behalf of the Institute of Race Relations,

London (Oxford: Oxford University Press, 1984). See also, Stephen Castles and Alastair Davidson (eds.) *Citizenship and Migration – Globalization and the Politics of Belonging* (London: Routledge, 2000); Mark J. Miller, *Foreign Workers in Western Europe – An Emerging Political Force* (New York: Praeger Pub. Text, 1982); and Saskia Sassen, *Guests and Aliens* (New York: New Press, 2000).

44 For a short summary of the debate over asylum, see the collection of articles by Judith Kumin, 'An Uncertain Direction', Stefan Teloken, 'It's Long Way to Harmonization', Melita Sunjic, 'Learning to Deal with Refugees', and by Ray Wilkinson, 'The Debate over Asylum', *Refugees*, 2 (113), 1998.

45 Pradip Kumar Bose (ed.), *Refugees in West Bengal – Institutional Practices and Contested Identities* (Calcutta: Calcutta Research Group, 2000).

46 On this a fascinating study, Joya Chatterjee, 'Rights or Charity? Government and Refugees – The Debate over Relief and Rehabilitation in West Bengal, 1947-1950' in Suvir Kaul (ed.), *Partitions of Memory* (Delhi: Oxford University Press), forthcoming). I am grateful to Joya for allowing me access to her essay.

47 It is significant to see that the principle of fixing responsibility was never applied on Israel or Britain in complete contrast to the way it was applied on Yugoslavia in the Kosovo case. See in this connection James Petras and Steve Vieux, 'Bosnia and the Revival of US Hegemony', *New Left Review*, No. 218, July August 1996, reprinted in included in *IRL – A Reader*, pp. 284-287; also, Mark Gibney, 'US Foreign Policy and the Creation of Refugee Flows' Howard Adelman (ed.), *Refugee Policy – Canada and the United States* (Toronto: York Lanes Press, 1991), reprinted in *IRL – A Reader*, pp. 288-296.

48 The same disconnection was evident in the settlement on the Chittagong Hill Tracts also, where return of refugees and return of their land was kept aside from the general framework of the settlement.

49 *Refugee Watch* carried successive reports on land alienation in southern Bhutan (Nos. 8, 9 & 12); see also detailed reports on the economic fate of the returnees in Sri Lanka (No 12), the fate of the returnees in the Chittagong Hill Tracts (Nos 8 & 10-11).

50 Edward Said, *The Politics of Dispossession – The Struggle for Palestinian Self-Determination, 1969-1994* (New York: Vintage Books, 1995), particularly 'introduction'.

51 Jacqueline Greatbatch, 'The Gender Difference – Feminist Critiques of Refugee Discourse' *International Journal of Refugee*, 1 (4), 1989, pp. 518-526.

52 *Consideration for Asylum Officers Adjudicating Asylum Claims from Women*, United States INS Gender Guidelines Memorandum, 26 May 1995, in *ILR – A Reader*, p. 42.

53 Unpublished paper on 'Refugee Women and the Fundamental Inadequacies in Institutional Responses in South Asia' presented at the IDPAD conference on forced population movements in South Asia, Chennai, 3-5 March 2001; see also by Paula Banerjee, 'Dislocating Women and Making the Nation', *Refugee Watch*, 17, December 2002.

54 Etienne Balibar, *Masses, Classes, Ideas: Studies on Politics and Philosophy Before and After Marx*, translated by James Swenson (London: Routledge, 1996), pp. 57-58.

55 Cassandra Balchin, 'United Against the UN: The UN Gender Mission Attitude Towards Afghan Women Refugees Within its Own Rank is Glaringly Hypocritical,' *Newsline*, April 1998, p. 95.

56 Paula Banerjee, 'Refugee Women and the Fundamental Inadequacies in Institutional Responses in South Asia'; see also, my defence of the experiential arguments in *The Marginal Nation*, particularly the concluding chapter titled, 'The Number Game'.

57 Joya Chatterjee, 'The Fashioning of a Frontier – The Radcliffe Line and Bengal's Border Landscape, 1947-52', *Modern Asian Studies* 33 (1), 1999.

58 Joya Chatterjee, 'Rights or Charity? Government and Refugees, The Debate over Relief and Rehabilitation in West Bengal, 1947-1950' in Suvir Kaul (ed.), *The Partitions of Memory – The Afterlife of the Division of India* (Delhi: Permanent Black, 2001), pp. 75-110.

59 Charles Tilly's rich concept of 'Transplanted networks' in Virginia Yans-Mclaughlin (ed.), *Immigration Reconsidered – History, Sociology and Politics* (New York: Oxford University Press, 1990), can be further worked on with the political insights gained from the researches into experiences of immigration into West Bengal and Assam from across the borders.

60 For example, David Newman and Anssi Passi, 'Fences and Neighbours in Postmodern World – Boundary Narratives in Political Geography', *Progress in Human Geography*, 22 (2), 1998, pp. 186-207; Oren Yaftachel, 'Israel: Metropolitan Integration or 'Fractured Regions'?', *Cities*, 14 (6), 1997, pp. 571-581; 'Ethnocracy and Its Discontents: Minorities, Protest and the Israeli Polity', *Critical Inquiry* 26 (4), Summer, 2000 (in press). I am personally thankful to Ansi Passi of the University of Oulu in Finland and Oren Yaftachel of Ben-Gurion University in Israel for allowing me access to these essays.

61 For details of these case laws, see my introductory essay, 'Power and Care – Building the New Indian State' in Ranabir Samaddar (ed.), *Power and Care – Practices of Asylum and Refugee Protection in India* (Sage, forthcoming, 2003); Tapan K. Bose has documented some of these in, 'Protection of Refugees in South Asia – Need for Legal Framework', SAFHR paper Series 6, January 2000.

62 The official European journal for EU cooperation with the African, Caribbean and the Pacific countries, *The Courier*, 187, July-August 2001, brought out a special dossier on migration, and in its opening lines of the dossier admitted that all streams of flow are mixing up, and said,

> Who are they? Refugees, asylum speakers, economic migrants, students or nomads? Should they be classified in this way or should all the reasons, which drove them to leave their countries of origin be taken into account? The departure of migrants and their arrival in refuge countries generates a wide range of problems relating to economic, demographic, and social, health and social issues. Although migration is not a recent phenomenon, management of it in today's age of globalisation is taking on a dimension that could be cause for concern. (p. 28).

I am grateful to Hubertus von Welck for allowing me access to this issue.

63 I borrow the term from the title of Gil Loescher, and John Scanlan, *Calculated Kindness – Refugees and America's Half-Open Door, 1945 – Present* (New York: Free Press, 1986).

64 Roland Barthes, *Empire of Signs*, trans. Richard Howard (New York: Hill and Wang, 1982), p. 8.

Chapter 10

Received Histories of War and Peace

The Geopolitical Context of Dialogue

The question, then, as we end our exploration of the dynamics of dialogue in all its forms is, how much and to what extent our received narratives of war and peace allow dialogic politics to emerge, and make the quest for justice a part of this dialogic politics. These received narratives of war and peace are the signs of a world where geopolitics, democracy, and dialogue for justice share the same space of history. These narratives also speak of the geopolitical context in which the dialogic politics for justice emerges in the most contentious form.

The geopolitical context in understanding the possibilities of the emergence of a dialogic order is significant. This context is made in five ways: (a) the way histories of war and peace reach us determines our geopolitical understanding of the world we live in and where our dialogic acts are shaped; (b) the way power is located; (c) the way world is re-structured periodically, sometimes marked by restoration as in recent time in 1989, reshaping the locations and character of geopolitical power and deeply affecting the capacity of nations, states, and peoples for political dialogue; (d) the way the decolonized masses have entered world history as subjects of international politics through re-negotiation of boundaries, borders, and the frontiers, making immigration one of the most profound issues of international politics (Already in the preceding chapter we have seen how immigration has impacted on humanitarianism as a political cult. Here we shall see how immigration has reoriented the geopolitics of our time, affecting the dialogic patterns of our politics.); (e) and finally, the stark contrast in the background of globalization between the irresponsibility of the restored power that searches for new forms of war and conquest and refuses any demand for dialogue, and the dialogic demand for justice.

Having surveyed patterns of dialogue as acts of contentious politics, forms of accommodation and understanding, the relation between dialogic politics and the strivings for minimal justice, the limits of constitutional dialogism in such endeavour, and how dialogue relates to war and peace, we must pay attention before we conclude to the broad context in which these dialogic practices take shape and are institutionalized. This is primarily a geopolitical context deeply influencing our capacity for dialogue and our notions of democracy, and to understand that context with all its complexities, we shall discuss in this chapter the above-mentioned five aspects. We begin this study of the context with reference to the received histories of war and peace.

The Burden of Received Histories

In 1985, impressed by elements of stability in the post-war international system the historian of cold war, John Lewis Gaddis called the post-war period as one of 'long peace'. In an interesting essay he argued that though the peacemakers fell out among themselves immediately after the war and built up the most unprecedented stock of weapons to destroy or at least keep each other in check, there was no war among them. There was no world war. Without a formal peace treaty, peace had survived. And even though small wars, hunger, pestilence, deprivation and domination remained and continued to characterize the world, there was no war like the second or the first great wars. Even though nothing survived of the wartime consultations between the great powers they had held back their guns. The post-war settlement survived 'twice as long as the far more carefully designed World War I settlement, has approximately equalled in longevity the great 19th century international systems of Metternich and Bismarck, and unlike those earlier systems after four decades of existence shows no perceptible signs of disintegration. It is, or, ought to be enough to make one think.'[1]

Gaddis of course did not leave it to the readers to think. With a flourish he concluded the essay with these words,

> The Cold War, with all of its rivalries, anxieties and unquestionable dangers, has produced the longest period of stability in relations among great powers that the world has known in this century; it now compares favourably as well with some of the longest periods of great power stability in all of modern history. We may argue among ourselves as to whether or not we can legitimately call this 'peace': it is not, I dare say, what most of us have in mind when we use that term. But I am not at all certain that the contemporaries of Metternich and Bismarck would have regarded their eras as 'peaceful' either, even though historians looking back on those eras today clearly do.
> What is to say therefore, how the historians a century from now – if there are any left by then – will look back on us? Is it not at least plausible that they will see our era, not as 'the Cold War' at all, but like those ages of Metternich and Bismarck, as a rare and fondly remembered 'Long Peace'? Wishful thinking? Speculation through a rose-tinted word processor? Perhaps. But would it not behove us to give at least as much attention to the question of how this might happen – to the elements in the contemporary international system that might make it happen – as we do to the fear that it may not?[2]

This was clearly an argument in the Hobbesean terms which would say that everybody's war against everybody else (that is, a condition of *total war*) would maintain peace as everybody would realize that everybody's war against everybody would destroy everybody. Evidently written in an age of Thatcherite belief in 'deterrence as peace', detente and SALT talks, when it was held that mutual acknowledgement of respective strengths and influences held the key to peace and working relationship, an age hardly fifteen years old and already seems so distant, 'cold war' in arguments like this transformed into 'long peace'.[3] Yet there will be no dearth of historians in another ten years who will tell us that peace visited us with the end of cold war. Winston Churchill had said that peace had returned in Europe with the end of what he called the 'second thirty years war'. It seems that our destiny is to be juggled between 'war' and 'peace' not only in a time frame but in a single incident frame also. All that we shall need is to use these two terms intermittently, interchangeably, depending on our past, the pasts we define for ourselves to start from.

But what happens to those for whom there is no world history, therefore no vantage tower of looking from the world point of view?[4] Those who are ravaged by wars like Biafra, Gulf, Vietnam? Those who remain 'local', bound by the geopolitics of remaining on the 'rim', constantly engaged in fending off the 'insurgency cross-fires'? It seems world history with its grand theory of globalization will have left out these peoples and regions, for their wars did not feature in a world conflagration, latest weapons were not used, international interventionist armies were not despatched, axis, alliances, dual and triple ententes were not formed, and the 'virtual transparency' in international diplomacy ruled out intrigues and thereby forced substantial chunks of humanity return to the barbaric wars of the old ages.[5] Are we then to conclude that 'long peace' is made of 'short wars', a reformulation of the old argument of Clausewitz, that peace is the condition of war? The end of cold war has made a paradigm shift, which everyone accepts. The point is, if we cannot hold on anymore to a 'balance mechanism dominated world system' as the explanatory device for today's wars and peace(s), what we are to fall back on?

Yet my purpose in this concluding chapter is not merely to argue that historians of war and peace have written their histories from a 'world-centric' view, a world configured according to their vision, thereby debarring many peoples and nations from this history and denying them any historical capacity to converse on the international themes, the purpose is to return to theme of geopolitics in a more fundamental way. Geopolitics not only configures the inside and the outside, a configuration that poses the greatest challenge to democracy in post-colonial conditions, it has one more function – its function in determining what will be their history of war and peace, and what will be our received histories of the same. After all, to lead their lives in the continuum of war and peace, post-colonial subjects not only deal with the threats of wars and actual wars and not only do they hope for peace, they live their lives under the histories of war and peace. We must therefore address the question of the conditions of war and peace in the way their histories present these to us, and in the way they sensitize us to dialogue with the realities of war and possibilities of reconciliation, realities that include histories of them. Whom the late historian Eric Wolf called *People without History*, are the new subjects of a post restoration global order. Therefore is the question, with which the book ends, what are the prospects for a dialogic order in the context of a globalization of anonymity, statelessness, of 'un-homing', loss of civility on which civilization depends, and the new weapons of control and subjugation? In the last ten years people have talked too much of global economics, for once let us talk of the politics of globalization, in the process of globalization of politics.

We should not forget in any case that this politics of globalization is built on a differential image of the new 'space-time relations', which as the claim goes have made territories and segments redundant, and have turned the original dispensation of geopolitical modernity – the Peace of Westphalia – a thing of antiquity. I have tried to show that this spatial-temporal resolution of current history – Gaddis' resolution is just one example of that – is based on attempts to iron out historically specific politics of war and peace with trans-historical explanations.[6] Just to give one example, the very peace that Gaddis celebrates occasioned massive wars against left wing opponents in an entire continent – South America – causing ten thousand deaths in just one country, Argentina, and whose records have just started coming out, and yet these were wars that caused relatively less disquiet, because they were based on 'social peace', and were thus relatively insignificant to the point of ignoring the internal dynamics of the union of monetarism and fascist rule.[7] To the post-colonial subject the ideas of great wars and long/great peace, therefore, appear in the context of globalization. For those who are not privileged to be the actors of 'world wars', and are consequently left out of world history for the most part of it, the problem of war and peace requires to be defined in a different way. The received definitions unfortunately do not give much.

The short wars of the last fifty years and particularly after 1989 – the year of restoration – have been gory, but more important have had world consequences. Yet the historiography, in world historical terms, of the hot wars in a 'period of cold war', that is the period of 'long peace', itself does not attach much importance to these short wars, not because these wars did not cause huge losses of life and resources, but these are thought to have had little consequences in terms of pattern of domination, great power rivalry, arms build up, arms control negotiations, and alliance patterns. If any, these seem to have reinforced them. Therefore, these are only footnotes to the script written in the main hall post war history writing, in which balance of power and systemic compulsions have hitherto held peace, howsoever precariously. Little wars thus remain bound to be represented in another image and it is not surprising that they are represented in a mirror of a great war. They have to be represented in world history in mirror of a great/long peace.

Take the example of the Vietnam War. Again here the loss of lives and resources, the use of chemical weapons, the new methods of aerial warfare are by now well documented and evoke little discussion (except for some needs for atonement or the need to exorcize the ghost of defeat once and for all), for Vietnam war is today acknowledged above all one of a scenario of killing fields. It is also acknowledged that it was one of the greatest national liberation wars, comparable with the liberation of China only, and it brought the age of de-colonization and national liberation wars to a close. But in terms of world history, i.e., in terms of the world pattern of competition and collaboration, rise and decline of powers, the Vietnam war is not given much importance, perhaps justifiably so. For, it is a myth that wars and victories by themselves empower a country. How war is executed, how it is sustained, how weaponry is deployed and used, how peace is made, how the strength behind all these is perceived, and above all how war and peace change relations with other countries, become crucial for the great power status that is ascribed to a country after a war. Seen in that light, the Vietnam War changed little. It did not create a system nor did it undo anything, it did not reduce anyone's status, nor could it enhance anyone's, in fact it did not affect the country in terms of power, which it should have had logically, and thus the war remained 'local' despite the measured colossal losses. Our history of war remains caught in their history of peace, their history of war traps our history of 'event-less-ness'. The cold war ended not with any peace treaty reviving the wartime arrangements between allies, but with the collapse of one of the adversaries. The greatest defeat in a war in the post 1945 era did not at all affect the losing power in a world perspective. Rather the party which was on the victorious side in the Vietnam war crashed out within 15 years of that famous victory. The compulsions of representation thus make a mockery of events. They are nothing, so long they are not of this 'world', so long they do not make 'world-history'.

How does one escape the compulsions of representation? Before reaching a tentative answer we can make a contrasting reference to the Gulf war. The Gulf war was a short war. It was not one of attrition, it showed how modern weaponry could be swiftly mobilized, deployed and used to telling effect. It quickly ended the battle cries of one of the parties who had promised a 'mother of all battles', and a 'thousand year war'. Yet world history judges the Gulf war more significant because it enhanced the world status of one party. It ushered in the so-called revolution in military affairs (RMA). War and peace in the Gulf assumes significance much beyond war and peace in that area. The implicit message of red republicanism in the Arab lands, the spectre of a Napoleonic adventurer upsetting a carefully constructed system of peace treaties in the region intended to bury the hatchet, the symbolized rise of a new concert of powers locked in a holy alliance against any adventurer, the emergence of a 'new South' engaged in a partnership with the 'new North' – a development which ensured the neutrality of everyone who would have mattered in the war – all these have made the dimension of power in the Gulf war and peace immensely significant much beyond the oft-cited factor of oil. Thus, both war and peace are judged by power and are represented in the scale of world history, the only privileged discursive field to have a measure of power.

Location of Power

The question of historically assigned locations in world history leads us to the second element of the context – the way power is located.

Power is measured in terms of influence. International relations theory has no hesitation in admitting that. But it does a quick trick on us when it tries to naturalize this by eschewing a genealogical exercise, as if 'spheres of influence' are natural as biospheres. Thus the closed circle of logic runs: if in the post-war era there has been no 'war' it is because spheres of influence worked well; if there were 'wars', these were collisions over spheres of influence; the fact that there was no 'war' showed that all countries were under one sphere or another; and if there were 'wars', that fact showed how spheres of influence were a reality of life. The law of naturalization works everywhere the same way. It works by avoiding a genealogical exercise.

This history is obviously unfair. First, it includes the nations not privileged to be world-actors in world history and world politics. And then the fact that they are there, in this world history and politics (after they have been included), is shown as evidence that there are privileged actors in the world whose acts influence others and force these 'others' to be drawn in the vortex of world history. A limited history of the emergence of spheres of influence in the Atlantic world and partly in the Pacific world is universalized and a provincial history takes on a global form. The history of the nations and countries, which have their specific dynamics of both war and peace, quite separate from the discourses of the 'Atlantic world' and the 'great Eurasian landmass' is colonized in a history that in reality is equally local and equally specific. It is specific to the clash of spheres of influence between the Atlantic powers and a power belonging to the Eurasian landmass. This is certainly a problem of historiography of the post-war world, but also a problem of rescuing the capacity of the nations not privileged to be world actors to make ironically their own wars and their own peace. Indeed the first problem, that is the problem related to historiography, generates the second one. If it is argued that 'for every thousand pages published on the causes of war there is less than one page on the causes of peace',[8] the problem that bounces back at us is: is it possible for the received discipline of history (particularly in its form as world history) which tends to assume a meta-form, to unearth the history of peace? For, in such history, little peace(s) are bound to remain encapsulated in histories of great powers, and conversely therefore little wars remain covered under the discourse of 'grand peace'.

The problem of disproportion from which studies of the history of peace suffer has therefore a deeper root. Power, influence, spheres, and area – the elements that make world history and create hierarchies in the grand format remain incompatible with a study, which for the present, remains only as the other self of this 'world'. Discursive experience has shown how a microscopic or for that matter telescopic examination cannot free a nation's or a country's history from being fated to be a 'part of the world'.[9] In this respect the revisionist history of the cold war has not been of much help either. It concentrated on the domestic roots of cold war policy of the United States and traced the interconnections between the domestic policy and policy abroad ('Labour at home, Russia abroad'); in many cases it showed how various Soviet initiatives in those early years were rebuffed or neglected, and unilateral-ism continued in a big way to be the basis of US cold war policy. This may have enriched our knowledge of the genesis of the cold war, but the epistemic basis of the cold war history has remained un-critiqued in such revisionist history writing. The defender of the cold war can look back on those years and may well admit the realist criticisms, but then can say, after all, the acknowledgement of the Soviet Union as a great power, the recognition of spheres of influence, and arms build up as policy of deterrence and as the 'zero option' – all these have prevented war and maintained peace.[10]

Yet it is obligatory to show how wars in the post-war era have had their own characteristics to such an acute degree that they cannot be subsumed under the success-story of the long peace. It will be of significance if we realize in that respect, how cold war was often an over-privileged category in determining the way we looked at the world and how the demise of the cold war has brought the world back to 'normalcy', where war and peace maintain their conditional character, i.e., their specificities.

The age of Metternich was shattered by the wars of 1848. Indeed historian Eric Hobsbawm places that age as the age of revolution, and the year of 1848 as the year when the age of revolution ended, and not the age of peace. The world of Bismarck, suitably an age of restoration and again not an age of peace, ended with the war that began in 1913. The post-war world from 1945 continuously experienced wars amidst long peace, and if we are to go by the estimate of the University of Michigan's 'correlates of war' project, of the 200 and odd wars at least 96 are notable as violent civil, interstate, and extra systemic conflicts.[11] The wars in the Gulf, Rwanda, Afghanistan, Algeria, Bosnia and Kosovo in particular, have drawn the cold war/the long peace to a close in a bizarre way. Going by the main argument of the cold war/long peace discourse, the end of 'long peace/cold war' in 1989-91 should have brought back wars, whereas this history of the cold war in the same breath would have us believe that victory in the cold war was achieved with the capitulation of one. The fact remains that the so-called world actors have only a limited ability to control these wars, they can at best restrain the parties sometimes and wars occur not always due to them, but in spite of them in as much peace prevails not always owing to them but notwithstanding them. In short the problematic in the study of peace is the link, the copula – the word (denominator of war and peace) and the world (the site of history). The task is to discover the link, to un-link, to allow peace to remain unanchored to a master discourse of grand war and grand peace.

But let us go back to the question of specific wars. The turmoil in Eastern Europe towards the end of the second war and in the immediate aftermath was certainly linked to the issue of sphere of influence, but not a product of that. In country after country like Greece, Poland, Czechoslovakia, the old imperial power that knew the region most intimately was found too weak to intervene, which it would have surely done had it the required strength. In colonies like India the final surge of the national freedom movement had started. The collapse of Britain and France, and Britain in particular, signalled the decline of the old imperial tradition and the beginning of de-colonization. The theatre was indeed the widest – from the borders of Oder-Niesse to the delta of Irabati, Mekong, and Yangtze. And it has taken almost the same period as of the long peace to complete – a process as fundamental as the emergence of super powers, appearance of nuclear weaponry, formation of blocs, and spheres of influence. It is not that the second process did not have effect on the first. Suffice it to mention that the very process of state formation in these 'new areas' in world politics was shaped in a substantive way by the great power strategy of partition to realize spheres of influence. Thus Germany was partitioned, so was Korea, then Vietnam; India, Yemen, Palestine, Cyprus – all these lands were subjected to partition by the great powers. Radha Kumar's researches on Bosnia have shown how partition has been for quite some time a managerial strategy of stabilizing a situation that seemingly is totally out of control, and in the process becomes the usher of a new round of instability.[12] The extent to which the states formed in these areas in the post-war period have been influenced by that mode of origin has not been historically measured and appreciated yet, not even in the case of Germany.[13]

In short, even though the end of the second war has a justified centralizing narrative, here too we have several histories of war, end of the war, and then peace. Indeed, whereas in some cases the actors of the war went by the supposedly discarded Wilsonianism, like resetting an international organization, following a policy of national self-determination and proclaiming a liberal world order, in other respects the actors were found to have proceeded not much beyond the nineteenth century concepts of balance of power, a committee of world policemen, and a series of alliances. In the most vital aspect of 'who controls Europe?' Europe behaved in an unprecedented manner. It not only welcomed the Marshall Plan, so compromised was it with threats of revolutions and the awareness of its own inability to maintain the earlier power and influence that it pleaded the United States to stay over, in the process reviving the nineteenth century notion of balance of power even by partitioning the continent. The history of cold war thus has within it many histories, some imprisoned in the past and some characteristic of the new age.[14]

In the same way the post-war wars have also their specific histories. First, partitions brought in its wake wars, some still continuing. The Kashmir war was fought three times between India and Pakistan. The dispute remains unsolved. Wars in Congo, Palestine, Lebanon, the Horn of Africa, Bosnia, are other instances. Second, low-key wars as form of diplomacy are very much the way in regions where state formation could not solve the nationality question. Third, republicanism wherever on ascendancy has sparked off wars. The unending rounds of war in Afghanistan are an evidence of the effects of that 'forced march of history'. Fourth, borders in parts of the globe have not only occasioned great games, but some major wars also as the war between China and Vietnam. Fifth, arms sale builds up the arsenal and the 'baroque arsenal' itself becomes a temptation to be used and tested. Sixth, there are covert wars fought by outside powers with mercenaries and also by trained émigrés as in Nicaragua, El Salvador, Angola, Laos, Burma and elsewhere. Seventh, the classic peasant revolutionary wars have also been a constant feature in the post-war world as in Mexico, though such wars are on decline. Eighth, there are the still un-measured (at best half-measured) wars (as in Latin America or Southeast Asia) against communist and other left-wing adversaries in these wars, that the cold war inspired and legitimated but had their internal and specific histories and resulted in deaths of several hundred thousands, which Carlos Nino has called the 'radical evils on trial' and Luis Mareno Ocampo has termed 'dirty wars'. Finally, the world witnesses the wars of self-determination in scattered places like Sri Lanka, Kurd region, Burma or Kashmir. It is instructive to remember that in most cases these types are not pure in reality. Ethnicity, border, partition, arms sale and a variety of other factors combine to produce wars.

Ironically with globalization as political policing weakens, wars become scattered. The policemen who are left to do the duty are often tired; war fatigue overwhelms them. A curious combination of isolationism and the lingering desire for world role marks the great powers at this time.[15] Only new political arrangements can take away some of their burdens. But that means first the post-war process has to exhaust itself. The long peace was the first half of that process. The scattered wars and peace(s) mark the second one. In this era therefore we shall see more security doctrines, more efforts to erect a system that can control the scattered wars and peace(s) while avoiding deployment commitments – security doctrines markedly different from that of the cold war/long peace years.

Structures of Restoration

Power is challenged; power is also restored at times. The structure of restoration affects the configuration of power, and influences thereby cycle of war and peace. Restoration in all ages is supposed to have brought peace. In some instances the restored peace creates conditions of new wars, in others it strengthens its own conditions and democracy. The point is: after the restoration of 1989 where do we stand?

Johan Galtung makes this point in a different way. Galtung argues that there is a 'peace structure', a 'war structure' and a 'non-war structure'. A 'peace structure' means a system of positive exchange and cooperation between the nations that satisfies five conditions, namely, symmetry, homology, interdependence (symbiosis), institution building, and dispersion (entropy). If the structure exhibits one or two features strongly, it will not be enough for peace. Thus between the big power and the dependent country there may be symbiosis as well as entropy. But since there is no symmetry, acute conflict may generate. On the other hand there may be symmetry and entropy, yet due to lack of other elements stated above there may be rigidity and consequently conflicts and war. The 'war structure' has also these elements with the 'only one major difference – instead of *positive* exchange and cooperation, there is *negative* exchange and destruction' (emphasis Galtung's). Thus in such scenario highly interdependent parties involve themselves in mutual destruction. Even with considerable institution-building war chances may accelerate rather than decelerate. In the 'non-war structure' the '*system is unstable, and may move towards the peace structure or towards the war structure*'[16] (emphasis Galtung's).

There may be agreement on the existence of a structure of peace conditions or a structure of war conditions, or for that matter a structure of neither-war-nor-peace condition. But the critical question to ask here would be, why are peace and war structures similar, that is to say, why are they potentially interchangeable? Galtung himself admits that there is no structural difference. In other words we cannot offer any irreducible structural explanation on the genesis of war/peace. The conditions of 1919-39 produced a war; these conditions were present in so many ways in the conditions of 1945-75. Yet there was no war. One explanation may be that a strong peace-consciousness developed early after 1945, and it was stronger than the peace-consciousness of 1935-39. It resisted many possibilities of war, more because this consciousness was present in various layers of society and in the layers of the political classes of different countries. But Galtung does not include consciousness as a structural element; to him this would be an extra-structural factor, indeed he speaks of it when he introduces the difference of 'positive'/'negative' as an extra-structural factor in distinguishing the two conditions. Ironically however, we may be tempted to taking resort to this extra-structural factor whenever we have to study closely a 'non-war-non-peace' structure. The proclivity of such a structure to slide into a war condition may be due to the presence or absence of certain extra-structural factor. Inevitably thus we are back to the issue of histories of wars, peace(s) and dialogues, and of historicizing them. We are presented with the difficult but absolutely essential task of finding out the history of the consciousness of peace. We can now see why the route of historical comparison in analysing the interchangeable nature of war and peace is the more significant one.

A comparison with events of the past in discussing the possibilities of peace in the post-restoration period, thus, may be gainful. The Napoleonic wars were concluded by the restoration of order, which was so suffocating that it needed a series of revolts for fresh winds to blow in Europe to certain extent. On the other hand, the Bismarckean order unwittingly allowed the labour movement and democratic charters to develop in Europe. The restoration of some old style regimes in East Europe after the war in 1945 was absurd and they quickly collapsed. There was effort to restore order in China after 1945. US intervention in support of restoration could not prevent the overthrow of the *ancien regime*; the revolutionary war succeeded. A new order emerged. It quickly attained stability having gained people's confidence and using its power skilfully. In Burma old order was restored and peace has not returned to that country since then. In the entire East Asian and the Pacific Rim, the Atlantic world led by the US buttressed its position after 1945 by constructing a defence perimeter around China. Corrupt regimes were held up. Wars therefore raged in that region for a long time. The US troops did not withdraw from the Korean peninsula, from the Philippines, from Okinawa in Japan, British troops stayed back in Malay, French troops in Indo-China.

Today we have a similar scattered picture before us. In the Balkans, the old order cannot come back; sometimes with intervention from outside the pre-1945 order is sought to be restored, but we do not know whether the experiment of setting up a state as part of restoring order and peace will be successful and whether without a state restoration peace restoration will be complete. Similarly in Somalia, Palestine, El Salvador, Algeria, Yemen, Lebanon, Rwanda, Congo, Afghanistan, Cambodia, and in some more countries – restoration of peace and order can no more be the restoration of the old; that is ruled out by the 'state of the state' itself; the state is ravaged by attrition, militarism, fratricide, corruption, elite gangsters and downright destruction by an occupant army – either by any of these factors or by all these combined. In those countries, introduction of a constitution and the erection of a state structure has become an essential part of the task of restoration. In 1848 declarations of constitutions signalled the challenge to the restored order; in 1919 order was restored on the basis of a liberal vision of a world and a new phase of constitutionalism – but this vision was so weak that forces of aggression and militarism went ahead to revise the settlement within ten years of it. In 1945 the liberal order could be restored at the heart of the Atlantic world only at the cost of partition, and the colonized world was invited to participate in that order through such grotesque form of de-colonization that the de-colonized nations still reel under its impact.

The task of peace restoration thus has landed into the deep, the heart of the question itself. The question of war and peace lies in the secret origins of the formation of state – shrouded in mystery, deep in slumber. Everyone talked of the states-system/s prevailing in the world or in parts thereof. They talked of treaties, strategies, alliances, diplomacies, and policies. State-system was to be strengthened, or reformed, or maintained, or changed. But what when the state itself becomes unsustainable? When state incessantly occasions war and is then eaten up by it? When peace is actualized even at the state level by private initiatives, i.e., by true public initiatives? When the history of the state becomes inadequate in chronicling the history of peace initiatives, of dialogues? When the task of restoring peace becomes inseparable from that of changing the state in order to restore it? It is then that we understand how this model of state and state restoration handed down by global history that is world history, has become so unworkable. We shall be able to see through the mysteries of war and peace – mysteries that succeed in making a period of the breakdown of the state appear as a period of long peace.[17]

It should not therefore possibly cause much surprise that the triumphal literature of late eighties and nineties characterized by that mood of 'globalism' does not fully reckon with the fact of the breakdown of the state in many parts of the globe. The crisis of the state as evinced in Bosnia, Afghanistan, Rwanda, Cambodia, parts of South Asia, and other places still appears as more of a managerial problem. Peacekeeping, peace enforcing, peace stabilizing and 'constitutional reinforcement of peace from outside' are considered part of an international managerial strategy for salvaging the state. But there the problem is defined as stabilizing the state against forces of 'ethnicity' which threaten and sometimes succeed in internationalizing an essentially local phenomenon. Therefore the international strategy is to quarantine the sickness, to closely monitor the situation, apply multilateral, multi-pronged therapy with various degrees of force, to cure the state of a problem which is seen as not intrinsic, but essentially one of internationalization of the local, the globalization of fragmentation.[18]

Yet there remains a sense of unease in this whole scenario. The triumphal literature emerging in the aftermath of cold war has developed around two main themes. The first theme claims, that with the failure and collapse of the socialist system, there is no ideological rival to liberal capitalism, by implication of the liberal system, no other aspiration backed by any kind of popular legitimacy. The second theme sees in the decline and departure of the parallel politics the emergence of a genuine world system whose last obstacles are in the process of being removed. States, particularly of the third world, cannot any more fantasize about national paths or national alternatives. Global economic institutions and international division of labour have finally made capitalism the site on which state is to be managed. Political programmes and economic practice are to recognize the necessity that state management, liberal values and economic openness within a world capitalist system have to go comfortably hand in hand.

The unease in this mood of triumph is reflected in the texture of the literature. Some say, that though the state is not under interrogation, the agenda of reconciling the state with the post-restoration world is hard and arduous. Some say that even in these sure times the state is severely threatened with massive migrations of population, ethnic extremities, hunger and pestilence in large parts of globe, unpredictable rogue behaviour, laxity of international laws, hordes of unemployed labourers, unruly currencies, and above all existence of a political class uneducated in the liberal doctrines of governance. Some say that globalization is threatened with clash of irreconcilable civilizations. Others finally point out that the global system, particularly the economic system, is itself unstable. The globe itself is not prepared for the massive forces of globalization. This uneven texture revealing the fears, suspicions, and uncertainties marking the triumphal moment makes us feel as if the world has re-entered the nineteenth century.[19]

Then too the states were not prepared for post-restoration world. Alliances, ententes, congresses, and treaties – all these marked the long century. The Metternichean age that seemed to Gaddis as the period of peace, we have already mentioned, appeared to others the 'age of revolution'.[20] By the time the states had been educated in the rules of a system, too much of imperatives and imponderables broke it up. Then too, the unanticipated factor of globalization of expansive trends played havoc with the carefully constructed edifice of states-system so much so that the truly monumental agenda in the Versailles deliberations could only be attempts to achieve 'national self-determination', 'home rule', 'settling borders' – the by now forgotten story of Wilsonianism.[21]

Immigration and the New Subjects of International Politics

The third world as a problem has reared its head again; it is again a problem for the successful completion of the Westphalian project of creating a global order and maintaining peace. Drawn in the vortex of several global events and developments like de-colonization, partitions, citizenship laws, emergence of minorities, ecological destructions, hardening of borders, immigration realities, and most important a geographical widening of capital and labour market, the career of the state in the third world today symbolizes the intractability of the problem of race, ethnicity, citizenship and alien-ness. States are crashing out of existence or are becoming shadows of past. States had to be formed *de novo* after 1919, again after 1945. In today's moment of triumph the same imperative has raised its head. Coming out of the sedate chambers of social anthropologists the subject of state-formation has become the issue of world politics today.[22]

One example that can reinforce the argument of historicity of wars and peace relates to the phenomenon of immigration in international politics today. There was a time when for its own formation a state required a subject population. However it was the territory, which more or less marked out this subject population.[23] Not so today. States require defining who are the citizens and by implication or by explicit formulation who is an alien. The category of immigration functions as a political marker of these two definitions. The sociological implication of this signification is immense. A fictive ethnic core, a differential *modus operandi* of statehood (that separates a state from others, separates its core from the periphery in its internal structure and defines its citizens as different from other groups) and the representation of hatred needed for the state to function – all these pertain to the fact that from a biological theory of subject population we have arrived at a sociological theory of the same with the consequence that a state can explicitly claim today the fact or the need for siting itself in a 'particular culture'. Immigration in this sense is not only an issue of labour economics but of the politics of the current time when state and nation have become almost un-differentiable, when the subject and the citizen have become almost synonymous, and the ideology of the state has in effect made a return to antiquity. In the era of state enterprises the old centuries when the cultural basis of some solidarity was taken as natural are perhaps on us. All this has immense consequence on the conflict pattern in the world today and the capacity of the international order to tackle the conflicts.

We can again look back to the lessons of history. Europe took nearly two centuries for its states to find their cultural basis. The Jewish question was finally solved only with extermination, their emigration, and the formation of a Jewish state. Still after the war when the states were to be erected anew, Christianity had to be repeatedly invoked for the stabilizing the states-system.[24] With immigration of Turkish, Arab, and Bosnian workers into the West European countries, we find Europe reaffirming its 'historic cultural basis'. Islam, Slavism, orthodoxy – these are the invasions against which states in Europe have raised guards. This situation confirms the 'presence of the past'.[25] We have here the reworking of the problematic of state-formation in a situation where the past story of the state refuses to vanish away like the ghost that hovers over the site that has witnessed repeated construction after periodic demolitions. That is why the question of state-formation, statelessness and prevention of conflicts has become irreducibly political. The project of nationalization of a society for imparting stability to the state is not complete if aliens are not marked out.

The irony is clear, clearer when we remember the circumstances of South Asia. The post-colonial nation form was to be the ultimate form of political institution. For the realization of this form, the process of constitution and development of nation form required the production of non-state people from time to time. Thus the entire process of decolonization including partitions, wars, and the deliberate acts of enforcing territoriality produced non-state persons in a country. But this process of production is so violent that it gives rise to anxiety, is this form going to be the ultimate form? Will not the massive flows of population give rise to other structures that rupture the fusion of the state and the nation? Indeed we find that the influx of people makes the process of nationalization ever delayed, ever incomplete, for every community produced by the process of state making is substantially imaginary, and therefore there always remains the possibility of newer communities being created out of the national process of ethnicization.[26]

This region was decolonized through the strategy of partition. However, as we have noted, as a key element in state-formation, partition has not been unique to the sub-continent, though in terms of 'globalization and fragmentation' their significance has been inadequately studied.[27] Crucial in such significance is the generation, through such process, of the continuing waves of migrants and refugees and their impact on state-formation and subsequent state-development. The emergence of modern India was through one of the biggest population transfers in human history. Between 1946 and 1951, nearly 9 million Hindus and Sikhs came to India and about 6 million Muslims went over to Pakistan. Of the said 9 million, 5 million came from what became West Pakistan and 4 million came from East Pakistan. Immigration into India, particularly eastern and northeast India has continued under the shadow of the exodus/influx of that time with of course new patterns but with very substantive continuities also.[28] Its impact on the state and politics at regional, national, and local levels, in terms of institutionalized and non-institutionalized domains, has been critical but again unfortunately inadequately studied. Though we have by now studies on the northeast, we are yet to generalize them as element of state-formation in South Asia. Think of the impact of the non-state persons such as the Chakmas in Arunachal and Mizoram on the politics of these states, the Bangladeshi aliens in Assam, the Sri Lankan Tamils in Tamil Nadu, Biharis in Bangladesh, the Bhutanese of Nepali origin in Nepal, the Afghans in Pakistan and Tajikistan, or the Bangladeshis in Karachi, Lahore and other parts of the Sind and Punjab.[29] The impact of the exodus and also the entry of various groups of people from and in Burma have been similarly acute in terms of the Burmese state and also the politics of South-east Asia.[30] These are all potential and actual issues of violent conflicts in the region. Ethnographic and anthropological studies have documented how population movement was a crucial factor in state-formation in past ages, related as it was with kin-linkages. What lends newness to the situation is the backdrop of citizenship against which alien-ness is defined, immigration is characterized, policies on 'near abroad' are formulated, the nation is periodically defined and announced, and the hierarchy internal to the subject population formulated. In this situation the old differences between *jus soli* (right of soil) and *jus sanguinis* (right of blood) do not mean much for both are equally operational in marking off the non-state persons and in fencing the (territorial) state. Non-state persons are non-citizens, often non-nationals, often they are unwanted where they are born and stay and also where they may be pushed back/forward. Citizen is an essential attribute of the state not only for defining the subject population, but also for defining the other end of a transactional relation whose one end is the state. That is why today the 'citizen' is often taken as the 'national', for *state* and *nation* as different entities do not signify much,[31] though at one time state had tried to adopt non-national forms. Religion, language, 'tribe', and territory – all these define minorities, and as such they define a particular category of citizenship and relate to the general problem of the national acceptance of the reality of immigration. Frankly speaking, in this respect the experiences of Europe and South Asia are not much different.[32] If we take into account the experiences of Germany and Italy as moderate examples, and Yugoslavia or the Caucasian states as rabid examples, the whole issue is not purely a 'third world' issue...The 'revenge of the nations' is visiting the states wide across the globe.[33] Hatred, violence, an exclusionist legal system of beneficiaries, a system of identifying the potential disruptors or groups of disruptors of statehood – each of these is required for the formation of a state. Seen in this light, the category of non-state persons shows that state-formation is not a one-time affair; it is an exercise in perpetuity in as much as state legitimating remains a permanent plebiscite.[34]

Migration is admittedly as old as human history, but as suggested earlier, international migration even of several decades and sometimes one or two centuries before creates implications for the revisionist strategy of the state and thus begins to exercise acute impact on the inter-state and international agenda. Thus Indians in Fiji or Uganda, Chinese in South-east Asia, Indian Tamils in Sri Lanka, Armenians in Azerbaijan, Turks in Germany, Palestinians in Jordan, Lebanon, and elsewhere in West Asia, are the classic non-state groups who pose a problem because they are perceived as a threat to the sovereignty to the state, for they are the minority, the perpetual reminder of state's fictive ethnicity. Involuntary migrations consequent to natural disasters, civil wars, political persecutions, repressions on minorities, total economic breakdowns involve millions of people. They are subjects of 'new slavery'. International system requires them, but is yet to devise a system of managing them. The international system needs migrant labour, invading armies of émigrés (Cuban émigrés in the USA or the Afghan mojahids in Peshawar), and sex-slaves. The hordes of involuntary migrants mix with streams of voluntary migrants. A widened information network, mobility of capital, the oil powered economic boom, re-division of global labour, proliferation of small arms and of other methods of low key insurgency, the growth in world population, the historical 'revenge of the nation' in the ethnic form – all these make immigration an important item in the agenda. Thus the state has become a matter of pressing concern in the list of global anxieties.

We all know that globalization in its first true incarnation that is colonialism, was the first impetus to migration. About 30 million people had arrived between 1861 and 1920 in the United States, a country of immigrants already responsible for the displacement and extermination of the indigenous population. Europe saw huge intra-Europe migration to the Baltic, the Black Sea and to the Central lands. The period between the Potato Famines, the Pogroms creating the fleeing Jew, the forced migration of an estimated 15 million slaves and the migration of indentured labour more widespread than slavery encompassing some 40 countries and involving 37 million people ending as late as 1941 is indeed a long one – a duration equal to that of colonialism. The political effects are only now being felt.[35] New patterns then emerged with changes in labour market and the general process of decolonization. Immigration studies began in right earnest. It is notable that immigration studies escaped the cold war framework in which the refugee question had remained.[36]

Here indeed is the point. Immigration studies and refugee studies represent a paradox in international relations theory. International relations theory, every student in this field knows, was bound by the reality and the rhetoric of cold war to the extent that the world which lay beyond the cold war and the great game of the post-45 era was virtually absent in this 'international'; the international was actually very much provincial. This corpus of knowledge known by the name of international relations was now trying to understand the phenomenon of immigration, which certainly related to cold war realities, but which in its complexities was beyond it, it was extra-cold war in its existence. Thus international relations theory remained a half way house. It became primarily a matter of political history, world history, and diplomatic history (though international political economy was an important element). It carries the marks of the knowledge of a cold war era that has given over to another age; at the same time that era has recoiled into another, of the past. The responses to the complex reality ranging from the neo-Nazi movements in Europe and elsewhere to the international institutions such as the UN organizations or the Peace-keeping forces, reveal the uncertain state of international relations as a form of knowledge, its anxious reaction to trends which had been long in human history but new in its significance in terms of politics – both of the state and the globe.[37]

Practitioners of international relations theory should not get angry over the fact that this discipline is found on many occasions, particularly in the post-colonial regions, as an emperor without clothes. The reason is three-fold.

First, the received body of knowledge of world relations is uncertain about the question so important to the post-colonial regions, namely, do contemporary borders exist primarily to control movement of people?

Second, in the context of the emergence of a vast army of footloose people on an unprecedented scale, it cannot answer, are border controls effective in stopping the entry of aliens?

Third, to the post-colonial subjects long deprived of standards of justice, this knowledge sometimes seems unconcerned with the issue, namely, are not these controls instruments of human rights abuses and injustices?

These issues remain outcaste to a discipline long occupied with great powers, great games, great outcomes and great destinies. This neglect has consequence in terms of political theory also. It is a consequence in form an irony. The neglect by political theory of the process and the practice of production of frontiers and borders has resulted in a barren landscape of ideas. The irony also is no less obvious. For, in the triumphal mood of post-1989 era, when it was claimed that globalization was doing away with borders, frontiers and borders returned with vengeance precisely because of the new questions posed by the politics of restoration. The issues that have come up are: Can citizenship be de-linked from nationality? Can there be citizen of a region? How will the politics of the new era, which depends much on borderless capital and bordered human resources, cope with borders that are 'here', therefore also 'there', a reality and yet something transcendental, a margin that is at once a result of the 'heartland' that this new politics creates? In such conjuncture, global history of war and peace finds itself embroiled in a mixed up situation of the real and the imaginary of borders. If peace in Europe, as I showed in the preceding pages, was possible – in the sense of an acute historiography – by wars elsewhere, such peace was creating borders, the only notable criterion of such borders being that these borders were vacillating. To have peace, walls had to be raised (*fortress Europe*), the programme of transferring liberal institutions to an international scale was slowed down, and populations and even some nations of Europe were made to 'await entry' into the peaceful world. Etienne Balibar, the French political philosopher, has commented with terseness,

> Thus borders are vacillating. This means that they are no longer localizable in an unequivocal fashion. It also means that they no longer allow a superimposition of the set of functions of sovereignty, administration, cultural control, taxation, and so on, and consequently a conferral on the territory, or better, on the duo of territory and population, of a simultaneously en-globing and univocal signification of 'presupposition' for all other social relations.[38]

In other words, as we see in the present history of the de-colonized region of South Asia, borders are no longer the shores of politics, but being produced *here* as produced *there* or because being produced *there*, borders are elements within the space of the political itself. Thus the border representing the war zones in Sri Lanka is to be found in Colombo, in structures of political-legal relations within the quarters of the state, within the Sri Lankan Diaspora – in spaces un-represented or un-representable by territory. Borders are therefore partitioning the world, rearranging distribution of space and territory, that is their function, the function of representing the local in the global, the function of turning every map into a 'world map', turning our spaces into parts of the world. This was true when the Durand line was drawn between British India and Afghanistan, as in 1948 when the Ceasefire line was drawn in Kashmir. Border is therefore a question of idealities – real because it is symbolic, vacillating because it is stabilizing. India is stabilized by having a contentious border up there in Kashmir, on its one side (LOC), below it where Jammu begins, and below everywhere, in every contentious field of politics of the nation.[39] Border in the post-restoration world will not be capable of securing identities, which is its main function – without producing the spiritual question that unsettles its capability of securing identity, namely its eternal interiority. Therefore, borders will belong increasingly to the world of fantasy, where 'ethnic cleansing' of one section of 'people' is in fact the suicide of all whose history is constituted by the repercussions of its own divisions. Immigration has heightened this function of border – *border to partition, partition to the death of secure identity*.[40] But this means also that we can turn the theme of Gaddis on its head, and ask the question of *difference* not in term of unqualified war and peace, or whose war and whose peace, but in terms of the capacity of reflecting on the project of governing oneself that this question of difference has represented.[41]

In other words, borders not merely present for us the issue of de-colonization, but the very colonization of history, not a specific question of a period (post-war, cold war), but an issue of generation which will always mean innovation of conceptions of historical correlations of nations – to put squarely, borders present the problematic of the relation of history to the nation-form. History, as I have tried to show through the book, presents the nation in a state of finality, so presents war and peace with its division and frontier as final. But this history so adept in presenting separation and exclusion becomes an extremely inadequate means to represent non-finality, non-wholes, and fractals. De-colonization has taken away from us the simplicity of isolation or unity by presenting to us the in-eradicable *remainder* of the nation-form, a form of exclusion and imbrications, individualities and complexities, borders and flows, in other words, presenting to us frontiers of encounters and engagements, new enchantments and disenchantments. What liberal politics of democracy fails to inquire into is that, how borders notwithstanding being the most illiberal of institutions have become essential to its theology by becoming the appropriate marker of a fragile equilibrium between functions of consensus and of conflict, representation of popular interest and control of those ideas and interests, citizenship and alien-hood, finally war and peace. It also fails to inquire, how globalization even in the form of cosmopolitanism fails to enchant large sections of humanity, how cosmopolitan demands on the nation-form (in fact it produces what Benedict Anderson calls 'long distance nationalism'[42]) can only create new territories, and how the strategy of capturing all that is around in the new cosmopolitan form will make the question even more insistent, *who speaks for the global, the cosmopolitan*?

What is equally significant in the context of a geopolitical colonization of history, especially of the post-colonial humanity, is a pan-glossing attitude to the history that is received, the TINA factor ('there is no alternative' argument).[43] The TINA factor overlooks the fact of history – the history of 1815, or 1851, or 1989, that globalization follows restoration, that wealth and poverty of nations become more glaring, and one of the most abiding human values, *responsibility*, is pushed to obscurity by the history of war and peace received in the age of globalization. Expectedly so, for this history must make us forget an equally significant historical question that is linked to responsibility, the politics of reparation – reparation of damage to body, mind, to the collective, culture, reproductive capacities – the issue of crime and forgiveness. Since the liberal-ethical politics does not acknowledge anything as 'margin', the silence to the indictment to *things belonging to the world* is not surprising, nor should be surprising the blindness of this world to the waves of disenchantment with the global. This disenchantment is in fact a resolve to disengage with high principles of liberalism, namely that, the individual is the ultimate theory of moral concern, that all individuals belong to the world of a universalized justice, and that same principles should apply universally to all lives and all cases. The disengagement is based on differential notions of principles, responsibilities, duties, and compensation. The disenchantment and consequent disengagement leads to a counter celebration, celebration of a different history, the history of self-rule and victim-hood, and celebration of alterity that critiques established identities.[44]

Indeed, as the idea of justice continues to capture imagination and create new images (is it not obvious that one captures imagination by creating image?) in an age that we know as one of globalization, *justice* itself loses its vestigial liberal connotations and takes on acute critical character. Thus, instead of the old discourse of the adverse effects of international trade and capital mobility on 'national economies', we have new discussions on the effects of the same on world's working poor, and the long-term effects on openness not of capital but of poverty to capital. Similarly, we now have concerns about lack of labour mobility in terms of its effects on the level of social justice. The debate on job versus income insecurity is now being turned upside down; we are no more content with thinking what adverse effect income security creates on job, but what adverse effect security of the metropolitan job-market has on income security of the working poor. Similarly, we have interrogations now of the given premises of the efficiency-equity trade offs; we have new experiences of the relation between credit market and rural poor, the dynamics of credit delivery, the dynamics of education and health, and most important, lessons on the relation between bargaining power and justice. In all these inquiries, we have much more vigorous link established between justice and democracy, and justice and quality of governance – in other words new inquiries about 'social justice in the global economy'.[45]

One must therefore, as against a seamless and flat history of democracy and globalization, or, global war and peace, think of the evolution of forms of global ideas and politics, their internal non-similarities, their ontological differences, and the continuing politics of assimilation. It means thinking anew of time and space as historical forms. This is a point we had made in chapter 3, and it remains to be seen how borders and boundaries while protecting liberal national institutions have made rapid thrusts of conservatism possible and have made violence the main mark in the history of liberalism in defence of sovereignty, government, territories –concerns of the chattering classes everywhere today. To repeat, one must therefore, as against this seamless history, turn attention to the overlays of betrayal, shames, responsibilities, enchantments, and forgiveness – to put it differently, the overlays that make a possible history of peace, that is to say a history that has the possibility to be written. War and peace appear in this possible history as frontier processes, a mixture of thickening and thinning, and a subject of mapping the history of friendship, promises, and betrayals.

The significant point to note in this discussion is that, among the defining features of the liberal politics has been the graduation of the state from the bondage of kinship, the constitution of the subject in form of individual citizen, the evolution of the state as one end of a transactional relation, and the uncertain merger of two forms of power – the centralized juridical power of the sovereign and the 'capillary' form. Yet the current pattern of conflicts shows, this graduation as suspect. The constitution of fictive ethnicity as the basis/core of the state, the configuration of minorities as immigrant communities, the invasions on territoriality which has been all along the prime attribute of modern state, show that the modern state may not be that modern; that while the state takes up the agenda of modernizing itself through nationalization, decentralization, constitutionalism and citizenship laws, and above all accepting and negotiating globalization, the state in fact while leading its people through such process is busy in reinventing kinship.[46] A revisionist state thus ends up by seeking forms of the past. The moment of triumph is in many ways a moment of admission of defeat. International relations at the end of the twentieth century featured by the globalization of fragmentation therefore looks so similar to the relations in the nineteenth century. Possibly it never left that age.

We may now paraphrase our arguments in terms of the themes of this book.

Till now we have argued that globalization and fragmentation are companions of each other today. The state of statelessness, of dispersed conflicts, of the imperative of enforcing the state to enforce peace in the post cold war world is a product of that complex situation. Such a state of international affairs is also demonstrative of the fact that international relations theory is inadequate to grasp this phenomenon. Thus while today's mood of triumph is disturbed by widespread statelessness, this state of statelessness reflects on the agenda of state building to the extent that we may say that we are in the age of state-formation again. In Afghanistan as much in Burma, Cambodia, Bosnia and Somalia, peace making is actually 'peace-enforcing', for peace is not achievable fundamentally through dialogue, it has to be 'enforced'. And wherever, dialogue was the key to political management of conflicts, it was part of a strategy. The reason is that the state in most of these cases has 'withered', there is no agency to provide the house to hold dialogues; and the state has now to be erected from outside so that peace can be enforced. Conflict theory dealt with asymmetries. Now conflict-prevention theory finds these asymmetries too real – so real that negotiation politics is often crumbling on its face. Peace has become 'elusive'.[47]

Yet we must not think that such a condition is suddenly on us. It has its own history. If the First World War showed how fragmented the world had been for the past hundred years, the Second War's end brought no decisive change in that situation. The contradictory trends of globalization and fragmentation were to prevent any resolution to the equilibrium-less world. In the beginning it seemed that individual states were being reconciled to the systemic world. Keynesianism, now limited in effects as individual (state) therapy, was supposed to be better as a global therapy. International economic order was growing and it was assumed that where Keynesianism would fail, monetarism would bring the rogue states to order. European imperial control would give way to a new charter of controls that would fine-tune the states-system. But the universalism of the Atlantic Charter had to domicile itself in the United States. Monetarism, Americanization, neo-imperial controls, and the fusion of Westernization and Americanization defeated the universalism of Atlantic Charter. In short, within Atlanticism lay the seeds of its failure, the very universalism was giving birth to its opposite. The contradiction was elemental. The realities of globalization characteristic of the cold war era, namely, trans-national corporations, territorial segmentation of production, financial integration, structured choices between orders and ideologies, the preponderance of the nation-state, and the fusion of globalization and Americanization, were seemingly absolute. Yet, immigration, labour flow, the emergence of the self in world politics demanding the right to determine its future, ethnicity, appearance of new nations and new regions, global recessions, monetarism and outright plunder, have ruled out any lasting impact of cold war and the way this 'war' has ended. The 'removal of the geopolitical constraints of the cold war'[48] removes the certainties of the preceding decades. It ushers in a regression to pre-existing times.[49]

Globalization, Barbarity, and the Dialogic Demand for Justice

However regression into pre-existing times, if that is what restoration is, is paradoxically linked, most of the time, to globalization, expansionism, and the legitimating ideology of universalism. This has happened partly to justify the act of restoration. But besides the need for legitimacy, there has been always a spring-effect, that is to say, the restored power always needed to follow an expansive policy – a kind of 'roll back' that Brzezinsky advocated in late forties, a policy that betrayed a sense of insecurity, of not being confident of the world. The sense of insecurity increased as the insecurity of the restored power afflicted the restoring power also. Therefore, raw force, barbarity, loss of civility, single minded will to destroy the adversary – all these have marked the global expansion of the victorious powers of both varieties, restored and the restorer. The spread of barbarity, the absolute evaporation of the culture of trust and reconciliation, and the concentration of force to erase the adversary are the characteristics of the new global age. This is facilitated, as the object of that force is not the vanquished power, which has left the stage in any case, but the anonymous people, the subjects without history, who do not have any more the state to protect them, but who raise the spectre of lawlessness. It is of that globalization that no one speaks of, and which this book has tried to discuss admittedly haltingly.

Michael Ignatieff in *Blood and Belonging* speaks of the almost semi-erotic gun culture of the checkpoints and contrasts it with the responsibility of guns in yester-times when violence reminded of the need to be responsible. You had the means of violence, so you had to responsible for you could not use them indiscriminately.[50] But that responsibility was gone with each period of restoration when the victorious power used violence indiscriminately, to show that the event of restoration was final. Thus the defeat of every anti-colonial uprising was followed by bloodbath. The Mutiny of 1857 in India, the Boxer uprising in China, or the uprisings in Vietnam had been drowned in the blood of genocide. Similarly, the victorious army after the Paris Commune in 1871, the state forces, hoodlums, and the gangsters in Indonesia in 1965-70, the fascists in Chile in 1974-75, and Lebanese rightist forces and the Israeli army in Beirut in 1983 – all pursued a policy of deliberate terror through indiscriminate use of violence. Then came chemical warfare, gas warfare, biological warfare, and above all production of more nuclear bombs that had been dropped once on the civilians of the two defenceless cities; all these showed how spread of barbarity accompanied waves of globalization. Hitler's methods, initially considered as exceptional, quickly gained acceptance wide across the globe – from French methods in Algeria, British methods in Malay, US methods in Vietnam, and Serbian methods in the Balkans to Pakistani methods in East Pakistan, and the games of the warlords in sub-Saharan Africa. Several countries not mentioned here have been no better. The issue is why and how are globalization and the spread of methods of genocide organically connected? It is here, it would be well to remember, local and the global are truly linked in a 'world history', or to be more precise, linked in a world of what Sankaran Krishna terms as 'mimetic histories'.[51]

Spread of technology and the scientific community, diffusion of methods of coercion and of counter-insurgency, privatization and eventual breakdown of the state, and the sinister air of virtual reality, all have played their role in universalizing terror, mass murder, and in the manufacture of global silence over such genocides accompanied by selective protests.[52] Clearly we are witnessing the eclipse of a sense of responsibility, a decrease of the area of shame, and a rise of values that are good on pride and not on shame.[53] The demise of ideology, we often forget, brings in the demise of the moral community also, howsoever the presence of such moral community might have conveyed at some time a sense of freedom-less-ness. In this era of virtual reality characterized by a lack of territorial constraints and mass public pressure, concentration of attack has ushered in revolution in 'military affairs' – sophisticated electronic and psychological warfare as shown in Kosovo – which has not been lived up to its claims of targeted annihilation, but certainly successful to a significant extent in manufacturing global silence.[54] The 'new wars' that we witness are products of globalization, which they in turn advance. We can advance four theses here.

(a) First, the civil wars of mutual claims of recognition and determination have done away with every sense of responsibility, restoration of trust, and a just reconciliation of claims. Precisely because old territorial forms of political living (territorial units such as a state, county, province, autonomous republics in a multinational state, nation-state) are considered inadequate or irrelevant by the political classes in many parts of the world, that the grab is up for new forms of territory as a matter of life and death struggle in the politics of recognition and determination. In this sense, new wars do not signify banal geopolitics.

(b) Second, the new wars on the basis of the RMA cannot be fought without a political consensus (G 7, new north-new south, Christian, western, Atlantic), and the RMA thus provokes globalization of confrontationist politics. It may not be preposterous to argue therefore that neutrality today is more difficult than it was in the days of cold war/long peace.[55]

(c) Third, a discriminatory history of responsibility encourages the global powers to wage new wars. That, responsibility too has a discriminatory history should not surprise us. Daniel Warner reminds us that when Max Weber was making the well-known distinction between an ethic of responsibility and an ethic of ultimate ends, using the figure of Martin Luther as the ultimate responsible man (in other words responsible to his own self ultimately), he was advocating irresponsibility towards others, who constituted their existence outside the universe (his person, his group, his state, or his followers) of the 'mature man' to whom responsibility remained limited to his universe only.[56] Warner says that even a communitarian perspective is not enough in forging a new ethic of responsibility. 'The very search for responsibility/community may itself be a part of a process of identity politics that denies new possibilities of openness. Understanding why we want to know who is responsible and the limits of that responsibility may be as important as, if not more important than, the process of determining responsibility that is so developed in international law...It is only when we appreciate the politics of identity and the closure that animate our desire to know that we can begin to glimpse new possibilities of responsibility/community outside of or beyond the limitations of that desire.'[57] In this sense, we can say that even the history of responsibility and reconciliation reaches the post-colonial subjects in a determined form.[58] As Mark Selden wrote in the report 'On Asian Wars, Reparations, Reconciliation', Germany unconditionally apologized for war crimes, the US did not own responsibility for war crimes in Vietnam, but went on forcing responsibility on other states for offences; none of the Latin American states has accepted responsibility for mass murders in seventies and eighties of the last century, and Japan has consistently refused to own blanket responsibility for rape and enslavement of nearly two hundred thousand women ('comfort women'), particularly after the Nanjing massacre.[59] This history of responsibility symbolized by a skewed war crimes tribunal at The Hague does not question the flagrant discrepancy, and is shown to be what it is by parallel trials such as the Women's International War Crimes Tribunal on Japan's Military Sexual Slavery that sat in Tokyo on 8-12 December 2000, or the trials by the citizens for deaths of thousands of people in India, Sri Lanka, and Pakistan and elsewhere in state-backed communal riots and frenzy of the security forces.[60] The discriminate nature of legal and moral justification for 'international' war crimes tribunals, set up by resolutions of the UN Security Council to punish crimes in only two areas in the world so far – the former Yugoslavia (for which a tribunal was set up in 1993) and Rwanda (1994) is evident. A U N press release issued last November quoted the President of both tribunals, Carla Del Ponte, complaining that the Yugoslav tribunal's 'historical credibility' was being 'undermined' by 'forced inaction on what had happened in Kosovo since June 1999'; and that the tribunal was still 'faced with difficulties, related primarily to the number of accused who remain at large'. As the Guardian correspondent, Paul Foot has pointed out in a dispatch, that she avoided any reference to the much more serious difficulties that had undermined the tribunals' historic credibility.[61] And these difficulties arose from the tribunals' limited scope, which had left them open to the charge that they were concerned only with war crimes committed in areas convenient to US foreign policy; and that the whole paraphernalia of courts and prosecutors set up by the UN Security Council represented yet another arm of the US state department. The most shocking example of the tribunals' impotence has been the state of Israel, which has continued to behave in the most flagrant contempt not only of the UN, seizing tracts of territory and exploiting and enslaving the inhabiting people, but also of the most basic laws and conventions covered by the statute of the international war crimes tribunals and written into international law by Security Council resolutions such as power to prosecute persons 'committing or ordering to be committed grave breaches of the Geneva conventions' including wilful killing, torture or inhuman treatment, wilfully causing great suffering or serious injury to body or health, and extensive destruction and appropriation of property not justified by military necessity and carried out unlawfully and wantonly. If there is any doubt about this, one has to consider only Article 3 of the Tribunals' statute that which refers to the 'wanton destruction of cities towns and villages or devastation not justified by military necessity', or Article 5 that empowers the Tribunal to prosecute persons responsible for crimes against humanity, including murder, if it is 'committed in armed conflict and directed against any civilian population'. Paul Foot has further commented on the most iniquitous regime of responsibility, as there is no need even for potential prosecutors to delve into Sharon's gruesome history – such as his role in the slaughter of over 60 Palestinians in the village of Qibya in 1953, documented by Israeli historian Benny Morris; or his responsibility as Israeli defence minister for the butchery of at least 2,000 people in the Lebanese refugee camps of Sabra and Shatila in 1982 (the climax of Sharon's invasion of Lebanon, denounced as illegal by the UN and even by Margaret Thatcher). Indeed, the daily breach by Israeli troops of the tribunals' own statute should pose a challenge to the tribunals' judges, prosecutors and administrators. The fact therefore is that, the emerging official discourse on responsibility and reconciliation remains completely insensitive to the truth that no healing is possible without reconciliation, no reconciliation possible without justice, and no justice is possible without some form of restitution.

(d) Fourth, the history of wars and militarism in the third world in the global mirror loses its own social and economic specificity. Exactly twenty years ago Irving Horowitz in a seminal work on third world militarism showed these specificities, and the specifics of democracy there. In his words, 'The third world is faced with a set of false options...Democracy is not addressed by politics or economics but has to do with health, education, welfare and the social uses of the public domain'.[62] Studying 'globalizing wars and integrating world systems', military nationalism and the state in the third world, issues of legitimacy and illegitimacy of third world rules, and 'the zero-sum economics and egalitarian politics', he further commented, 'The lesson is clear – While one cannot predetermine dictatorial or democratic outcomes, neither should one prejudge the possibility of outcomes simply on the basis of the military origins of many Third World Revolutions'.[63] Written at the heydays of three worlds theory, Horowitz still rings true in the post-1989 age of democratic transition and restitution.

All in all, this is a curious history of globalization; and this curious history of globalization that predicates the contentious politics of dialogue should not escape our eyes. Community, state, nation, region, civilization, and the globe, all have been dumped into the box that the current history of globalization is. They may jostle for space. But that is fine with today's global order. Similarly, the 'internationalism of Gillette's chairman' and the one 'expounded by Tagore or Gandhi' both find accommodation in this current history.[64] This history is producing regions at a furious pace that may work as containers of all these entities. Asia was never a whole to justify its name. It was either east or south or the near east or central or the rim named as Asia-Pacific.[65] Similarly, Europe had never been Europe – the two treaties of Locarno and Rapallo exemplified the agony of trying to make sense of the many Europes in one Europe, an agony that still continues.[66] It all depends on the gaze. The production of political regions to bring the disparate political entities into some order and tie them on a template has become global. In reading this curious history where a sudden expansion of finance market, and increase of mobility of capital, information and skill are invoked to assert the supremacy of the monetarist economics and the end of the age of political confrontation (with the possible exception of the clash between civilizations, but that is not a political clash, but a civilization-clash, so the argument would run), we shall do well to remember that behind our back, almost escaping our gaze, a new global politics is shaping up. The hazy contours of that politics show the signs of a new moral community that tries to find a balance between the self and the outside, attempts to acquire a global character that is evinced in the huge public protests against the policies of the World Trade Organization (WTO), the World Bank-IMF (International Monetary Fund), the World Economic Forum, and the formation of an all-American Free Trade Zone (of course excluding 'dictators and communist countries') speaks out against a 'runaway world', criticizes inhumane labour conditions, risky technologies, violations of human rights, trans-national corporations, and emphasizes a new ethics marked by a respect for plurality, accommodation, sense of responsibility, and a notion of just reconciliation.[67]

There is no doubt that the current phase of global politics is pushing for a new dialogue for justice. It does not mean however, that this politics will be no less confrontationist than the politics of the cold war. If the protests at Genoa are an indication, rising before us are the new questions, new dialogues, new confrontations. Even though the grand statesmen of the West assembled in Genoa on 20 July this year would not have liked the images of conflict that their summit left in its wake, the media that they cannot do without had a satisfying day with some help from the city's *carabinieri*. Carlo Giuliani, a young political activist, was shot dead at the venue of the G-8 summit; hundreds were injured. The worldwide audience of TV onlookers since three years ago in Seattle, and then successively in Prague and Gothenburg, could not but be bemused at how this gathering of the cultured representatives of the trans-Atlantic ruling class, assembled to discuss ways and means of reducing poverty, had provoked such vehement opposition, leaving one dead, still others mutinous, and people across the globe derisive of the pomp and circumstance that accompanied the summit.

The protesters with the grassroots trade unions (SLAI-Cobas) were in a combative mood and well prepared to make the 20 July 'rebellion in Genoa' a success. Among those present in the ancient mercantilist capital were delegations from Yugoslavia, Mexico, Turkey, Greece, Sardinia and elsewhere. As expected, the demonstrators represented various views and orientations – unions, political activists (including anarchists and communists), development NGOs, religious and humanitarian groups, human right activists and a whole lot of others attracted to Genoa by the assembly of Big Men. The slogans ranged from the demand to withdraw NATO from Yugoslavia and the Balkans, to the dissolution of Hague 'inquisition', to support for the Palestinian *Intifada,* and opposition to the embargo against Iraq and Cuba. However, the general theme that cut across the disparate groups was in the arena of global economics and politics – the need for 'globalization from below', a reduction of the current Third World debt burden, and the control of international capital flows by 'civil society' rather than by the international corporations and the governments at their beck and call.

Just a week after the gathering of the big men in Genoa, a much-less-publicized anti-imperialist Camp was organized in Assisi, the mediaeval town of monks and mendicants, also in Italy. It was dedicated to developing political perspectives for the anti-globalization movement, which the organizers felt was at a decisive stage following the G-8 meeting and its well-publicized fallout. The violence in Genoa were seen as a signal to the movement – how to continue the protests against globalization, establish 'civic control' over the emerging agenda in the wake of globalization, and stand firm to confront the forces that are pushing the one-way economic globalization. The continent most devastated by capitalism was present in Genoa as it was in Assisi: Lumumbist groups from Zaire-Congo whose country is being torn apart by Western intervention; people like Dr. Bashir Kurfi, an anti-imperialist intellectual from Nigeria; and activists from Sierra Leone, Senegal, Chad, Guinea...The delegates from West Asia questioned the West's management of peace in the context of the second *Intifada*, and sought lessons from the collapse of the Oslo agreement. Other activists pointed to how the Western powers were courting Turkey for the sake of its military machine even while in Ankara's jails political prisoners were engaged in fasts unto death. In Assisi, the revolutionary women from Afghanistan explained their struggle against the Pakistan-backed Taliban, while the Filipino members of the 'Migrante' suggested the political possibilities available to Asian migrants in the West. The Mexicans warned of how the struggle for state power could be forgotten when 'civil society' issues take over the agenda, as it had in their country in these times of *neo-Zapatism*.[68]

Just as globalization has many faces, so too does the protest in its opposition. The rebellion in Genoa thus raises a problematic. By embodying contradictory phenomena, it asks of politics a whole series of questions that used traditionally to be part of the elite domain. If the meeting of G-8 was a manifestation of international democracy at work, how could such pomp and glory presume to represent the majority of people on this planet? How could those who claim to care for the poor and the victims allow 'humanitarian bombs' and transnational murders? Was it possible for the manifold nature of the Genoa protest to be rewritten in a coherent standard political format, making it more powerful? That is what was requiring an answer, beyond the agitation and stone throwing.

There were many images available at Genoa, starting with the ambition of the trans-Atlantic club and the failure of its rhetoric. Then there was the desperate effort of the protesters to increase their ranks through global means, and the mobs' resolve to spit on the moral indignation expressed against the agitation by the political leaders of that club. The unreality of the agenda of the super-rich to fight poverty was mirrored in the death of the protestor on the Genoa pavement. It was a hallucinatory world all right. Just as it happened in the case of Raskolnikov's murder of his landlady, in Dostoyevsky's *Crime and Punishment*, the killing of the youth in Genoa has left its clues everywhere. That lone, but symbolic, death now guarantees that henceforth the protests will continue wherever the symbols of global power present themselves – demanding that the global powers must talk with affected people on decisions that affect their lives. The G-8 may hereafter move their annual hermitage into the deepest forest of Canada or the furthest corner of Tasmania, but it does not look like they can easily meet amidst the glass towers and wide boulevards any longer. The docile welcome of leaders that used to happen in the past will be a memory, with the protestors forcing them to remember the wealth that they represent – and not the world that they would presume to.

But an assembly of emperors away from the bustle and din of crowd – will it serve the purpose? To ask this is to ask, *what* precisely is the purpose? Again, in Dostoyevsky's hands punishment and crime were not events in sequel; they were one. Crime was punishment. The moment of commitment of the crime was the moment when punishment had begun. The long shadows outside the palaces and glass hotels of Seattle, Gothenburg and Genoa are there to stay. The protestors had wanted the moments of assembly to become moments of derision and disbelief and now they have become that. An assembly deep in the forest or in a faraway island cannot attract glory. Montesquieu had reminded us more than two centuries ago, how the pursuit of glory was important for kingdoms. Modern republics, as kingdoms of today, pursue policies that promise glory, for glory brings in its wake power, wealth, and satisfaction. Hence the dilemma, how to pursue glory without murder, pomp without derision, riches without poverty, and democracy without coercion?

More than to those assembled at Assisi, this will remain a question to those who had met at Genoa.

The fault-lines of this new interrogative politics are not yet clear. But there is no doubt that in this politics we shall see the engagement of the principle of justice[69] with that of claims and rights, of the ethic of responsibility with the reality of power unconstrained by responsibility, of the politics of dialogue with that of coercion, of our history of wars and peace with their history of the same, an engagement of our statelessness with their states. Above all, the politics of dialogue is pushing forward the demand now being made on States and powers to assume responsibility – responsibility of the state to guarantee human rights, international criminal responsibility, demand on the great powers to accept the reality of the incipient forms of justice, and the responsibility to avoid unilateral actions, and engage in dialogue The early marks of these engagements are visible in the forms of new political realities in many parts of the world. They give lie to the claims of this curious history, that globalization promotes trade, trade promotes development, development produces interdependence, and interdependence produces peace. We can use one phrase that sums up the cinema, *liberal peace.*[70]

Notes

1 John Lewis Gaddis, 'The Long Peace – Elements of Stability in the Postwar International System' in Gaddis, *The Long Peace – Inquiries into the History of the Cold War* (New York: Oxford University Press, 1987), p. 216.

2 Ibid. p. 245.

3 Contrary to what many international relations experts think, 'peace through deterrence' is not a new idea; it is as old as the beginning of the realist school. See, Warn Bullogh and Raoul Naroll, *Deterrence in History* (New York: State University of New York Press, 1969). Apart from the philosophical basis that Hobbes provided, to which I referred in the beginning (chapter 2), the implicit arguments of the modern deterrence school bear an extra-ordinary similarity with the views that German sociologists were advocating several decades ago in course of their theory of what they had termed as 'realpolitic'. Thus, Ratzenhofer said,

The idea of perpetual peace is a relic of a class of ideas which supposes the state to have arisen through Rousseau's social contract...It is true that the human race may sometimes approach the ideal condition of peace...but we are still far removed from this stage, and the peace idea propagated by *women of both sexes* is fallacious of and injurious to its advocates' (emphasis mine) – Ratzenhofer cited by James Alfred Aho, *German Realpolitik and American Sociology – An Enquiry into the Sources and Political Significance of the Sociology of Conflict* (London: Associated University Press, 1975), p. 46.

Aho notes the 'German slough of romantic despondency' from which the 'sociology of conflict begins' and citing the work of Gumplowicz he also shows how the German sociologists made conflict the reason of sociology's existence (pp. 14 and 39). The writings of the German sociologist Georg Simmel were most influential. In *Conflict and the Web of Group Affiliation* (1908), trans. Kurt H. Wolff (New York: Free Press of Glencoe, 1955) he wrote of conflict as an integrative factor vital for association, hostility as a primary factor in human nature, and competition as socializing, objective, and civilizing. (pp. 13-84).

4 This was in his mind when the former UN Secretary General Boutros-Ghali during his visit to Sarajevo in 1992 had said, 'Bosnia is a rich man's war. I understand your frustration, but you have a situation here that is better than ten other places in the world...I can give you a list.' – cited in Richard Holbrooke, *To End a War* (New York: The Modern Library, 1998), p. 175. Human misery is not comparable to be made banal, particularly misery in a war that took about two hundred thousand lives. But it remains undeniable that some misery attracts rich man's attention while others do not, but whether that is to good effect or not is another matter.

5 Yet it must be remembered that in the Gulf as much in Vietnam it was a multinational army, the loss in terms of human life was also colossal if not as much, new weaponry deployed with as much rapidity, peace efforts by third parties as desperately and as carefully made.

6 Justin Rosenberg, *The Follies of Globalisation Theory* (London: Verso, 2000), p. 61.

7 Luis Moreno Ocampo, 'Beyond Punishment – Justice in the Wake of Massive Crimes in Argentina', *Journal of International Affairs*, 52 (2), Spring 1999, pp. 669-689.

8 Geoffrey Blainey, *The Causes of War* (New York: Free Press, 1977 paperback, and then 1988), cited in *The Long Peace*, p. 217.

9 On the genealogy of area consciousness I have written elsewhere in details in *Whose Asia is it Anyway? Region and the Nation in South Asia* (Calcutta: Pearl Publishers, 1996).

10 Examples of such revisionist accounts, Joyce Kolko and Gabriel Kolko, *The Limits of Power – The World and The United States Foreign Policy, 1945-1954* (New York:

Pantheon Books, 1972); Thomas G. Paterson, *Soviet American Confrontation – Postwar Reconstruction and the Origins of the Cold War* (Baltimore: Johns Hopkins, 1973).

11 Cited in 'The Long Peace', p. 218. It is also significant to note in this connection that 'between September 1945 and 1988, 105 states intervened one or more times in 131 other states and non-self governing territories, and these actions represented at least 639 interventions' and at least 269 armed conflicts started between 1945 and 1988. – Herbert K. Tillema, *International Armed Conflict since 1945 – A Bibliographic Handbook of Wars and Military Interventions* (Boulder, Colorado: Westview Press, 1991), p. 6. See also in this connection Lewis F. Richardson, *Statistics of Deadly Quarrels* (Pittsburg: Boxwood Press, 1960). In South Asia, for example, the conflicts like those in Kashmir, Sind, Sri Lanka, the CHT, and Afghanistan have shown with remarkable consistency both internal and external linkages. One insurgency analyst has chosen to describe the situation in terms of 'crossfire'. See Subir Bhaumik, *India's North East – Insurgency Crossfire* (New Delhi: Lancers, 1996).

12 Radha Kumar, *Divide and Fall? Bosnia in the Annals of Partition* (London: Verso, 1997); see also the review essay by Arvind N. Das, 'The End of Geography', *Biblio*, March-April, 1998, pp. 8-9; on this theme further, Clive J. Christie, 'Partition, Separatism and National Identity – A Reassessment', *Political Quarterly*, 63 (1), January-March 1992, pp. 68-72. However Christie's argument about the need to separate two kinds of partition, ideological and identity generated, can be accepted only in a broad format of decolonization where the simultaneous processes of partition, decolonization and new state formation were ushered in by war and the decline of the old imperial tradition.

13 Indeed only with the reappearance of 'one' Europe, we now see what the partition of Europe did to the continent. The least that we can say is that with old rivalries reappearing and the states system in Central and Eastern Europe being redrawn, Europe often reminds us of an older stage in state formation. The impact of the post 1945 arrangement of spheres of influence in East Europe in particular does not always seem very profound. For a relevant discussion, see Geza Jeszensky, 'More Bosnias? National and Ethnic Tensions in the Post-Communist World', *East European Quarterly*, 31 (3), September 1997; and Sarah A. Kent, 'Writing the Yugoslav Wars – English Language Books on Bosnia (1992-96) and the Challenges of Analysing Contemporary History', *American Historical Review*, 102 (4), October 1997.

14 One has to only see some of the debates in the European circles of political thinking reflected in the pages of *The European Journal of International Affairs* to have an idea of how Europe thinks today of the 'Americanisation of the continent'. For example, 'An Interview with Mark Eyskens', the previous Prime Minister of Belgium, 7, Winter 1990; 'A Conversation with Joachim Fest', the co-editor of *Frankfurter Allgemeine Zeitung*, Ibid.; Giuseppe Sacco, the well-known columnist, 'Europe and the World', 10 (4), 1990; the whole section on 'De Gaulle and his Century', 9 (3), 1990; 'Britain in Today's World – An Interview with Michael Heseltine', a prominent British conservative politician, 6, Autumn 1989; and the two essays in the same number, William Safran, 'Is France becoming America?' and Marie-France Toinet, 'Convergence in Disguise'.

15 On war fatigue of great powers and the curious combination of post-Vietnam mood of isolationism and a lingering desire of world role, one of the most revealing accounts is that on the US role in the Balkan crisis, particularly in the Dayton negotiations, by the chief US negotiator, Richard Holbrooke, *To End a War* (New York: The Modern Library, 1998), Chapter 15, pp. 215-227.

16 Johan Galtung, 'Regional Security Commissions – A Proposal' in Galtung and Sverre Lodgaard (eds.), *Co-operation in Europe* (Oslo and Assen: Universitetsforlaget and International Peace Research Institute, 1970), pp. 73-83.

17 The mystery seems to be most to the celebrated campaigners of peace. Of the nine Nobel Peace Prize recipients in 1971-80 only two were concerned with human rights violations and popular protests: Amnesty International (1977) and Adolfo Perez Esquivel (1980). The rest, typical of the worries of a world where only the presence or absence of nuclear weapons counted and the typical conflict-resolution mumbo-jumbo of the cold war age were – to mention a few: Willy Brandt (1971), Henry Kissinger (1973), Eisaku Sato (1974), Andrei Sakharov (1975) and Menachem Begin and Mohamed Anwar Al-Sadat (1978). Again in 1981-1990, Lech Walesa (1983), International Physicians For The Prevention of Nuclear War (1985), XIVth Dalai Lama (1989) and the most singularly incapable of addressing the question of war and peace in the third world, Mikhail Sergeyevich Gorbachov (1990). The Nobel Peace Lectures are also instructive as they show how peace movement in the two decades had only the Atlantic world in mind when they spoke of 'global' dangers. See the 2 volumes of Tore Frangsmyr (ed.), *Nobel Lectures – Peace, 1971-1980 and 1981-1990* (Singapore: Scientific Press, 1997).
The myopia is addressed partly in Mary Durfee and James N Rosneau, 'Playing Catch-Up: International Relations Theory and Poverty', *Millennium: Journal of International Studies*, 25 (3), 1996, pp. 521-545.

18 On the issue of peace-keeping, peace-enforcing and helping the state to stand on its feet there is a growing literature which begins of course with the evaluation of the Military Staff Committee of the UN. See, for example, Jonathan Soffer, 'All for One or All for All – The UN Military Staff Committee and the Contradictions within American Internationalism', *Diplomatic History*, 21 (1), Winter 1997, pp. 45-69; Eva Bertram, 'Reinventing Governments – The Promise and Perils of United Nations Peace Building', *Journal of Conflict Resolution*, 39 (3), September 1995, pp. 387-418; D. Holiday and W. Stanley, 'Building the Peace – Preliminary Lessons from El Salvador', *Journal of International Affairs*, 46 (2), 1993, pp. 415-438.

19 The triumphalist literature is extremely varied. However the reader can get an idea of the uneven texture from Francis Fukuyama, *The End of History and the Last Man* (London: Penguin, 1992); J.L. Gaddis, 'International Relations Theory and the End of Cold War', *International Security*, 17 (3), 1992-93; Paul Kennedy, *Preparing for the Twentieth Century* (New York: Random House, 1993); W.C. Wohlforth, 'Realism and the End of the Cold War', *International Security*, 19 (3), 1994-95; A. Varsori (ed.), *Europe 1945-1990s – The End of an Era?* (London: Macmillan, 1995); Thomas J. McCormick, 'Troubled Triumphalism – Cold War Veterans Confront a Post-Cold war World', *Diplomatic History*, 21 (3), 1997; S.P. Huntington, 'The Clash of Civilizations?', *Foreign Affairs*, 72 (3), 1993; Huntington, 'If not Civilizations, What? Paradigms of the Post Cold War World', *Foreign Affairs*, 72 (5), 1993.
The return to nineteenth century in today's discourse of war and peace has to be on some other occasion discussed in greater detail. Not only historians like Gaddis recall the last century while discussing peace, ironically references to cold war also bring back the preceding century. Theodore Draper in the essay, 'Prophets of the Cold War' in his *Present History – On Nuclear War, Detente, and Other Controversies* (New York: Random House, 1983), pp. 371-399, recounts the theme as it had featured in the nineteenth century, and discusses in that context the famous remark by Alexis de Tocqueville made more than hundred years ago about Russo-American control of the globe.

20 Gaddis, 'The Long Peace'; Eric Hobsbawm, *The Age of Revolution: Europe 1789-1848* (London: Weidenfeld and Nicholson, 1975).

21 On the impact of Wilsonianism, M.P. Leffler, *The Elusive Quest – America's Pursuit of European Stability and French Security 1919-1933* (Chapel Hill: University of North Carolina Press, 1979); for an analysis of Wilsonian 'liberal' policies aimed against Bolshevism, David S. Fogelsong, *America's Secret War against Bolshevism – U.S. Intervention in the Russian Civil War* (Chapel Hill: University of North Carolina Press, 1995).

Anthropologists will always remind us that while the ubiquity of the theme of 'return' in current literature of conflict and peace perplexes the international relations researchers and policy managers, modes of conflict resolution may not be that different today from those who do not share the 'international'. On this see, Alean Al-Krenawi and John R. Calgary, 'Conflict Resolution Through a Traditional Ritual Among the Bedouin Arabs of the Negev', *Ethnology*, 38 (2), Spring 1999, pp. 163-174; Pradip Bose, 'The Anthropology of Reconciliation' in Helmut Reifeld and Ranabir Samaddar (eds.), *Peace as Process* (Delhi: Manohar Publishers, 2001).

22 It will be however unfair to claim that there is a complete absence of literature about the question of forming the state anew. For example, some like Christopher Clapham on the basis of the experiences of the state in sub-Saharan Africa speak of 'Degrees of Statehood', and not of one state in 'Degrees of Statehood', *Review of International Studies*, 24, 1998, pp. 143-157; see also Richard Devetak, 'Incomplete States: Theories and Practices of Statecraft' in John MacMillan and Andrew Linklater (eds.), *Boundaries in Question – New Directions in International Relations* (London: Pinter Publishers, 1995), pp. 19-40.

23 On this a recent and thorough guide to the international relations literature is Malcolm Anderson, *Frontiers – Territory and State Formation in the Modern World* (Cambridge: Polity Press, 1997).

24 On the role of the Church in stabilizing post-War Europe, the entire collection of essays in the *European Journal of International Affairs*, 12 (2), 1992, is rewarding.

25 Etienne Balibar, 'Racism and Nationalism' in Etienne Balibar and Immanuel Wallerstein, *Race, Nation, Class – Ambiguous Identities* (London: Verso, 1991), pp. 38-45.

26 Some of the new works on South Asian Diaspora bring out bring out the emerging nature of non-national groupings and bring into light the phenomenon of immigration predicating nationalism and ethnicity. See for example the detailed essay by Papiya Ghosh, 'Who Went Where and How are They Doing? Pakistanis and Indians outside South Asia', *Refugee Watch*, 12, December 2000, pp. 24-33.

27 Clive J. Christie's argument in 'Partition, Separatism and National Identity – A Reassessment' about the need to separate two kinds of partition, ideological and identity generated, can be accepted only in a broad format of decolonization where the simultaneous processes of partition, decolonization and new state- formation were ushered in by war and the decline of old imperial tradition.

28 I have discussed these in my book, *The Marginal Nation – Transborder Migration from Bangladesh to West Bengal* (Delhi: Sage Pub, 1999); on this see also the essays in Ranabir Samaddar (ed.), *Reflection on Partition in the East* (Delhi and Calcutta: Vikas Pub. and Calcutta Research Group, 1997).

29 On the entire Chakma question including its impact on the Northeast, Subir Bhaumik, Meghna Guhathakurta and Sabyasachi Basu Ray Chaudhury (eds.), *Living on the Edge – Essays on the Chittagong Hill Tracts* (Kathmandu and Calcutta: South Asia Forum for Human Rights and Calcutta Research Group, 1997); for a wide ranging survey of the conditions of uprooted peoples of South Asia readers will certainly gain from, Tapan Bose and Rita Manchanda (eds.), *States, Citizens and Outsiders – The Uprooted Peoples of South Asia* (Kathmandu: South Asia Forum for Human Rights, 1997).

30 For an understanding of this we remain indebted to Bertil Lintner for his monumental work, *Burma in Revolt – Opium and Insurgency since 1948* (Bangkok: White Lotus, 1994).

31 T.K. Oommen, for example, notes the conflation, but considers such conflation a mistake. My argument has been however the opposite as I try to show that the reality of massive immigration and its links to statehood makes such conflation an understandable exercise. See his introductory essay, 'Conceptualising the Linkage between Citizenship and National Identity' in Oommen (ed.), *Citizenship and National Identity – From Colonialism to Globalism* (Delhi: Sage, 1997), pp. 13-51.

32 On this the reader may find the papers contributed to the Strasbourg Conference on human rights hosted by the Council of Europe in 1995 helpful. See, Jurgen Axer (ed.), *The Human Rights Community and Conflict Resolution in South Asia – The Applicability of European Examples* (Brussels and New Delhi: Friedrich Naumann Stiftung, 1996).

33 A. Minc cited in Alfonso Alfonsi, 'Citizenship and National Identity – The Emerging Stirrings in Western Europe' in *Citizenship and National Identity – From Colonialism to Globalism*, p. 64; on Europe see also, D. Thranhardt (ed.), *Europe – A New Immigration Continent* (Lit, Munster, 1992); also David Carment and Patrick James, 'Internal Constraints and Interstate Ethnic Conflict – Towards a Crisis-based Assessment of Irredentism', *Journal of Conflict Resolution*, 39 (1), March 1995, pp. 82-109.

34 It is significant to note that to one of the close observers of the migration scene the issue of trans-border population movement does not appear as a problematic in the international relations theory, it is at best a problem for the state to manage, particularly in terms of security – see Myron Weiner's essay, 'Security, Stability and International Migration' in Sean M. Lynn-Jones and Steven E. Miller (eds.), *Global Dangers – Changing Dimensions of International Security* (Cambridge, MA: The MIT Press, 1995), pp. 183-218. However, a perusal of the international instruments and other legal texts will bring out the point. See for example, Convention relating to the Status of Stateless Persons, 28 September 1954, UN Treat Series 5158, Vol. 360, *Official Records of the Economic and Social Council, Seventeenth Session, Supplement 1*, E/2596; Convention on the Reduction of Statelessness, UN Document A/CONF, 9/15, 1961, *Official Records of the General Assembly, Ninth Session, Supplement 21*, A/2890; Convention for the Reduction of Cases of Statelessness, adopted by the International Commission on Civil Status in Berne, 13 September 1773; and finally Special Protocol Concerning Statelessness, adopted by the Hague Conference of Plenipotentiaries, League of Nations, 12 April 1930, entry into force 0n 10 December 1973.

35 For a broad survey, Varginia Yans-MacLaughlin (ed.), *Immigration Reconsidered – History, Sociology and Politics* (New York: Oxford University Press, 1990), see 'Introduction'.

36 Gil Loescher, and John Scanlan, *Calculated Kindness – Refugees and America's Half-Open Door, 1945-Present* (New York: Free Press, 1986).

37 This uncertainty is evident in any common study in the field of international relations today. For example, Sita Bali, 'Migration and Refugees' in Brian White, Richard Little and Michael Smith (eds.), *Issues in World Politics* (London: Macmillan, 1997), chap. 10; today it is inconceivable that any such study will leave out the 'new issues', a stark contrast to the power-studies and system-studies of yesteryears; but these new studies still carry the traces of an uncertain comprehension.

38 Etienne Balibar, 'The Borders of Europe', trans. J. Swenson, in Pheng Cheah and Bruce Robbins (eds.), *Cosmopolitics – Thinking and Feeling Beyond the Nation* (Minneapolis: University of Minnesota Press, 1998), p. 219.

39 I discuss this process in details in *A Biography of the Indian Nation, 1947-97* (Delhi: Sage Publications 2001).

40 Anita Sengupta in a remarkable essay, 'Imperatives of National Territorial Delimitation and the fate of Bukhara, 1917-1924', *Central Asian Survey*, 19 (3-4), 2000, pp. 397-418, shows how the modern imperatives of national territorial delimitation ended the old spiritual existence of Bukhara, which was neither Uzbek nor Tajik, and yet, how this delimitation of boundaries failed to provide security of identity, and Bukhara's mid-way nature came to symbolize the political destiny of bounded entities in Central Asia. For further details on the politics of border-making in modern Central Asia, see Anita Sengupta, 'Territorial Delimitation in Central Asia, 1917-27' in Ranabir Samaddar (eds.), *Space, Territory and the State – New Readings in International Politics* (Hyderabad: Orient Longman, 2002).

41 John Lewis Gaddis however never found any reason for reflection on his method of differentiating epochs of war and peace. He has been quite sure that historical knowledge is enough for such making such distinction. In fact, in an essay, 'History, Theory and Common Ground', *International Security*, 22 (1), Summer 19997, pp. 75-85, and subsequently published in a revised form as 'In Defense of Particular Generalization – Rewriting Cold War History, Rethinking International Relations Theory' in Colin Elman and Miriam Fendius Elman (eds.), *Bridges and Boundaries – Historians, Political Scientists and the Study of International Relations* (Cambridge, MA: The MIT Press, 2001), pp. 301-326, he remarked that historians were doing well without what he termed as 'grand patterns of history', and historian's 'distrust of theory' in the sense of speculation that did not trace processes over time. Thus while he said that historians need not be 'closet theorists', his own reflections on cold war history (*We Now Know – Rethinking Cold War History*, New York: Oxford University Press, 1997) did not consider the possibility of a theoretical lacuna latent in 'grand theorizing the cold war' in the sense of universalizing a strictly local phenomenon.

42 Benedict Anderson, *The Spectre of Comparisons – Nationalism, Southeast Asia and the World* (London: Verso, 1998), ch.3, 'Long-Distance Nationalism', pp. 58-74. Ben Anderson characterizes long-distance nationalism in this way, '...First of all, it is the product of capitalism's remorseless, accelerating transformation of all human societies. Second, it creates a serious politics that is at the same time radically unaccountable. The participant rarely pays taxes in the country in which he does his politics; he is not answerable to its judicial system; he probably does not cast even an absentee ballot in its elections because he is a citizen in a different place; he need not fear prison, torture or death, nor need his immediate family. But well and safely positioned in the First World, he can send money and guns, circulate propaganda, and build intercontinental computer information circuits, all of which can have incalculable consequences in the zones of their ultimate destinations. Third, his politics, unlike those of activists for global human rights or environmental causes, are neither intermittent nor serendipitous. They are deeply rooted in consciousness that his exile is self-chosen and that the nationalism he claims on e-mail is also the ground on which an embattled ethnic identity is to be fashioned in the ethnicized nation-state that he remained determined to inhabit. That same metropole that marginalizes and stigmatizes him simultaneously enables him to play, in a flash, on the other side of the planet, national hero'. (p. 74).

43 Students of Indian politics are familiar with this term, most used in comments on elections.

44 One of the significant writings where the disenchantment with the received history of the global is discussed as a complicated process is Achille Mbembe, 'African Modes of Self-Writing', *Identity, Culture & Politics – An Afro-Asian Dialogue*, 2 (1), pp. 1-39.

45 I borrow this phrase from the title of Pranab Bardhan's essay, 'Social Justice in the Global Economy', *Economic and Political Weekly*, 36 (5-6), 3-10 February, 2001, pp. 467-480, where some of these issues are discussed at length. However unlike Bardhan, I would place the issue of justice in the context of the global regime of power, and not in the context of the scope that the pursuit of justice can secure, Bardhan thinks, within the structure of the current phase of globalization. To me it seems important to stress the limits of these micro-searches for justice within the structure of globalization, because while these searches carry instructive lessons, their limits show the imperative for conceptualizing a new politics of justice – thinking and feeling beyond the nation-form and globalization, both of which have become banal in terms of ideas.

46 Two narrative accounts which suggest the continuity I speak of here have influenced my view: Neal Ascherson, *Black Sea* (London: Jonathan Cape, 1995,) and the set of essays on riots in modern history in Paul R. Brass (ed.), *Riots and Pogroms* (London: Macmillan, 1996).

47 William Zartman admits that the asymmetries make any politics of negotiations to end civil wars uncertain. The case studies presented by him including ones that are around non-state persons show that these conflicts with inbuilt asymmetries make the state once again an issue of international relations. See, I. William Zartman (ed.), *Elusive Peace – Negotiating an End to Civil Wars* (Washington D.C.: Brookings Institution, 1995), pp. 7-12.

48 Indeed the historiography of cold war brings out the still inconclusive debate as to the relative importance of geo-political considerations and cultural factors in the making of cold war and suggests that uncertainties were there possibly still then and contributed greatly to the genesis of the cold war. Massive immigration today has brought back the uncertainties, which had been successfully covered by the certainties of the cold war decades. For an idea of the ambiguities that lay in the genesis of the cold war the best guide may be Michael J. Hogan (ed.), *America in the World – The Historiography of American Foreign Relations since 1941* (New York: Cambridge University Press, 1995); also by Hogan, *The End of the Cold War – Its Meaning and Implications* (New York: Cambridge University Press, 1992).

49 On globalization and fragmentation, I have immensely gained from the succinct historical overview of the two interlinked processes in Ian Clark, *Globalization and Fragmentation – International Relations in the Twentieth Century* (New York: Oxford University Press, 1997).

50 Michael Ignatieff, *Blood and Belonging* (London: Routledge, 1993).

51 Sankaran Krishna uses the term in his *Postcolonial Insecurities – India, Sri Lanka, and the Question of Nationhood* (Delhi: Oxford University Press, 2000), chapter 1, p. 3.

52 See for example, Noam Chomsky, *World Orders – Old and New* (London: Pluto Press, 1998). Andrew Bell-Fialkoff admirably brings out the epidemic nature of practices of genocide in *Ethnic Cleansing* (New York: St. Martin Griffin, 1999).

53 Ben Anderson drew my attention to this aspect of globalization in a conference on 'Cosmopolitanism and the Nation State' organized by the Prince Claus Fund of The Hague and the Asian Development Research Institute of Patna, Patna, 22-26 February 2001.

54 Eric Hobsbawm in his insightful lecture on 'Barbarism – A User's Guide' does not note the political reasons and consequence of the globalization of genocide. See Hobsbawm, *On History* (London: Abacus, 1997), pp. 334-350.

It is also important to remember the *almost* reality of this reality and not to lose sight of the fact that this is still not the real. For example, the media may seem to make and then represent the real. But the CNN factor may itself be illusory, which is what virtual reality partly is. The RMA operates in such air-conditioned atmosphere. See the important essay by Peter Viggo Jakobsen, 'Focus on the CNN Effect Misses the Point – The Real Media Impact on Conflict Management is Invisible and Indirect', *Journal of Peace Research*, 37 (2), March 2000, pp. 131-144.

55 Here I draw ideas from the critical geopolitical writings of recent times. See for a summary, Klaus Dodds, 'Political Geography II – Some Thoughts on banality, New Wars and the Geopolitical Tradition', *Progress in Human Geography*, 24 (1), 119-129. This is a three-part essay. Interested readers may read the other two, 'Political Geography I – The Globalisation of World Politics' (1999) and 'Political Geography III – Critical Geopolitics after Ten Years'. I am grateful to Klaus Dodds for allowing me access to this massive summary of critical thought in geopolitics full with insights.

56 Daniel Warner, 'Searching for Responsibility/Community in International Relations' in David Campbell and Michael J. Shapiro (eds.), *Moral Spaces – Rethinking Ethics and World Politics* (Minneapolis: University of Minnesota Press, 1999); Max Weber's argument appears in his lecture, 'Politics as a Vocation' in Hans Gerth and C. Wright Mills (eds.), *From Max Weber – Essays in Sociology* (New York: Oxford University Press, 1946), pp. 77-129.

57 Daniel Warner, 'Searching for Responsibility/Community in International Relations', p. 2.

58 Kevin T. Jackson in *Charting Global Responsibilities – Legal Philosophy and Human Rights* (Lanham: University Press of America, 1994) points out the existing inadequacy of both legal philosophy and international humanitarian law in enforcing responsibility. He suggests an 'integrity' approach to international law – a perspective that would 'take seriously the notion of civil and human rights in the international arena and the institutional obligations they impose on individuals, nation-states, and corporations' and would be capable of providing a 'provisional philosophical paradigm' within which we can 'understand the special legal responsibilities facing the global community'. (Preface). He also draws our attention to the need to study the form that a philosophy of international human rights law must take in order to become 'applicable on international actors (transnational actors, nation-states, individuals) qua addresses or duty-holders of that law'. (Preface).

59 Mark Selden, 'On Asian Wars, Reparations, Reconciliation', Economic and Political Weekly', 36 (1), 6 January 2001, pp. 25-26. Also see Cathy Lisa Schneider, 'Violence, Identity and Spaces of Contention in Chile, Argentina and Columbia', *Social Research*, 67 (3), Fall 2000, pp. 773-802.

60 There reports have been brought out civil liberty committees, for example in India by PUDR (Peoples' Union for Democratic Rights), PUCL (Peoples' Union for Civil Liberties), Committee for Independent Initiative on Kashmir, APDR (Association for the Protection of Democratic Rights), and most notably APCLC (Andhra Pradesh Civil Liberty Committee) whose pioneering reports on fake encounters and murders of political activists in Andhra Pradesh gained wide attention, and saved the lives of many.

61 *The Guardian*, 'A Tribunal for Israel' by Paul Foot, August 7, 2001.

62 Irving Louis Horowitz, *Beyond Empire and Revolution – Militarization and Consolidation in the Third World* (New York: Oxford University Press, 1982), p. 263.

63 Ibid. p. 183.

64 See the two essays by Valerian Rodrigues, 'Whose Call Do We Hearken (in South Asia)? Community, Civilization, State, Nation, Region or the Globe?' and Rakhahari Chatterji, 'Globalization, Culture and The Nation-State' in B. Ramesh Babu (ed.), *Globalization and the South Asian State* (Delhi: South Asian Publishers, 1998), pp. 139-169.

65 I have written on this in details in my *Whose Asia is it Anyway? Region and the Nation in South Asia* (Calcutta: Maulana Abul Kalam Azad Institute of Asian Studies and Pearl Publications, 1996).

66 Eric Hobsbawm, 'The Curious History of Europe' in *On History*, pp. 287-301.

67 There is, however, no reason to think that the current wave of anti-globalization protests is without predicaments, not the least of which is the capacity of global institutions to absorb the protests and modify themselves accordingly. See comments by Raimo Vayrymen, 'Anti-Globalisation Movements at the Crossroads', The Joan B. Kroc Institute for International Peace Studies, University of Notre Dame, Policy Brief (4), November 2000.

68 For details of the 'rebellion in Genoa' some websites are available, one being www.mondialiasation.org. The reader can also see my article, 'Crime and Punishment in Genoa', *Himal* (Kathmandu), August 2001.

69 In invoking justice here I am aware of course that justice is not a monolith, and not uniform in practices. Precisely therefore is the significance of the practices of justice in the current age – in democratization, in transitional regimes, in peace and reconciliation, in institutional practices at both municipal and international levels, in realization of rights, economic development, and in consideration to past and future generations. One has to only see the diverse forms of justice currently thought and practiced, as chronicled in the collection of essays on justice in *Journal of International Affairs*, 52 (2), Spring 1999. See also for an up to date summary of issues and discussions on justice, Richard Falk, *Human Rights Horizons – The Pursuit of Justice in a Globalising World* (London: Routledge, 2000).

70 For a more guarded presentation of such thesis, see the essay by Havard Hegre, 'Development and the Liberal Peace – What Does It take to be a Trading State?' *Journal of Peace Research*, 37 (1), January 2000, pp. 5-30. On the political factors predicating such a neat sequence R. William Ayres, 'A World Flying Apart? Violent Nationalist Conflicts and the End of the Cold War', Ibid., pp. 105-117. See also on this theme, Solomon William Polachek, 'Conflict and Trade – An Economic Approach to Political International Interactions' (pp. 89-120) and Jeffrey H. Bergstrand, 'On Modeling the Impact of Arms Reduction in World Trade' (pp. 121-142) in Walter Isard and Charles H. Anderson (eds.), *Economics of Arms Reduction and the Peace Process* (Amsterdam: Elsevier Science Publishers, 1992).

Chapter 11

Epilogue: Ten Principles

Dialogic Acts

Readers will have noticed that besides the preceding chapter, where I try to draw explicitly inferences from the current geopolitical context in which the dialogic politics of justice is shaping up, in other chapters too in narrating various practices of dialogue, I have drawn in detail the geo-political and ethno-political contexts in which dialogic acts take place or are barred, because these contexts suggest how these dialogic practices gather their forms amidst war and peace of our time. I have tried to demonstrate how the ethno-political and geo-political contexts sit heavily on the way in which dialogues are carried out.[1] I have not attempted here, as the readers have discovered, any grand or a coherent theory of dialogue, but a sheer descriptive account of *dialogic situations and dialogic relations*, to indicate what dialogue signifies to the politics of our time, and the questions it poses for a possible politics of justice. Hence the importance of relational accounts in studying practices – the discursive and institutional forms of the dialogic acts, the dialogic situations, the rough contours of the dialogic universe. Therefore, the question with which this book ends is, what is this act, which makes dialogue a dialogue, in other words, what is this with which a dialogue constitutes itself?

The word, *act*, as we know is a carrier of double meaning – in the sense of activity it is production, and in the sense of form, an institution. In order to produce, we must stand apart from the means with which we produce in order to act on them. We give ourselves to what we are producing and what we want to produce, by not giving into it – that is to say, a total conversion of our alienation or distancing into our identity as an actor, that is our political subject-hood and individuality. It means that in practicing and forming dialogue, as political subjects we are away from it, and then into it. We are not inherently dialogists just in the way we are not naturally unilateral beings. We are dialogic because relations compel dialogues – we have differences, quarrels, reasons to accommodate, appreciate, and these are, as I have emphasized in the foregoing pages, relational acts.

What is the truth of these relational acts as we find them in situations of war, peace, diplomacies, and the strivings for rights and justice, and what do these relational acts mean in terms of the truth of dialogue? That being my inquiry here, I have reflected on forms and institutions of dialogue, in particular on what political rationality has made into one of the most established forms of dialogue or conversation today, namely constitution and constitutional forms of accommodation and conversation. In several chapters and in several contexts in this book, I raised the issues of the limits of constitutional deliberations and constitutional forms of accommodation in order to emphasize the limits of legal and juridical conversations and dialogue, and the trajectory dialogic politics has to take towards ensuring minimal justice.

Let me give an instance of the limits that constitution and law-bound dialogues have on the political issues of our time. This will make clear why I am emphasizing the need to study dialogic situations in the broadest possible political context. Is the right of self-determination the source of conflict or a contribution to conflict prevention? The dilemma evident in search for a constitutional answer to a fundamentally political question is nothing but an agony of what law does to our questions of political subject-hood. In all ex-colonial countries of South Asia where constitution-making and legal rationality was not a matter of philosophy but direct acts of conflict-moderation, resolution, and prevention, dialogue is at once an object (a demand, an issue, a theme of quarrel) and a subject (an act, a procedure, a process that constitutes into a self). Law makes us citizens and subjects, so that we can dialogue in our polity, yet law places limits on our capacity and power to dialogue in order to re-make our political society. Our role as political subjects founds the basic law through among things animated conversations that is, the constitution; yet, it is constitution that defines our political subjectivity and the power to dialogue. In this logical circle we have the self-foundation of law – the aporia in terms of dialogic politics.

Constitution, as I have repeatedly attempted to outline, lays the guidelines of negotiating conflicting relations and building up unities. But constitution, precisely because these are fundamental rules, make negotiation difficult, at times impossible. How can we have a fundamental negotiation? If it is fundamental, it means we cannot negotiate, we can negate. Constitution by creating a political body, in that footstep we can include international law also, by appealing to the body of internationally recognized (that is legal) actors recognizes its subjects who are then objects to be constitutionally governed. In this logical closure, law derives its greatest strength, and dialogic politics faces its greatest challenge. It is this juridical finality of the conflict-resolution procedure, which dialogic acts at times tear apart, for while often being subject to constitution and law, dialogue presents truths beyond constitution, and suggests new ways of constituting that unity of the subject and the object. Politics while vexed, and at times paralysed with the duality of the juridical truth and the dialogic truth, becomes under pressure of the dialogic acts what it is, namely an act that gives birth to law and constitutions, and immediately proceeds thereafter to subvert it with new ideas and conversations.

Therefore Kant was only deluding himself when he wrote that, 'the idea of a cosmopolitan right...a complement of the unwritten code of law – constitutional as well as international code of law (is) necessary for the public rights of mankind in general and thus for the realization of perpetual peace. For only by endeavouring to fulfil the conditions laid down by this cosmopolitan law can we flatter ourselves that we are gradually approaching the ideal'.[2] But he was hinting at the same time at a significant truth in that same third definitive article, that perpetual peace is perpetual effort, and that 'mutual interest' and 'means of mediation' lead to 'a concord in a state of peace', in other words, there are practices of peace, perpetual practices. The regulatory function of law is not enough. It is also not enough to say that empirical practices of conversation will survive by translating themselves into juridical knowledge. For if law exists as I said on a self-foundation, that is to say, it does not need apparently a politics and history, dialogic practices as relational acts in politics, makes this self-foundation weak, unstable, with what stands on this foundation, that is law, always appealing to an agency outside, that is politics, to help it out, to restore social faith in it. In this structure of 'in' and 'out', we can rediscover the fundamentals of the politics of dialogue. Neither war, nor the supposed opposite, constitution, can suppress it for long.

All this can be re-told in another way, namely that, dialogue already inscribed in a juridical order, rediscovers itself by discovering limits of dialogue in its juridical existence. Thus constitution does not resolve issues but provokes issues by provoking dialogue to step out of bounds. That is the act of dialogue – its amenability to organization, governing, government, and yet reaching beyond to the irreducible conversation on the street, of the masses. This is the closest that we can possibly come to its truth as an act, the truth of its possibility and deficit. This is what the post-colonial experience of war, partition, self-determination, accommodation, forced migration, care, protection, and peace efforts, signifies – the unending possibility of dialogic acts and the dialogic deficit. Studying dialogic practices as situations and practical experiences is therefore crucial. It will lead us to the acts of the street, the politics of the corporeal action as different from communicative action, the collective desire and acts of the masses to dialogue, ally, unite, befriend, understand, accommodate, and co-exist – less a matter of ideology, more a matter of urge, practicality, compulsion, hence existence.

Dialogue is thus less a matter of ideal than of contention. This is what makes it the wild spring, beyond training, familiarization courses on meetings, conventions, assemblies, and drafting of treaties and laws, capsules of conversational techniques, and etiquettes of diplomats and conflict resolution specialists. That is why the monarchs of today want to civilize dialogue, teach the masses how to co-exist (in constitutional form, within constitutional bounds) because dialogue on the street is an object of fear, *the fear of the unanticipated and the unknown alliances, understandings, negotiating capacities, and realities*. If Marx seems almost alone in recognizing that the rational foundation of political reason is singularly incapable of addressing the issues of real, physical acts of control, domination, subjugation, rule, violation of relations, and the disuniting and disembodying of selves, there is nothing to complain. For, the abyss over which the state, public sphere, civil society and all other official registers uncertainly hover, is the world where the mono-logic existence of authority is being incessantly challenged, where dialogue is challenging the power of unilateral legislation and the codes thereof, where as Marx suggested in so many of his writings, the deepest privacy of power is contested by innovative acts of the masses of which political conversation is one of the most pronounced form.

Three moments in that act, which is actually a process – the moment of dialoguing (on) what we know; then comes the second moment, the moment of dialoguing (on) what we do not know; and then the obituary moment, dialoguing (on) what we do not want, or do not want to do, the unlearning, so that we can know the new a little, we can become little more wise, to arrive at crossroads where anew we begin to dialogue (on) what we have known. Our existence has challenged authority in this way.

Even in this geo-politically sanctioned world characterized by geo-politically sanctioned histories of war and peace, and similarly sanctioned forms of dialogue, it will remain so.

Ten Principles of the Relational Act of Dialoguing

I have admitted, this is partly a speculative account of the politics of dialogue based on current history of war and peace in this region. It is neither a general theory nor a continuous history of the principles of dialogue. Rather than drawing out the noblest principles in which dialogic urges may be wrapped, I have offered a quick gloss on current historical experiences of South Asia. Instead of basing my case on well-drawn comparisons of South Asian experiences amongst themselves and with those belonging to outside, I have preferred a broad brush to point out certain high features of the dialogic acts in the current age. I shall therefore appeal my readers to treat this book exactly as it is: a historically grounded sketch needing further research, confirmation, modification and necessary revision. And I shall have to admit that this putative account of the dialogic world reveals a contentious model. The model speaks of contending rights, their limits, the trans-valuation of several aspects of a dialogic situation and the process of focalization inbuilt in a dialogic act, the absence of any inherent virtue in a dialogic act, and the existence of dialogue as a field of governing relations. In the model that emerges from the pages of this book, dialogue is thus posed as a totally dialectical interrelationship of subjects.[3] It suggests qualitative interaction, logic of polyphony, and is different from relating to persons as objects, which suggests a purely functional relation. In this subject-to-subject relation, the contradictions of South Asian, if not global, politics come into relief. The subjects cannot be removed from dialogic politics, nor can they become the master of it. The politics of dialogue is thus a politics of engagement in an alienated world. Contending all values, it becomes a value in itself.

We can describe the contentious situation of dialogue in terms of certain principles. These principles suggest a model that emerges from the book. I shall speak of ten principles.

First, while we can ground the question, where a dialogic situation comes from – in philosophy, morals, metaphysics, and rational choice – this model instead puts this question in terms of political history. The model speaks of dialogue as a relational act.

Second, in treating dialogue as historical products, outcomes of contentious situations, my argument is that in creation of democracy, resisting arbitrariness of power, framing rules of rights and justice, struggling for constitutionalism and then stepping beyond it, dialogue on the street has a major role to play.

Third, dialogues in their political formation resemble the right to engage and be heard, and such a right belongs to a category – the category of a people forming a political community with all contentions cracking the community repeatedly from within and without. Dialogues mark the cracks and not unity.

Fourth, dialogues form the field of governing people and their relations. It can advance democracy, or can be a governmental technique, particularly by becoming a tool for advancing democracy. A dialogic relation can be both a mark of democracy, and its deficit, or to put it slightly differently, mark of democracy's deficit and supplement.

Fifth, this happens, because while dialogue is a complex and its nature is plural, a dialogic situation is characterized by trans-valuation of all other aspects and a consequent focalization in the situation. It is this process that turns the dialogic situation into a governmental field.

Sixth, dialogue as opposed to war and as the route to peace politically aims at making relations stable by means of law and constitutional forms; yet constitution stymies dialogue by making rules fundamental, in other words by obstructing newer accommodative acts and forms of accommodation.

Seventh, dialogue emerges from two phenomena – (a) formation of states and constitutions, and (b) the geopolitical realities of partition, war and peace marked by the reproduction of these conditions. Though these two are viewed often separately, a model that looks at dialogue as the product of confluence of the two refreshes our understanding about the possibilities and limits of dialogic practices and acts.

Eighth, expansion of notions of justice, rights, and citizenship invariably expands the dialogic universe, which now welcomes in its presence issues and entities hitherto considered out of bounds of dialogue, such as alien-hood, methods of sharing sovereignty, re-claiming and re-judging the past, justice as fairness, responsibility, or partitioned selves.

Ninth, a dialogic act is never about reaching truth, but about determining the relation of the dialogists (and their desire to dialogue) with truth.

Tenth, and this follows from the earlier feature, the politics of dialogue proposes an alternative, not by proposing an alternative ideology or ethics, but by proposing an unending series of possibility, alternative realities and arguments within what is happening, within what we construe as realities of war and peace. A dialogic politics is therefore in many ways a critical politics.

This model, expressed in the above mentioned ten principles, and the two figures, is empiricist (though I would not prefer to call it so), because it draws regularities from varying experiences of post-colonial realities, and does not claim for a universal theory of dialogue, but for historically embedded account of conversational acts. The model is also partly speculative, for as yet we do not have a well-documented history of dialogues to base this model on. This model is contentious, because it takes geopolitical realities into account. And, finally, it is designed openly as an account of the present. Yet the plausibility of such an account derives from the fact that besides ideologies, cultural inclinations, and consensual attitudes, dialogue springs from politically contentious situations. Indeed, its mandate originates from a desire to contend the consensus that has been hitherto one of the most effective ways to govern people. The model also has the possibility of accommodating different forms and differential practices of dialogue within it. One of the essential advantages of this model, I shall end by saying, is that it tells us that the story is far from finished, and that we shall have to keep on wandering about the trajectories that dialogic politics can take. It will need a great deal more than we currently know to discern the deep lines of that trajectory that has brought dialogic politics in this post-colonial region to its present stage, and the diverse possibilities of its journey. We cannot say with certainty that the particular type of dialogic regime, which the present global politics is attempting to impose on the world will become a stable reality for some time, but certainly a sense of reality demands that we study closely the contentions, dialogues, and their patterns, so magisterially laid bare before us by the political events of our time.

Configurations of the Dialogic Acts

(Figure 1)

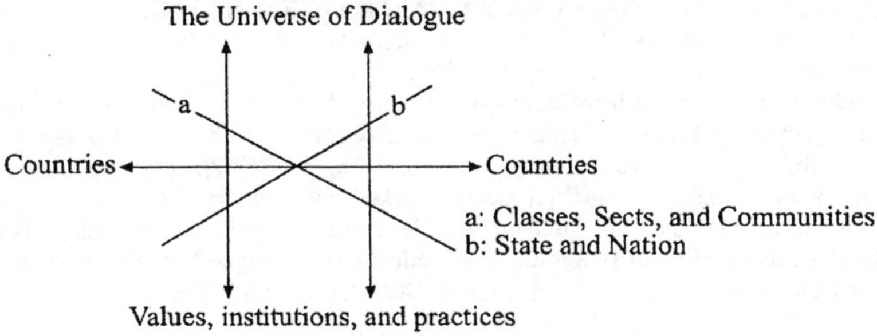

(Figure 2)

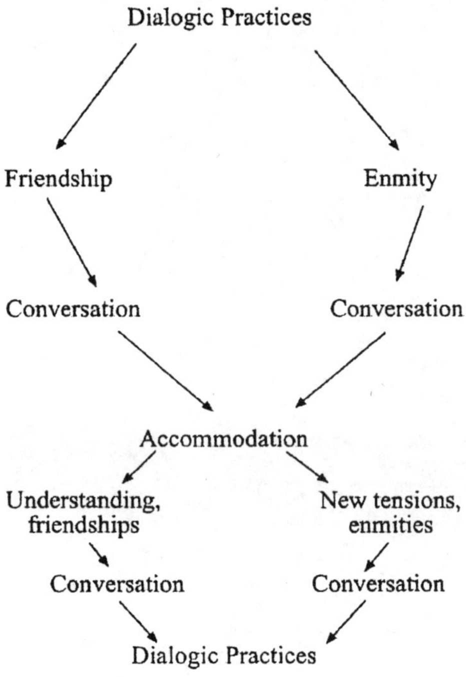

Notes

1 Itty Abraham in an insightful discussion on the Bandung Conference shows how the Afro-Asian dialogue at Bandung was linked to state, place, and identity. See his essay, 'State, Place, Identity – Two Stories in the Making of Region' in K. Sivaramakrishnan and Arun Agarwal (eds.), *Regional Modernities in Stories and Practices of Development* (Stanford University Press, 2002).

2 Immanuel Kant in the 'Third Definitive Article of Perpetual Peace in his *Perpetual Peace* reprinted in David P. Barash (ed.), *Approaches to Peace – A Reader in Peace Studies* (New York: Oxford University Press, 2000), pp. 125-126.

3 The issue of a dialogic situation being marked by 'subjects' and not 'objects' is forcefully argued by Ludmila Bulavka and Alexander Buzgalin in their essay, 'The Next Hundred Years of Mikhail Bakhtin – The Dialectic of Dialogue Versus the Metaphysics of Post-Modernism', *Counter-Hegemony*, 3, May 2000, pp. 25-33.

Bibliography

Primary Reports

'Annual Report of the Parliamentary Commissioner of Hungary for National and Ethnic Minority Rights, 1 January – 31 December 1998', Parliamentary Commissioner's Office, Budapest 1999; same report for 1999, Budapest, (2000).

'Autonomy and Minority groups – A Legal Right in International Law?', E/CN.4/Sub.2/AC.5/ 2001/CRP.5, 5 March 2001.

'Autonomy in the 21st Century – Through the Theoretical Binoculars', E/CN.4/Sub.2/AC.5/2001/ CRP.1, 5 March 2001.

'Examples of Autonomy in Finland – The Territorial Autonomy of the Aland Islands and the Cultural Autonomy of the Indigenous Saami People', Report to the UN Commission on Human Rights, E/CN.4/Sub.2/AC.5/2001/WP.5, 8 May 2001.

'Final Text of the Commentary to the Declaration on the Rights of Persons Belonging to National or Ethnic, Religious and Linguistic Minorities', E/CN.4/Sub.2/AC.5/2001/2, 2 April 2001.

'Fourteen Dead on the Border – For What?' a report by David Bacon, Oakland, California (6/15/01) on: portsideMod@netscape.net, 20 June 2001.

'Handbook for Applying the Guiding Principles on Internal Displacement', Manual of the Brookings Institution Project on Internal Displacement, 1999.

'National Minorities in the Slovak Republic', Ministry of Foreign Affairs of the Slovak Republic, Bratislava, (2001).

'Participation of Minorities in Decision Making', E/CN.4/Sub.2/AC.5/2001/CRP.6 – 30 March 2001.

'Report of the Two Day Workshop of the CHA (Consortium of Humanitarian Agencies) on Districts Meet in Colombo, 3-4 April' (mimeo), 2001.

'Study on the Use of the Autonomy Approaches in the Russian Federation', report to the Commission on Human Rights, E/CN.4/Sub.2/AC.5/20001/WP.3 – 23 March 2001.

'The Peace Process in Northern Ireland', E/CN.4/Sub.2/AC.5/2001/CRP.3, 5 March 2001.

'War and Peace in Sri Lanka – Policy Options for Canada', South Asia Partnership, Canada, Ottawa, 22 March 2000 (mimeo).

Report of the Working Group on Minorities on its Sixth Session, Geneva, 22-26 May 2000 E/CN.4/Sub.2/2000/27, 30 June 2000.

Collection of International Instruments and Other Legal Texts Concerning Refugees and Displaced Persons (Geneva: Geneva: Division of International Protection of the Office of the United Nations High Commissioner for Refugees, 1995).

Consideration for Asylum Officers Adjudicating Asylum Claims from Women, United States INS Gender Guidelines Memorandum, 26 May 1995.

Constituent Assembly Debates.

Convention for the Reduction of Cases of Statelessness, adopted by the International Commission on Civil Status in Berne, 13 September 1973.

Convention on the Reduction of Statelessness, UN Document A/CONF, 9/15, 1961, *Official Records of the General Assembly, Ninth Session, Supplement 21, A/2890.*

Convention relating to the Status of Stateless Persons, 28 September 1954, UN Treaty Series 5158, Vol. 360, *Official Records of the Economic and Social Council, Seventeenth Session, Supplement 1*, E/2596.

Rajya Sabha Debates, Vol. XXII, 26 April 1958.

Report of the Committee of Concerned Citizens, 1997-2000 (hereafter Committee Report), published by S.R. Sankaran, Convenor of the Committee of Concerned Citizens, Hyderabad, March, (2000).

Report of the International Colloquy on the Guiding Principles on Internal Displacement (Washington, D.C.: The Brookings Institution Project on Internal Displacement, 2000).

Report of the Representative of the Secretary General Mr. Francis M. Deng, submitted pursuant to the Commission on Human Rights Resolution 1999/47 – Internally Displaced Persons, Economic and Social Council, E/CN 4/2000/83, 26 January, 2000.

Report of the Working Group on Minorities on its Fifth Session, Geneva, 25-31 May 1999, E/CN.4/Sub.2/1999/21, 24 June(1999).

Report of the Working Group on Minorities on its Fourth Session, Geneva, 25-29 May 1998, E/CN.4/Sub.2/1998/18, 6 July, (1998).

UN Docs A/AC. 96/SR.371, paragraph 107, SR 372, paragraph 60; SR 373, paragraphs 34, 78.

Books and Journals

Adelman, Howard, ed. (1995), *Refugee Policy – Canada and the United States* (Toronto: York Lanes Press).

Ahmar, Moonis (1998), 'Ethnicity and Constitutional Reforms – A Pakistani Perspective' in Iftekharuzzaman (ed.), *Ethnicity and Constitutional Reforms in South Asia* (Dhaka: University Press Limited).

Ahmed, Ishtiaq (1997) 'Exit, Voice and Citizenship' in Tomas Hammar, Grete Brochmann, Kristof Tamas and Thomas Faist (eds.), *International Migration, Immobility and Development – Multidisciplinary Perspectives* (New York: Berg).

Ahmed, Syed Jaffar (1990) *Federalism and Pakistan – A Constitutional Study* (Karachi: Oxford University Press).

Aho, James Alfred (1975) *German Realpolitik and American Sociology – An Enquiry into the Sources and Political Significance of the Sociology of Conflict* (London: Associated University Press).

Akgonenc, Oya (1998) 'A Precarious Peace in Bosnia-Herzegovina – The Dayton Accord and its Prospect for Success' in Stefano Bianchini and Robert Craig Nation (eds.), *The Yugoslav Conflict and its Implications for International Relations* (Ravenna: A. Longo Editore).

Ali, Mehrunnisa (1990) 'Jinnah's Perception of Pakistan's Foreign Policy and the Kashmir Issue', *Pakistan Horizon*, 43 (2).

Al-Krenawi, Alean and John R. Calgary, (1999) 'Conflict Resolution Through a Traditional Ritual Among the Bedouin Arabs of the Negev', *Ethnology*, 38 (2), Spring.

Allen, Tim and Jean Seaton, eds. (1999) *The Media of Conflict: War Reporting and Representations of Ethnic Violence* (London: Zed Books).

Anderson, Benedict (1991) *Imagined Communities – Reflections on the Origin and Spread of Nationalism* (London: New Left Books).

Anderson, Benedict (1998) *The Spectre of Comparisons – Nationalism, Southeast Asia and the World* (London: Verso).

Anderson, Malcolm (1997) *Frontiers – Territory and State Formation in the Modern World* (Cambridge: Polity Press).

Andrew Linklater (eds.), *Boundaries in Question – New Directions in International Relations* (London: Pinter Publishers).

Annual Surveys of the UNHCR; www.unhcr.ch.

Arendt, Hannah (1954) *Between Past and Future – Six Exercises in Political Thought* (London: Faber and Faber).

Armson, Cynthia J, ed. (1999) *Comparative Peace Processes in Latin America* (Washington, D.C. and Stanford, California: Woodrow Wilson Center Press, and Stanford University Press).

Arrow, Kenneth J. (1992) 'The Basic Economics of Arms Reduction' in Walter Isard and Charles H. Anderson (eds.), *Economics of Arms Reduction and the Peace Process* (Amsterdam: Elsevier Science Publishers).

Ascherson, Neal (1995) *Black Sea* (London: Jonathan Cape).

Ashley, Richard K. (1987) 'The Geopolitics of Geopolitical Space – Towards a Critical Theory of International Politics', *Alternatives*, 12.

Ashley, Richard K. (1984) 'The Poverty of Neo-realism', *International Organisation*, 38.

Ashrawi, Hanan (1995) *This Side of Peace* (New York: Simon & Schuster).

Asmal, K. and others, (1996) *Reconciliation through Truth – A Reckoning of Apartheid's Criminal Governance* (Cape Town: David Philip Publishers).

Axer, Jurgen, ed. (1996) *The Human Rights Community and Conflict Resolution in South Asia – The Applicability of European Examples* (Brussels and New Delhi: Friedrich Naumann Stiftung).

Ayres, R. William (2000) 'A World Flying Apart? Violent Nationalist Conflicts and the End of the Cold War', *Journal of Peace Research*, 37 (1), January.

Babu, B. Ramesh, ed. (1998) *Globalisation and the South Asian State* (Delhi: South Asian Publishers).

Bailey, F.G. (1957) *Caste and the Economic Frontier* (Manchester: Manchester University Press).

Bajpai K. *et al.*, (1995) *Brasstacks and Beyond: Perception and Management of Crisis* (New Delhi: Manohar Publications).

Bajpai, Kanti (1999) 'CBMs: Contexts, Achievements, Functions' in D. Banerjee (ed.), *Confidence Building Measures in South Asia* (Colombo: Regional Centre for Strategic Studies).

Balibar, Etienne (1993) 'Spinoza, The Anti-Orwell – The Fear of the Masses' in Etienne Balibar, *Classes, Masses, Ideas – Studies on Politics and Philosophy before and after Marx* (London: Routledge).

Balibar, Etienne (1993) *Classes, Masses, Ideas – Studies on Politics and Philosophy before and after Marx* (London: Routledge).

Balibar, Etienne (1996) 'Is European Citizenship Possible', *Public Culture*, 8 (2), Winter.

Balibar, Etienne (1998) 'The Borders of Europe' in Pheng Chea and Bruce Robbins (eds.), *Cosmopolitics – Thinking and Feeling beyond the Nation* (Minneapolis: Minnesota University Press).

Balibar, Etienne (1998) *Spinoza and Politics*, trans. Peter Snowdown (London: Verso).

Balibar, Etienne (1999) 'Algeria, France: One nation or Two?' in Joan Copjec and Michael Sorkin (eds.), *Giving Ground – The Politics of Propinquity* (London: Routledge).

Balibar, Etienne (2000) 'What makes a People a People? Rousseau and Kant' in Mike Hill and Warren Montag (eds.), *Masses, Classes, and the Public Sphere* (London: Verso).

Balibar, Etienne and Immanuel Wallerstein, (1991) *Race, Nation, Class – Ambiguous Identities* (London: Verso).

Ball, Terence, ed. (1987) *Idioms of Inquiry – Critique and Renewal in Political Science* (Albany, N.Y.: State University of New York).

Banerjee, Paula (2001) 'The Tenuous Control over the Line of Control' in Helmut Reifeld and Ranabir Samaddar (eds.), *Peace as Value* (Delhi: Manohar).

Banerjee, Paula (2002) 'Dislocating Women and Making the Nation', *Refugee Watch*, 17, December.

Baral, Lok Raj (1998) 'Ethnicity and Constitutional Reforms in Nepal' in Iftekharuzzaman (ed.), *Ethnicity and Constitional Reform in Asia* (Dhaka: University Press Limited).

Bardhan, Pranab (2001) 'Social Justice in the Global Economy', *Economic and Political Weekly*, 36 (5-6), 3-10 February.

Barthes, Roland (1982) *Empire of Signs*, trans. Richard Howards (New York: Hill and Wang).

Basu Ray Chaudhuri, Sabyasachi and Shahid Fiaz, eds. (1999) *The Kargil War – From the News Files*, SAFHR Paper 6 (Kathamndu: South Asia Forum for Human Rights).

Bell, Christine (1999) 'A Comparative Perspective on Peace Initiatives', *Accord – An International Review of Peace Initiatives*, 7.

Bell-Fialkoff, Andrew (1999) *Ethnic Cleansing* (New York: St. Martin's Griffin).

Bendahamane, Diane and John McDonald Jr. (1986) *Prospective on Negotiation – Four Case Studies and Interpretations* (Washington, D.C.: Foreign Service Institute).

Benhabib, Seyla (ed.), (1995) *Democracy and Difference* (Princeton: Princeton University Press).

Berghahn, Volker R. and Hanna Schissler (eds.), (1987) *Perceptions of History: An Analysis of School Textbooks* (Oxford: Oxford University Press).

Bergstrand, Jeffrey H. (1992) 'On Modeling the Impact of Arms Reduction in World Trade' in Walter Isard and Charles H. Anderson (eds.), *Economics of Arms Reduction and the Peace Process* (Amsterdam: Elsevier Science Publishers).

Bertram, Eva (1995) 'Reinventing Governments – The Promise and Perils of United Nations Peace Building', *Journal of Conflict Resolution*, 39 (3), September.

Bertrand, Didier (2000) 'The Forced Repatriation of Tamil Asylum Seekers from Europe', *Refugee Watch*, 12, December.

Beyani, Chaloka (1995) 'State Responsibility for the Prevention and Resolution of Forced Population Displacements in International Law', *International Journal of Refugee Law*, special issue.

Bhargava, Rajeev (2000) 'The Moral Justification of Truth Commissions' in Charles Villa-Vicencio and Wilhelm Verwoerd, *Looking Back Reaching Forward – Reflections on the Truth and Reconciliation Commission of South Africa* (London and Cape Town: Zed Press and University of Cape Town Press).

Bhargava, Rajeev, ed. (1998) *Secularism and its Critics* (Delhi: Oxford University Press).

Bhargava, Rajeev, Amiya Kumar Bagchi and R. Sudarshan (eds.), (1999) *Multiculturalism, Liberalism and Democracy* (Delhi: Oxford University Press).

Bhattachan, Krishna B. (2000) 'Possible Ethnic Revolution or Insurgency in a Predatory Unitary State Hindu State, Nepal' in Dhruba Kumar (ed.), *Domestic Conflict and Crisis of Governability in Nepal* (Kathmandu: Centre for Nepal and Asian Studies), pp. 135-162.

Bhattacharjea, Ajit (1997) *Countdown to Partition – The Final Days* (Delhi: HarperCollins Publishers India).

Bhaumik, Subir (1996) *India's Northeast – Insurgency Crossfire* (Delhi: Lancer Publishing House).

Bhaumik, Subir (1999) 'The Accord That Never Was: A Critique of the 1975 Shillong Accord', in Jehan Perera (ed.), *Peace Process in Nagaland and Chittagong Hill Tracts – An Audit Report*, SAFHR Paper Series 5, Kathmandu.

Bhaumik, Subir, Meghna Guhathakurta and Sabyasachi Basu Ray Chaudhury, eds. (1997) *Living on the Edge – Essays on the Chittagong Hill Tracts* (Kathmandu and Calcutta: South Asia Forum for Human Rights and Calcutta Research Group).

Bianchini, Stefano (1999) 'The Changing Notion of Self-Determination – Global Repurcussions for the International Community' in Henry Huttenbach and Francesco Privitera (eds.), *Self-Determination from Versailles to Dayton – Its Historical Legacy* (Ravenna: Longo Editore, 1999).

Bianchini, Stefano and George Schopflin, eds. (1998) *State Building in the Balkans – Dilemmas on the Eve of the 21st Century* (Ravenna: Longo Editore).

Bianchini, Stefano and Robert Craig Nation, eds. (1998) *The Yugoslav Conflict and its Implications for International Relations* (Ravenna: A. Longo Editore).

Bilgrami, Akeel (1994) 'Two Concepts of Secularism – Reason, Modernity and the Archimedean Ideal', *Economic and Political Weekly*, 9 July.

Blainey, Geoffrey (1977) *The Causes of War* (New York: Free Press).

Blair, Harry W. (1996) 'Democracy, Equity and Common Property Resource Management in the Indian Subcontinent', *Development and* Change, 27.

Bose, Pradip K. (1992) 'Text of Language, Language of Text – The Post-modern Dilemma', *Margins*, February 2000.

Bose, Pradip K. (2001) 'Anthropology of Reconciliation' in Ranabir Samaddar and Helmut Reifeld (eds.), *Peace as Process – Reconciliation and Conflict Resolution in South Asia* (Delhi: Manohar).

Bose, Pradip Kumar (1999) 'Trust and the Refugee Experience', *Refugee Watch*, 7, June.

Bose, Pradip Kumar, ed. (2000) *Refugees in West Bengal – Institutional Practices and Contested Identities* (Calcutta: Calcutta Research Group).

Bose, Sumantra (1997) *The Challenge in Kashmir – Democracy, Self-Determination and a Just Peace* (New Delhi: Sage Publications).

Bose, Sumantra (2001) 'Kashmir at the Crossroads – Problems and Possibilities', *Security Dialogue*, 32 (1), March.

Bose, Tapan and Rita Manchanda, eds. (1997) *States, Citizens and Outsiders – The Uprooted Peoples of South Asia* (Kathmandu: South Asia Forum for Human Rights).

Bose, Tapan K. (2000) 'Protection of Refugees in South Asia – Need for Legal Framework', SAFHR paper Series 6, January.

Boulding, Kenneth E. (1962) *Conflict and Defense – A General Theory* (New York: Harper and Row).

Boulding, Kenneth E. (1978) *Stable Peace* (Austin: University of Texas Press).

Bradnock, Robert W. (1998) 'Regional Geopolitics in a Globalising World – Kashmir in a Geopolitical Perspective', *Geopolitics*, 3 (2), Autumn.

Brass, Paul R., ed., (1996) *Riots and Pogroms* (London: Macmillan).

Brecht, Bertolt (2001) 'The Threepenny Lawsuit' in Marc Silberman (ed. and trans.), *Bertolt Brecht on Film and Radio* (London: Methuen).

Brock-Utne, Birgit (1985) *Educating for Peace – A Feminist Perspective* (New York: Pergamon Press).

Brubaker, Roger (1996) *Nationalism Reframed* (Cambridge, MA: Cambridge University Press).

Bullogh, Warn and Raoul Naroll, (1969) *Deterrence in History* (New York: State University of New York Press).

Butalia, Urbshi (1998) *The Other Side of Silence – Voices from the Parititon of India* (New Delhi: Viking).

Campbell, David and Michael J. Shapiro (eds.), (1999) *Moral Spaces – Rethinking Ethics and World Politics* (Minneapolis: University of Minnesota Press).

Canovan, Margaret (2000) 'Patriotism is not Enough', *British Journal of Political Science*, 30.

Capotorti, F., (1993) *Study on the Rights of Persons Belonging to Ethnic, Religious, and Linguistic Minorities* (UN Sub-commission on Prevention of Discrimination and Protection of Minorities, UN sales no E 78, XIV.1).

Care, Geoffrey (2000) 'Transnational Refugee Recognitions – A Judicial Perspective', unpublished paper, presented at the International Conference on Forced Migration, Displacement and Conflict Resolution, University of Jadavpur, Calcutta, 20-22 April.

Carment, David and Patrick James (1995) 'Internal Constraints and Interstate Ethnic Conflict – Towards a Crisis-based Assessment of Irredentism', *Journal of Conflict Resolution*, 39 (1), March.

Castles, Stephen and Alastair Davidson (eds.) (2000) *Citizenship and Migration – Globalisation and the Politics of Belonging* (London: Routledge).

Castles, Stephen and Godula Kossack, (1984) *Immigrant Workers and Class Structure in Western Europe*, (Oxford: Oxford University Press, and Institute of Race Relations, London).

Castles, Stephen and Mark J. Miller, (1998) *The Age of Migration – International Population Movements in the Modern World*, second edition (London: Macmillan Press).

Caws, Peter, ed. (1989) *The Causes of Quarrel* (Boston: Beacon Press).

Chaddha, Maya (1997) *Ethnicity, Security and Separatism in India* (New York: Columbia University Press).

Chadha Behera, Navnita, Paul M. Evans and Gowher Rizvi (1997) *Beyond Boundaries: A Report on the State of Non-official Dialogues on Peace, Security and Cooperation in South Asia* (Ontario: University of Toronto – York University Joint Centre for Asia Pacific Studies).

Chandoke, Neera (1999) *Beyond Secularism – The Rights of Minorities* (Delhi: Oxford University Press).

Chandrakanthan, A.J.V. (1998) 'The Negotiations Between President Premadasa and the LTTE – A Retrospective Assessment' in Kumar Rupesinghe (ed.), *Negotiating Peace in Sri Lanka – Efforts, Failures and Lessons*.

Chari, P. R (2000) 'India's Defence Expenditure: Can It Be Reduced,' *Defence Expenditure In South Asia: India and Pakistan*, RCSS Policy Studies 12, June.

Charney, Craig (1999) 'Civil Society, Political Violence, and Democratic Transition and the Peace Process in South Africa, 1990-1994', *Comparative Studies in Society and History*, 41 (1), January.

Chasie, Charles (1999) *The Naga Imbroglio* (Kohima: Standard Publishers).

Chatterjee, Joya 'Rights or Charity? Government and Refugees, The Debate over Relief and Rehabilitation in West Bengal, 1947-1950' in Suvir Kaul (ed.), *The Partitions of Memory – The Afterlife of the Division of India* (Delhi: Permanent Black, 2001).

Chatterjee, Joya (1999) 'The Fashioning of a Frontier – The Radcliffe Line and Bengal's Border Landscape, 1947-52', *Modern Asian Studies*, 33 (1).

Chatterji, Rakhahari (1998) 'Globalization, Culture and The Nation-State' in B. Ramesh Babu (ed.), *Globalization and the South Asian State* (Delhi: South Asian Publishers).

Chea, Pheng and Bruce Robbins (eds.), (1998) *Cosmopolitics – Thinking and Feeling beyond the Nation* (Minneapolis: Minnesota University Press).

Chimni, B.S. (ed.), (2000) *International Refugee Law – A Reader* (Delhi: Sage).

Chomsky, Noam (1978) 'An Exception to the Rules', *Inquiry*, 17 April.

Chomsky, Noam (1998) *World Orders – Old and New* (London: Pluto Press).

Christie, Clive J. (1992) 'Partition, Separatism and National Identity – A Reassessment', *Political Quarterly*, 63 (1), January-March.

Churchill, Ward (1999) *Struggle for the Land – Native North American Resistance to Genocide, Ecocide and Colonization*, revised edition (Winnipeg: Arbeiter Ring Publishing).

Clapham, Christopher (1998) 'Degrees of Statehood', *Review of International Studies*, 24.

Clapp, Robin and Tihana Majcen, (2000) 'Deportation of Rejected Asylum Seekers from the European Union – Avoidable Deaths?', *Human Rights Tribune* 7 (4), December.

Clark, Ian (1997) *Globalization and Fragmentation – International Relations in the Twentieth Century* (New York: Oxford University Press).

Clifford, James and George E. Marcus, eds. (1986) *Writing Culture: The Poetics and Politics of Ethnography* (Berkeley: University of California Press).

Cohen, Roberta and Francis M. Deng, (1998) *Masses in Flight – The Global Crisis of Internal Displacement* (Washington, D.C.: Brookings Institution Press).

Cohen, Roberta and Francis M. Deng, (1998) *The Forsaken People – Case Studies of the Internally Displaced* (Washington, D.C.: Brookings Institution Press).

Cordier, Andrew W. and Wilder Foote, eds. (1965) *The Quest for Peace – The Dag Hammarskjold Memorial Lectures* (New York: Columbia University Press).

Creighton, Colin and Martin Shaw, eds. (1987) *The Sociology of War and Peace* (London: The Macmillan Press).

Cross, Susan and Robert Rosenthal (1999) 'Three Models of Conflict Resolution – Effects on Inter-group Expectations and Attitudes', *Journal of Social Issues*, 55 (3), Fall.

Crow, Ben, Alan Lindquist, and David Wilson, (1995) *Sharing the Ganges – The Politics and Technology of River Development* (Delhi: Sage Publications).

Dahl, Robert (1989) *Democracy and its Critics* (New Haven: Yale University Press).

Dahrendorf, Ralf (1988) *The Modern Social Conflict – An Essay on the Politics of Liberty* (New York: Weidenfeld and Nicholson).

Daniel, E. Valentine and John Ch. Knudsen (eds.), (1995) *Mistrusting Refugees* (California: University of California Press).

Das, Arvind N. (1998) 'The End of Geography', *Biblio*, March-April.

Datta, P.S. (1995) *Ethnic Peace Accords in India* (Delhi: Vikas Publications).

Day, Graham and Robbins, David (1987) 'Activists for Peace – The Social Basis for Peace Movement' in Colin Creighton and Martin Shaw (eds.), *The Sociology of War and Peace* (London: The McMillan Press).

De Lima, Vivian and Carmencita Karagoag, eds. (1995) *Peace, Disarmament and Symbiosis in the Asia-Pacific*, proceedings of the Conference on Peace, Disarmament and Symbiosis in the Asia-Pacific (Quezon City, Philippines: Solidaridad Foundation).

Derrida, Jacques (1992) *The Other Heading – Reflections on Today's Europe*, trans. by Pascale-Anne Brault and Michael B. Naas (Bloomington and Indianapolis: Indiana University Press).

Derrida, Jacques (1992) *The Politics of Friendship*, (London: Verso).

Derrida, Jacques (1997) Discussion with students of the University of Sussex, 1 December, on *The Politics of Friendship*, www.sussex.ac.uk/Units/frenchthought/derrida.htm.

Dhavan, R., R. Sudarshan, and S. Khurshid, eds. (1985) *Judges and the Judicial Power – Essays in Honour of Justice Krishna Iyer* (Bombay: Oxford University Press).

Diamond, Jared (1997) *Guns, Germs, and Steel* (New York, W.W. Norton & Company).

Diehl, P. and G. Goertz (1998) 'Territorial Changes and Militarized Conflict', *Journal of Conflict Resolution*, 32, March.

Dingwaney, Anuradha and Carol Maier, eds. (1996) *Between Languages and Cultures: Translation and Cross-cultural Texts* (Delhi: Oxford University Press).
Dodds, Klaus (1999) 'Political Geography I – The Globalisation of World Politics' *Progress in Human Geography*, 24 (1).
Dodds, Klaus (2000) 'Political Geography II – Some Thoughts on banality, New Wars and the Geopolitical Tradition', *Progress in Human Geography*, 24 (2).
Dodds, Klaus (2001) 'Political Geography III – Critical Geopolitics after Ten Years', *Progress in Human Geography*, 25 (1).
Douzinas, Costas and Ronnie Warrington, (1995) 'A Well-founded Fear of Justice – Law and Ethics in Postmodernity' in Jerry Leonard (ed.), *A Reader in (Post) Modern Critical Theory* (New York: State University of New York Press).
Dower, N (1998) *The World Ethics – The New Agenda* (Edinburgh, Edinburgh University Press, 1998).
Draper, Theodore (1983) *Present History – On Nuclear War, Detente, and Other Controversies* (New York: Random House).
Durfee, Mary and James N Rosneau, (1996) 'Playing Catch-Up: International Relations Theory and Poverty', *Millennium: Journal of International Studies*, 25 (3).
Echevarri-Gent, J. (1993) *The State and the Poor – Public Policy and Political Development in India and the United States* (Berkeley: University of California Press).
Eco, Umberto (1994) 'Tolerance and the Intolerable', *Index on Censorship*, 1 (2), May-June.
Elman, Colin and Miriam Fendius Elman, eds., (2001) *Bridges and Boundaries – Historians, Political Scientists and the Study of International Relations* (Cambridge, MA: The MIT Press).
Emmanuel, S.J. (1998) 'Kumaratunga-Prabhakaran Talks – A Northern View' in Kumar Rupesinghe (ed.), *Negotiating Peace in Sri Lanka – Efforts, Failures and Lessons* (London: International Alert).
Engineer, Ashgar Ali (2002) 'Da'wah or Dialogue?', *Islam and Modern Age*, June.
Enloe, Cynthia (1998) 'All the Men are Militias, All the Women are Victims – The Politics of Masculinity and Femininity in Nationalist Wars' in Lois Ann Lorentzen and Jennifer Turpin (eds.), *The Women and War Reader* (New York: New York University Press).
Etzioni, Amitai (1962) *The Hard Way to Peace – A New Strategy* (New York: Collier Books).
Falk, Richard (1978) 'The Moral Argument as Apologia', *The Nation*, 25 March.
Falk, Richard (2000) *Human Rights Horizons – The Pursuit of Justice in a Globalizing World* (London: Routledge).
Fiaz, Shahid and Ranabir Samaddar, (2001) 'Peace Audit report 2 – Peace process in Sri Lanka', SAFHR Paper 8, Kathmandu, April, also www.safhr.org.
Fiaz, Shahid and Sabyasachi Basu Ray Chaudhuri (eds.), (2000) *The Ten Week War in Kargil: A Dossier*, SAFHR Papers 6, Kathmandu.
Finn Laursen, (1999) 'Civil Society and European Integration', *The Annals* (Issue on Civil Society and Democratization), 565, September.
Fogelsong, David S. (1995) *America's Secret War against Bolshevism – U.S. Intervention in the Russian Civil War* (Chapel Hill: University of North Carolina Press).
Forms of Autonomy – The Practice of Autonomous and Self-Government Arrangements, Report of the International Committee of Lawyers for Tibet, 1999 (mimeo).
Forsythe, David P. (1993) *Human Rights and Peace* (Lincoln: University of Nebraska Press).
Foucault, Michel (1988) 'Technologies of the Self' in Luther H. Martin, Huck Gutman and Patrick H. Hutton (eds.), *Technologies of the Self – A Seminar with Michel Foucault* (Amherst: University of Massachusetts Press).

Foucault, Michel (1998) *The Essential Works*, Vol. 2, *Aesthetics, Method, and Epistemology*, ed. James D. Faubion, trans. Robert Hurley and others (London: Allen Lane, The Penguin Press).

Foucault, Michel (2001) 'Governmentality' in James Fabion (ed.), *Michel Foucault – Essential Works 3, Power* (London: Allen Lane, Penguin).

Franceschet, Antonio (1999) 'The Ethical Foundations of Liberal Internationalism', *International Journal*, LIV (3), Summer.

Frangsmyr, Tore, ed. (1997) *Nobel Lectures – Peace, 1971-1980 and 1981-1990*, 2 vols. (Singapore: Scientific Press).

Fukuyama, Francis (1992) *The End of History and the Last Man* (London: Penguin).

Fukuyama, Francis (1995) *Trust – The Social Virtues and the Creation of Prosperity* (New York: Free Press).

Fukuyama, Francis (1999) *The Great Disruption – Human Nature and the Reconstitution of Social Order* (New York: The Free Press).

Gaddis, John Lewis (1987) *The Long Peace – Inquiries into the History of the Cold War* (New York: Oxford University Press).

Gaddis, John Lewis (1992-93) 'International Relations Theory and the End of Cold War', *International Security*, 17 (3).

Gaddis, John Lewis (1997) 'History, Theory and Common Ground', *International Security*, 22 (1), Summer.

Gaddis, John Lewis (1997) *We Now Know – Rethinking Cold War History* (New York: Oxford University Press).

Gaddis, John Lewis (2001) 'In Defense of Particular Generalization – Rewriting Cold War History, Rethinking International Relations Theory' in Colin Elman and Miriam Fendius Elman (eds.), *Bridges and Boundaries – Historians, Political Scientists and the Study of International Relations* (Cambridge, MA: The MIT Press).

Gain, Philip and Shishir Moral, (1998) 'Land Rights, Land Use and Ethnic Minorities of Bangladesh', a report of the SEHD, Dhaka.

Galanter, Marc (1963) 'Law and Caste in Modern India', *Asian Survey*, 3.

Galanter, Marc (1997) *Law and Society in Modern India* (Delhi: Oxford University Press).

Galtung, Johan (1970) 'Regional Security Commissions – A Proposal' in Johan Galtung and Sverre Lodgaard (eds.), *Co-operation in Europe* (Oslo and Assen: Universitetsforlaget and International Peace Research Institute).

Galtung, Johan (1985) 'Twenty-five Years of Peace Research – Ten Challenges and Some Responses', *Journal of Peace Research*, 2 (2).

Galtung, Johan (1995) 'Geopolitics after the Cold War – An Essay in Agenda Theory' in Vivian De Lima and Carmencita Karagoag (eds.), *Peace, Disarmament and Symbiosis in the Asia-Pacific*, proceedings of the Conference on Peace, Disarmament and Symbiosis in the Asia-Pacific (Quezon City, Philippines: Solidaridad Foundation).

Galtung, Johan and Sverre Lodgaard, eds. (1970) *Co-operation in Europe* (Oslo and Assen: Universitetsforlaget and International Peace Research Institute).

Gambetta, Diego (1988) *Trust – Making and Breaking Cooperative Relations* (Oxford: Blackwell).

Ganguly, Sumit (1986) *Origins of War in South Asia: Indo-Pakistani Conflicts since 1947* (Boulder: Westview).

Ganguly, Sumit (1997) 'Past Attempts at Mediation and Conflict Prevention in South Asia: Would CSBMs Have Made A Difference?' in Sumit Ganguly and Ted Green (eds.), *Mending Fences: Confidence and Security Building Measures in South Asia* (Delhi, Manohar Publications).

Ganguly, Sumit (1999) *The Crisis in Kashmir: Portents of War, Hopes of Peace* (Cambridge: Cambridge University Press).

Ganguly, Sumit and Ted Green (eds.), (1997) *Mending Fences: Confidence and Security Building Measures in South Asia* (Delhi, Manohar Publications).

Geertz, Clifford (1980) *Negara – The Theater State in Nineteenth-Century Bali* (Princeton, NJ: Princeton University Press).

Ghai, Yash (2001) *Public Participation and Minorities*, a report of the Minority Rights Group International, London.

Ghosh, Papiya (2000) 'Who Went Where, and how are they doing? Pakistanis and Indians outside South Asia', *Refugee Watch*, 12, December.

Ghosh, Sudhir (1967) *Gandhi's Emissary* (Boston: Houghton Mifflin Company).

Gibney, Mark (1991) 'US Foreign Policy and the Creation of Refugee Flows' in Howard Adelman (ed.), *Refugee Policy – Canada and the United States* (Toronto: York Lanes Press).

Giddens, Anthony (1988) *The Consequences of Modernity* (Cambridge: Polity Press).

Gilbert, Geoff (2001) 'Autonomy and Minority Groups – A Legal Right in International Law?', paper for the Seventh Session of the UN Working Group on Minorities, E/CN.4/Sub.2/AC.5/2001/CRP.5, 5 March.

Gopal, S. (1989) *Jawaharlal Nehru: A Biography*, Vol. 3, (Delhi: Oxford University Press).

Gow, James (2000) 'A Revolution in International Affairs?', *Security Dialogue*, 31 (3), September.

Greatbatch, Jacqueline (1989) 'The Gender Difference – Feminist Critiques of Refugee Discourse' *International Journal of Refugee Studies*, 1 (4).

Grinberg, Lev (2000) 'Imagined democracy, Imagined Peace – Reframing Israeli Politics', paper presented in the second Humphrey International Conference on 'Challenges to Democracy – Peripheries as a Vantage Point', Ben Gurion University, Be'er Sheva, 19-21 May.

Grover, R (ed.), (1995) *The Story of Kashmir: Yesterday and Today* (Delhi: South Asia Books).

Guha, Sumit (1998) 'Power and law in Past and Present' in Satish Saberwal and Heiko Sievers (eds.), *Rules, Laws, Constitutions* (Delhi: Sage).

Gutmann, Amy (ed.), (1994) *Multiculturalism* (Princeton, N.J.: Princeton University Press).

Gutmann, Amy and Dennis F. Thomson, (1996) *Democracy and Disagreement* (Cambridge: Belknap Press of the Harvard University Press).

Gyawali, Deepak and Ajaya Dixit, (1999) 'Mahakali Impasse and Indo-Nepal Water Conflict', *Economic and Political Weekly*, 27 February.

Habermas, Jurgen (1976) *The Legitimation Crisis*, trans. Thomas McCarthy (Cambridge: Polity Press).

Habermas, Jurgen (1996) *Between Facts and Norms – Contributions to a Discourse Theory of Law and Democracy* (Cambridge, MA: The MIT Press).

Habermas, Jurgen (1996) *Theory and Practice*, trans. John Viertel (Cambridge: Polity Press).

Hadenius, Axel, ed. (1997) *Democracy's Victory and Crisis* (Cambridge: Cambridge University Press).

Hampson, Fen Osler (1996) *Nurturing Peace – Why Peace Settlements Succeed or Fail* (Washington, D.C.: United States Institute of Peace).

Harrington, Michael (1987) *The Next Left – The History of A Future* (London: I.B. Tauris).

Hasan, K. Sarwar (ed), (1966) *Documents on the Foreign Relations of Pakistan: The Kashmir Question* (Karachi: Government of Pakistan).

Hasan, Mushirul (1998) *Legacy of a Divided Nation – India's Muslims since Independence* (Delhi: Oxford University Press).

Hasan, Mushirul (ed.), (1993) *India's Partition – Process, Strategy, and Mobilization* (Delhi: Oxford University Press).

Hasan, Mushirul (ed.), (1995) *India Partitioned – The Other Face of Freedom*, 2 Volumes (Delhi: Roli Books).

Hasan, Mushirul, ed. (1993) *India's Partition – Process, Strategy, and Mobilization* (Delhi: Oxford University Press).

Hathaway, James C. (1991) *The Law of the Refugee Status* (Toronto: Butterworths).

Hayner, Priscilla B. (1994) 'Fifteen Truth Commissions, 1974-1994 – A Comparative Study', *Human Rights Quarterly*, 16.

Hayner, Priscilla B. (1996) 'Commissioning the Truth – Further Research Questions', *Third World Quarterly*, 17 (1).

Hazarika, Sanjoy (2000) *Rites of Passage – Border Crossings, Imagined Homelands, India's East and Bangladesh* (Delhi: Penguin).

Hegre, Havard (2000) 'Development and the Liberal Peace – What Does It take to be a Trading State?' *Journal of Peace Research*, 37 (1), January.

Henrici, Peter in (1989) 'Two Types of Philosophical Approach to the Problem of War and Peace' in Peter Caws (ed.), *The Causes of Quarrel* (Boston: Beacon Press).

Hermann, Tamar (2001) 'Reconstituting Boundaries as a means of Peacemaking: The Israeli-Palestinian Case', unpublished paper, February.

Hewitt, Vernon (1995) *Reclaiming the Past: The Search for Political and Cultural Unity in Contemporary Jammu and Kashmir* (London).

Hill, Mike and Warren Montag (eds.), (2000) *Masses, Classes, and the Public Sphere* (London, Verso).

Hoare, Quintin and Geoffrey Nowell Smith, eds. (1996) *Selections from the Prison Notebooks of Antonio Gramsci*, Indian edition (Madras: Orient Longman).

Hobbes, Thomas (1972) *Leviathan*, reprint, Ed, C.B. Macpherson (Harmondsworth: Penguin Books).

Hobsbawm, Eric (1975) *The Age of Revolution: Europe 1789-1848* (London: Weidenfeld and Nicholson).

Hobsbawm, Eric (1997) *On History* (London: Abacus).

Hogan, Michael (1992) *The End of the Cold War – Its Meaning and Implications* (New York: Cambridge University Press).

Hogan, Michael J, ed. (1995) *America in the World – The Historiography of American Foreign Relations since 1941* (New York: Cambridge University Press).

Holbrooke, Richard (1998) *To End a War* (New York: The Modern Library).

Holiday, D. and W. Stanley, (1993) 'Building the Peace – Preliminary Lessons from El Salvador', *Journal of International Affairs*, 46 (2).

Horam, M. (1988) *Naga Insurgency: The Last Thirty Years* (New Delhi: Cosmo).

Horowitz, Irving Louis (1963) *Games, Strategies and Peace* (Philadelphia: American Friends Service Committee, 1962).

Horowitz, Irving Louis (1963) *The War Games – Studies of the New Civilian Militarists* (New York: Ballantine).

Horowitz, Irving Louis (1973) *War and Peace in Contemporary Social and Philosophical Theory* (London: Souvenir Press).

Horowitz, Irving Louis (1982) *Beyond Empire and Revolution – Militarization and Consolidation in the Third World* (New York: Oxford University Press).

Huntington, S.P. (1993) 'If not Civilizations, What? Paradigms of the Post Cold War World', *Foreign Affairs*, 72 (5).

Huntington, S.P. (1993) 'The Clash of Civilizations?' *Foreign Affairs*, 72 (3).

Hutchings, Kimberly (1999) asks in Hutchings, *International Political Theory* (London: Sage).

Huttenbach, Henry (1999) 'The Decline of European Politics – Then and Now' in Huttenbach, Henry and Francesco Privitera, (eds.) *Self-Determination from Versailles to Dayton – Its Historical Legacy* (Ravenna: Longo Editore, 1999.

Huttenbach, Henry and Francesco Privitera, eds. (1999) *Self-Determination from Versailles to Dayton – Its Historical Legacy* (Ravenna: Longo Editore).

Iftekharuzzaman (ed.), (1998) *Ethnicity and Constitional Reform in Asia* (Dhaka: University Press Limited).

Ignatieff, Michael (1993) *Blood and Belonging* (London: Routledge).

Ignatieff, Michael (2000) *The Rights Revolution* (Toronto: Anansi Publishing).

Ihonvhere Julius O. 'Engaging the Leviathan (2001) Constitutionalism and the New Politics in Africa', Annual Minority Rights Address, organized by the International Centre for Ethnic Studies (Colombo), Geneva, 18 May.

Ihonvhere, Julius O (2000) *Towards the New Constitutionalism in Africa* (London: Centre for Democracy and Development).

Iralu, Kaka D. (2000) *Nagaland and India, Blood and the Tears – A Historical Account of the 52 Years Indo-Naga War and the Story of Those Who were Never Allowed to Tell It* (no publisher's name, Kohima).

Isard, Walter (1992) *Understanding Conflict and Science of Peace* (Cambridge, MA: Blackwell).

Isard, Walter and Charles H. Anderson (eds.), (1992) *Economics of Arms Reduction and the Peace Process* (Amsterdam: Elsevier Science Publishers).

Isard, Walter and Charles H. Anderson, eds. (1992) *Economics of Arms Reduction and the Peace Process* (Amsterdam: Elsevier Science Publishers).

Ishiyarna, John T. (1999) 'Representational Mechanisms and Ethnopolitics – Evidence from Transitional Democracies in Eastern Europe', *East European Quarterly*, 33 (2), Summer.

Ivekovic, Ivan (1998) 'Identity – Usual Bias, Political Manipulations and Historical Forgeries – The Yugoslav Drama' in Stefano Bianchini and George Schopflin (eds.), *State Building in the Balkans – Dilemmas on the Eve of the 21st Century* (Ravenna: Longo Editore, 1998).

Ivekovic, Ivan (1999) 'Contemporary Ethnic Conflicts in a Comparative Perspective' in Ursula Oswald-Spring, *Peace Studies from a Global Perspective* (Delhi: Madhyam Books).

Jackson, Kevin T. (1994) *Charting Global Responsibilities – Legal Philosophy and Human Rights* (Lanham: University Press of America).

Jakobsen, Peter Viggo (2000) 'Focus on the CNN Effect Misses the Point – The Real Media Impact on Conflict Management is Invisible and Indirect', *Journal of Peace Research*, 37 (2), March.

Jaleel, Muzamil (1999) 'It Was Not Our War' in Sankarshan Thakur and others, *Guns and Yellow Roses: Essays on the Kargil War* (New Delhi: Viking).

Jeszensky, Geza (1997) 'More Bosnias? National and Ethnic Tensions in the Post-Communist World', *East European Quarterly*, 31 (3), September.

Jha, Prem Sankar (1996) *Kashmir 1947: Rival Versions of History* (Delhi: Oxford University Press).

Joseph, Sarah (1998) *Interrogating Culture – Critical Perspectives on Contemporary Social Theory* (New Delhi: Sage).

Kahn, Victoria (1999) 'The Duty to Love – Passion and Objectivity in Early Modern Political Theory', *Representations*, 68, Fall.

Kalin, Walter (2000) *Guiding Principles on Internal Displacement – Annotations*, Studies in Transnational Legal Policy, 32 (Washington, D.C.: The American Society of International law and the Brookings Institution).

Kapur, Harsh (1994) *India's Foreign Policy – 1947-92: Shadows and Substance* (New Delhi/Thousand Oaks/London: Sage Publications).
Kaul, Suvir, ed. (2001) *The Partitions of Memory – The Afterlife of the Division of India* (Delhi: Permanent Black).
Kay Hailbronner, (1988) 'Non-refoulement and 'Humanitarian' Refugees – Customary International Law or Wishful Legal Thinking?' in David Martin (ed.), *The New Asylum Seekers – Refugee Laws in the 1980s* (Dordrecht: Martinus Nijhoff Publishers).
Keane, John (1998) *Civil Society – Old Images, New Visions* (Cambridge: Polity Press).
Kelman, Herbert (1999) 'The Interdependence of Israeli and Palestinian National Identities – The Role of the Other in Existential Conflicts', *Journal of Social Issues*, 55 (3), Fall.
Kemp, Adriana (1998) 'From Politics of Location to Politics of Signification – The Construction of Political Territory in Israel's First Years', *Journal of Area Studies*, 12.
Kemp, Adriana (1999) 'The Frontier Idiom on Borders and Territorial Politics in Post-1967 Israel', *Geography Research Forum*, 19.
Kennedy, Charles (1993) 'Managing Ethnic Conflict – The Case of Pakistan', *Regional Politics and Policy*, 1 (3), Spring.
Kennedy, David 'International Refugee Protection', *Human Rights Quarterly*, 8 (1).
Kennedy, Paul (1993) *Preparing for the Twentieth Century* (New York: Randon House).
Kent, Sarah A. (1997) 'Writing the Yugoslav Wars – English Language Books on Bosnia (1992-96) and the Challenges of Analysing Contemporary History', *American Historical Review*, 102 (4), October.
Keohane, R and S. Hoffman, eds. (1991), *The New European Community – Decision-Making and Institutional Change* (Boulder: Westview Press).
Kitz, N.J. (ed.), (1995) *Transitional Justice – How Emerging Democracies Reckon with Former Regimes*, Vol. 2 (Washington D.C.: US Institute of Peace Press).
Klintworth, Gary (1992) 'Chinese Perspectives on India as a Great Power,' in R. Babbage and Sandy Gordon (eds.), *India's Strategic Future: Regional State or Global Power* (New York).
Knight, Jack (1999) 'Constitutionalism and Deliberative Democracy' in Stephen Macedo (ed.), *Deliberative Politics – Essays on Democracy and Disagreement* (New York: Oxford University Press).
Kodikara, S.U. (1985) 'The Separatist Eelam Movement in Sri Lanka – An Overview', *India Quarterly*, 37, 1987.
Kolko, Joyce and Gabriel Kolko, (1972) *The Limits of Power – The World and The United States Foreign Policy, 1945-1954* (New York: Pantheon Books).
Krishna, Sankaran (2000) *Postcolonial Insecurities – India, Sri Lanka, and the Question of Nationhood* (Delhi: Oxford University Press).
Krugman, Paul (1999) *The Return of Depression Economics* (London: Allen, Lane The Penguin Press).
Kuenne, Robert E. (1992) 'Disarmament Negotiations as an Exercise in Mature Rivalry' in Walter Isard and Charles H. Anderson (eds.), *Economics of Arms Reduction and the Peace Process* (Amsterdam: Elsevier Science Publishers).
Kumar, Dhruba (ed.), (2000) *Domestic Conflict and Crisis of Governability in Nepal* (Kathmandu: Centre for Nepal and Asian Studies).
Kumar, Radha (1997) *Divide and Fall? Bosnia in the Annals of Partition* (London: Verso).
Kumar, Radha (2001) 'Lessons Learnt – The Options Ahead', *Transeuropeannes*, 19-20.
Kumar, Ram Narayan (2001) *A Complex Denial*, SAFHR Paper 9, Kathmandu.
Kumin, Judith (1998) 'An Uncertain Direction', *Refugees*, 2 (113).
Kushnck, Louis and James Jennings, eds. (1999) *An Introduction to Poverty – The Role of Race, Power, and Politics* (New York: New York University Press).
Kymlicka, Will (1989) *Liberalism, Communitarianism, and Culture* (Oxford: Clarendon).

Lakey, George (1987) *Powerful Peacemaking – A Strategy for a Living Revolution* (Philadelphia: New Society).

Lawoti, Mahendra (2001) 'Exclusionary Democratization – The Politics of Minority Domination in Nepal', (unpublished) paper presented to the 'Dialogue on Democracy and Pluralism in South Asia', Jawaharlal Nehru University, New Delhi, 14-17 March.

Leffler, M.P. (1979) *The Elusive Quest – America's Pursuit of European Stability and French Security 1919-1933* (Chapel Hill: University of North Carolina Press).

Lintner, Bertil (1997) *Burma in Revolt – Opium and Insurgency since 1948* (Bangkok: White Lotus, 1994).

Linz, Juan and A. Stepan (1996) *Problems of Democratic Transition and Consolidation* (Baltimore: Johns Hopkins University Press).

Lockhart, Charles (1979) *Bargaining in International Conflicts* (New York: Columbia University Press).

Loescher, Gil and John Scanlan, (1986) *Calculated Kindness – Refugees and America's Half-Open Door, 1945 – Present* (New York: Free Press).

Long, David (1995) 'The Harvard School of Liberal International Theory', *Millennium – Journal of International Studies*, 24, Winter.

Lopez, George A., ed.(1989) *Peace Studies – Past and Future*, special number of *The Annals of the American Academy of Political and Social Science*, Vol. 504, July.

Lorentzen, Lois Ann and Jennifer Turpin, eds. (1998) *The Women and War Reader* (New York: New York University Press).

Lubbers, Ruud (2002) 'Asylum for All', *Harvard International Review*, Volume 24, Spring.

Luff, Donna (1999) 'Dialogues Across the Divides – Moments of Rapport', *Sociology*, 33 (4), November.

Lustick, I. S. (1993), *Unsettled States, Disputed Lands – Britain and Ireland, France and Algeria, Israel and West Bank-Gaza* (London: Cornell University Press).

Lynn-Jones, Sean M. and Steven E. Miller (eds.), (1995) *Global Dangers – Changing Dimensions of International Security* (Cambridge, MA: The MIT Press).

Lyotard, Jean Francois (1978) *The Differend – Phrases in Dispute* (Manchester: Manchester University Press).

Macedo, Stephen, ed. (1999), *Deliberative Politics – Essays on Democracy and Disagreement* (New York: Oxford University Press).

MacIntyre, Alasdair (1984) 'Is Patriotism a Virtue?' – E.H. Lindley Memorial Lecture (Lawrence: University of Kansas).

MacMillan, John and Andrew Linklater, eds. (1995) *Boundaries in Question – New Directions in International Relations* (London: Pinter Publishers).

Madan, T.N. (1997) *Modern Myths and Locked Minds* (Delhi: Oxford University Press).

Madison, James (1981) 'Universal Peace' in M. Meyers (ed.), *Sources of the Political Thought of James Madison* (Hanover: Brandeis University Press).

Magliana, Melissa (2000) *The Autonomous Province of South Tyrol – A Model of Self-Governance?* (Bolzano: European Academy of Bolzano).

Mahajan, Gurpreet (ed.), (1998) *Democracy, Difference and Social Justice* (Delhi: Oxford University Press, 1998).

Mahajan, Sucheta (2000) *Independence and Partition – The Erosion of Colonial Power in India* (New Delhi: Sage Publications).

Mamdani, Mahmood (2001) *When Victims Become Killers – Colonialism, Nativism, and the Genocide in Rwanda* (Princeton, NJ: Princeton University Press).

Mann, Michael (1987) 'War and Social Theory – Into Battle with Classes, Nations and States' in Colin Creighton and Martin Shaw (eds.), *The Sociology of War and Peace* (London: The McMillan Press).

Mann, Michael (1999) 'The Dark Side of Democracy – The Modern Tradition of Ethnic and Political Cleansing', *New Left Review*, 235, May-June.

Manogaram, C. (1987) *Ethnic Conflict and Reconciliation in Sri Lanka* (Honolulu: University of Hawai Press).

Mansbridge, Jane (1995) 'Using Power/Fighting Power – The Polity' in Seyla Benhabib (ed.), *Democracy and Difference* (Princeton: Princeton University Press).

Mansbridge, Jane (1999) 'Everyday Talk in a Deliberative System' in Stephen Macedo (ed.), *Deliberative Politics – Essays on Democracy and Disagreement* (New York: Oxford University Press).

Manual for Early Warning and Early Response, (1998) issued by FEWER (Forum for Early Warning and Early Response), London.

Mao Zedong on Diplomacy (Beijing: Foreign Language Press, 1998).

Martin, David (ed.), (1988) *The New Asylum Seekers – Refugee Laws in the 1980s* (Dordrecht: Martinus Nijhoff Publishers).

Martin, Luther H, Huck Gutman and Patrick H. Hutton (eds.), *Technologies of the Self – A Seminar with Michel Foucault* (Amherst: University of Massachusetts Press).

Massey, James (1999) *Minorities in Democracy – The Indian Experience* (New Delhi: Monohar).

Mattausch, John (1987) 'The Sociology of CND' in Colin Creighton and Martin Shaw (eds.), *The Sociology of War and Peace* (London: The Macmillan Press).

Mbembe, Achille (2001) 'African Modes of Self-Writing', *Identity, Culture & Politics – An Afro-Asian Dialogue*, 2 (1).

Mccormick, Thomas J. (1997) 'Troubled Triumphalism – Cold War Veterans Confront a Post-Cold War World', *Diplomatic History*, 21 (3).

McGregor, Andrew (1999) 'Quagmire in West Africa – Nigerian Peacekeeping in Sierra Leone, 1997-1998', *International Journal*, LIV (3), Summer.

Mekenkamp, Monique, Paul van Tongeren, and Hans van de Veen, eds. (2002) *Searching for Peace in Central and South Asia* (Boulder, Colorado: Lynne Rienner Publishers).

Menon, Ritu and Kamla Bhasin, (1998) Borders & Boundaries – Women in India's Partition (Delhi: Kali for Women).

Menon, V.P. (1956) *The Story of the Integration of the Indian States* (Bombay: Orient Longmans).

Mgxashe, Mxolisi (2000) 'Reconciliation, A Call to Action', in Charles Villa-Vicencio and Wilhelm Verwoerd, *Looking Back Reaching Forward – Reflections on the Truth and Reconciliation Commission of South Africa* (London and Cape Town: Zed Press and University of Cape Town Press).

Miller, Mark J. (1982) *Foreign Workers in Western Europe – An Emerging Political Force* (New York: Praeger Pub. Text).

Miller, Steven A. (1996) *Debating the Democratic Peace* (Cambridge, MA: The MIT Press).

Misra, Udayon (1988) *North East India – Quest for Indentity* (Guwahati: Omsons).

Moghadam, Valentine (1993) *Modernizing Women – Gender and Social Change in the Middle East* (Cairo: The American University in Cairo Press).

Montag, Warren (2000) 'The Pressure of the Street – Habermas' Fear of the Masses' in Mike Hill and Warren Montag (eds.), *Masses, Classes, and the Public Sphere* (London, Verso).

Mouffe, Chantal (1998) 'The Political Community: Universitas or Societas?' reprinted in Gurpreet Mahajan (ed.), *Democracy, Difference and Social Justice* (Delhi: Oxford University Press, 1998).

Mouffe, Chantal (1999) 'Deliberative Democracy or Agonistic Pluralism?', *Social Research*, 66 (1), Fall.

Mouffe, Chantal (2000) *The Democratic Paradox* (London: Verso).

Munck, Ronaldo and Purnaka L. de Silva, eds. (2000) *Postmodern Insurgencies – Political Violence, Identity Formation and Peace Making in Comparative Perspective* (London and New York: McMillan and St. Martin's Press).

Naga Resistance and Peace Process, a compendium of documents relating to peace process in Nagaland (Delhi: Other Media Communications, 2001).

Nandy, Ashis (1983) *The Intimate Enemy: Loss and Recovery of Self under Colonialism*, (Delhi: Oxford University Press).

Nandy, Ashis (1988) 'The Politics of Secularism and the Recovery of Religious Tolerance', *Alternatives*, 13 (2), April.

Nandy, Ashis (2001) 'Defining a New Cosmopolitanism – Towards a Dialogue of Asian Cultures', *Identity Culture & Politics – An Afro-Asian Dialogue*, 2 (1), January.

Navlakha, Gautam (1997) 'Armed Forces Special Powers Act', *Economic and Political Weekly*, 32 (6).

Navlakha, Gautum (2000) 'Kashmir: An Idea Which Cannot Be Suppressed,' *Economic and Political Weekly*, January 29.

Nehru, B.K. (1997) *Nice Guys Finish Second* (New Delhi: Viking).

Newman, David and Anssi Passi (1998) 'Fences and Neighbours in Postmodern World – Boundary Narratives in Political Geography', *Progress in Human Geography*, 22 (2).

Nicholson, Frances and Patrick Twomey, eds. (1999) *Refugee Rights and Realities – Evolving International Concepts and Regimes* (Cambridge: Cambridge University Press).

Nino, Carlos (1996) *Radical Evil on Trial* (New Haven: Yale University Press).

Nye, J, R. Keohane and S. Hoffman, S., eds. (1993) *After the Cold War – International Institutions and State Strategies in Europe, 1989-1991* (Cambridge, MA: Harvard University Press).

O'Leary, Brendan Ian S. Lustick, and Thomas Callaghy, eds. (2001) *Right-sizing the State – The Politics of Moving Borders* (Oxford: Oxford University Press).

Oakshott, Michael (1998) 'On Civil Association' reprinted in Gurpreet Mahajan (ed.), *Democracy, Difference and Social Justice* (Delhi: Oxford University Press, 1998).

Ocampo, Luis Moreno (1999) 'Beyond Punishment – Justice in the Wake of Massive Crimes in Argentina', *Journal of International Affairs*, 52 (2), Spring.

O'Connor, James (1984) *Accumulation Crisis* (New York: Blackwell).

Oomen, T. K. ed. (1998), *Citizenship and National Identity – From Colonialism to Globalism* (Delhi: Sage).

Pandey, Gyanendra (2001) *Remembering Partition – Violence, Nationalism and History in India* (Cambridge: Cambridge University Press).

Pankhurst, Donna (1999) 'Issues of Justice and Reconciliation in Complex Political Emergencies – Conceptualizing Reconciliation, Justice and Peace', *Third World Quarterly*, 20 (1).

Parekh, Bhikhu (2000) *Rethinking Multiculturalism – Cultural Diversity and Political Theory* (Hampshire: Palgrave).

Paris, Roland (1997) 'Peace building and the Limits of Liberal Internationalism', *International Security*, 22 (2), Fall.

Paterson, Thomas G. (1973) *Soviet American Confrontation – Postwar Reconstuction and the Origins of the Cold War* (Baltimore: Johns Hopkins).

Patnaik, Arun (1997) 'The Concept of Individual Freedom', *Indian Journal of Human Rights*, 1 (1).

Penrose, Jan (2000) 'The Limitations of Nationalist Democracy – The Treatment of Marginal groups as a Measure of State Legitimacy', *Hagar – International Social Science Review*, 1 (2).

Perera Jehan, ed. (1999) *Report of a Peace Audit on Nagaland and the Chittagong Hill Tracts*, SAFHR Paper 4, Kathmandu.

Petras, James and Steve Vieux, (1996) 'Bosnia and the Revival of US Hegemony', *New Left Review*, No. 218, July-August.

Pettygrew, J (1995) *The Sikhs of Punjab – Unheard Voices of State and Guerrilla Violence* (London: Zed Books).

Philipson, Liz (2001) *Breaking Recurring Themes in the Cycles of War and Peace in Sri Lanka*, www.c-r.org.

Pierson, Christopher (1996) *The Modern State* (London: Routledge).

Polachek, Solomon William (1992) 'Conflict and Trade – An Economic Approach to Political International Interactions' in Walter Isard and Charles H. Anderson (eds.), *Economics of Arms Reduction and the Peace Process* (Amsterdam: Elsevier Science Publishers).

Praag, Michael C. Walt van, (1998) 'The Implementation of the Right to Self-Determination as a Contribution to Conflict Prevention', a UNESCO report, UNESCO Centre of Catalonia, Barcelona.

Prodromou, Elizabeth H. (1996) 'Paradigms, Power and Identity: Rediscovering Orthodoxy and Regionalizing Europe', *European Journal of Political Research*, 30(2), September, 1996. Michael Rogin, 'The Two Declarations of American Independence', *Representations*, 55, Summer.

Putnam, Robert D. (1993) *Making Democracy Work – Civic Traditions in Modern Italy* (Princeton, NJ: Princeton University Press).

Putnam, Robert D. (1995) 'Bowling Alone – America's Declining Capital', *Journal of Democracy*, 6.

Putnam, Robert D. (1996) 'The Strange Disappearance of Civic America', *American Prospect*, 24.

Quester, George H. (1989) 'International Security Criticisms of Peace Research' in George A. Lopez (ed.), *Peace Studies – Past and Future*, special number of *The Annals of the American Academy of Political and Social Science*, Vol. 504, July.

Rabinow, Paul (1996) 'Representations are Social Facts – Modernity and Post-modernity in Anthropology' in Paul Rabinow, *Essays on the Anthropology of Reason* (Princeton: Princeton University Press).

Racine, Jean Luc (2001) 'The Uncertain Triangle: India, China and Pakistan – The Regional and International Dimensions', unpublished essay.

Radan, Peter (1999) 'Yugoslavia's Internal Borders as International Borders – A question of Appropriateness', *East European Quarterly*, 33 (2), Summer.

Rajan, Nalini (1998) *Secularism, Democracy, Justice – Implications of Rawlsian Principles in India* (New Delhi: Sage).

Ranade, Sudhanshu (2000) 'The Kashmir Gambit', *Economic and Political Weekly*, 35 (52-53), December 30.

Rapaport, Anatol (1970) 'Conflict Resolution in the Light of Game Theory and Beyond' in Paul G. Swingle (ed.), *The Structure of Conflict* (New York: Academic Press).

Rawls, John (1993) *A Theory of Justice* (New York: Cambridge University Press, 1972) and *Political Liberalism* (New York: Columbia University Press).

Rawls, John (1999) 'Laws of Peoples' in Samuel Freeman (ed.), *Collected Papers of John Rawls* (Cambridge, MA: Harvard University Press).

Read, Edward (ed.), (1968) *Beyond Coexistence – The Requirements of Peace* (New York: Grossman Publishers).

Regan, Richard J. (1986) *Just War* (Washington, DC: Catholic University of America Press).

Reifeld, Helmut and Ranabir Samaddar, eds. (2001) *Peace as Process* (Delhi: Manohar Publishers).
Ribeiro, Julio (1998) *Bullet for Bullet – My Life as a Police Officer*, (Delhi: Penguin Books).
Richardson, James L. (1997) 'Contending Liberalism – Past and Present', *European Journal of International Relations*, 3.
Richardson, Lewis F. (1960) *Statistics of Deadly Quarrels* (Pittsburg: Boxwood Press).
Rodrigues, Valerian (1998) 'Whose Call Do We Hearken (in South Asia)? Community, Civilization, State, Nation, Region or the Globe?' in B. Ramesh Babu (ed.), *Globalisation and the South Asian State* (Delhi: South Asian Publishers).
Roht-Arriaza, Naomi (1999) 'Institutions of International Justice', *Journal of International Affairs*, 52 (2), Spring.
Rosenberg, Justin (2000) *The Follies of Globalisation Theory* (London: Verso).
Rueschemeyer, D, E. Stephens, and J. Stephens, (1992) *Capitalist Development and Democracy* (Chicago: University of Chicago).
Rupesinghe, Kumar (ed.), (1998) *Negotiating Peace in Sri Lanka – Efforts, Failures and Lessons* (London: International Alert).
Rupesinghe, Kumar (with Sanam Anderlini), (1998) *Civil Wars, Civil Peace* (London: Pluto Press).
Rushdie, Salman (1992) *Imaginary Homelands – Essays and Criticism, 1981-91* (London: Granta Books).
Saberwal, Satish and Heiko Sievers, eds. (1998) *Rules, Laws, Constitutions* (Delhi: Sage).
Sacco, Giuseppe (1990) 'Europe and the World', *The European Journal of International Affairs*, 10 (4).
Safran, William (1990) 'Is France becoming America?' and Marie-France Toinet, 'Convergence in Disguise', *The European Journal of International Affairs*, 10 (4).
Said, Edward (1995) *The Politics of Dispossession – The Struggle for Palestinian Self-Determination, 1969-1994* (New York: Vintage Books).
Said, Edward W. (1992) *The Question of Palestine* (New York: Vintage Books).
Samad, Yunus (1995) *A Nation in Turmoil* (Delhi: Sage Publications).
Samaddar, Ranabir (1996) *Whose Asia Is It Anyway – Region and the Nation in South Asia* (Calcutta: M. A. K. Azad Institute of Asian Studies/Pearl Publishers).
Samaddar, Ranabir (1997) 'Flowing Waters and the Nationalist Metaphors', *Studies in Conflict and Terrorism*, 20 (2), April-June.
Samaddar, Ranabir (1998) 'The Failed Dialectic of Territoriality and Security, and the Imperatives of Dialogue', *International Studies*, 35 (1).
Samaddar, Ranabir (1999) *The Marginal Nation – Transborder Migration from Bangladesh to West Bengal* (Delhi: Sage Publications).
Samaddar, Ranabir (2000) 'Leaders and Publics – Stories in the Time of Transition', *Indian Economic and Social History Review*, 37 (4), October-December.
Samaddar, Ranabir (2001) *A Biography of the Indian Nation 1947-97* (Delhi: Sage Publications).
Samaddar, Ranabir (2002) *Paradoxes of the Nationalist Time – Political Essays on Bangladesh* (Dhaka: University Press Limited).
Samaddar, Ranabir (2003) 'Power and Care – Building the New Indian State' in Ranabir Samaddar (ed.), *Refugees and the State* (Delhi: Sage).
Samaddar, Ranabir, ed. (1997) *Reflection on Partition in the East* (Delhi: Vikas Publications).
Samaddar, Ranabir, eds. (2002) *Space, Territory and the State – New Readings in International Politics* (Hyderabad: Orient Longman).
Samaddar, Ranabir (ed.), (2003) *Refugee and the State* (Delhi: Sage).

Sandel, Michael J. (1992) *Liberalism and the Limits of Justice* (Cambridge: Cambridge University Press).

Sanders, Lynn (1997) 'Against Deliberation', *Political Theory*, June.

Sarkar, Sumit (1994) 'The Anti-Secularist Critique of Hindutva: Problems of a Shared Discursive Space', *Germinal*, 1.

Sassen, Saskia (2000) *Guests and Aliens* (New York: New Press).

Schaeffer, Robert K. (1999) 'Severed States – Dilemmas of Democracy in a Divided World' (New York: Rowman & Littlefield Publishers).

Schneider, Cathy Lisa (2000) 'Violence, Identity and Spaces of Contention in Chile, Argentina and Columbia', *Social Research*, 67 (3), Fall.

Seaton, Jean (1999) 'Why Do We Think the Serbs Do It? The New Ethnic Wars and the Media', *Political Quarterly*, Summer.

Seiglie, Carlos (1992) 'Determinants of Military Expenditure' in Walter Isard and Charles H. Anderson (eds.), *Economics of Arms Reduction and the Peace Process* (Amsterdam: Elsevier Science Publishers).

Selden, (2001) 'On Asian Wars, Reparations, Reconciliation', *Economic and Political Weekly*, 36 (1), 6 January.

Selected Works of Mao Ze Dong, Volume IV (Beijing: Foreign Language Publishing House, (1989).

Sengupta, Anita (2002) 'Territorial Delimitation in Central Asia, 1917-27' in Ranabir Samaddar (ed.), *Space, Territory and the State – New Readings in International Politics* (Hyderabad: Orient Longman).

Sengupta, Anita, (2000) 'Imperatives of National Territorial Delimitation and the fate of Bukhara, 1917-1924', *Central Asian Survey*, 19 (3-4).

Sense, Paul D. (1999) 'Democracy and Maturity – Deciphering Conditional Effects on Levels of Dispute Intensity', *International Studies Quarterly*, 43 (3), September.

Seshu, Geeta (1999) 'Media and Kargil: Information Blitz with Dummy Missiles', *Economic and Political Weekly*, 9 October.

Sharp, Gene (1973) *The Politics of Non-violent Action* (Boston: Porter Sergeant).

Silberman, Marc (ed. and trans.), (2001) *Bertolt Brecht on Film and Radio* (London: Methuen).

Simmel, Georg (1955) *Conflict and the Web of Group Affiliation* (1908), trans. Kurt H. Wolff (New York: Free Press of Glencoe).

Singh, Gurharpal (2000) *Ethnic Conflict in India – A Case Study of Punjab* (Basingstoke: Macmillan Press).

Singh, Gurharpal (2000) *Ethnic Conflict in India – A Case-Study of Punjab* (Basingstoke: Macmillan Press).

Skocpol, Theda (1985) 'Bringing the State Back In: Strategies of Analysis in Current Research' in Peter Evens, Dietrich Rueschmeyer and Theda Skocpol (eds.), *Bringing the State Back In* (Cambridge: Cambridge University Press).

Skocpolm Theda, ed. (1998) *Democracy, Revolution, and History* (Ithaca: Cornell University Press).

Smith, D. E. ed. (1966) *South Asian Politics and Religion* (Princeton, N.J.: Princeton University Press).

Soffer, Jonathan (1997) 'All for One or All for All – The UN Military Staff Committee and the Contradictions within American Internationalism', *Diplomatic History*, 21 (1), Winter.

Stoessinger, John George (1963) *The Refugee and the World Community* (Minneapolis: The University of Minnesota Press).

Strangleman, Tim (1999) speaks of these in his study of railway industry, 'The Nostalgia of Organizations and Organization of Nostalgia – Past and Present in the Contemporary Railway Industry', *Sociology*, 33 (4), November.

Sunjic, Melita (1998) 'Learning to Deal with Refugees', *Refugees*, 2 (113).

Swingle, Paul G, ed. (1970) *The Structure of Conflict* (New York: Academic Press).

Swingle, Paul G. (1970) 'Dangerous Games' in Paul G. Swingle (ed.), *The Structure of Conflict* (New York: Academic Press).

Syse, Henrik (2000) 'Ethics, Sovereignty, and Self-Defense – A Rejoinder', *Security Dialogue*, 31 (4), December.

Tai, Yong Tan and Gyanesh Kudasia, (2000) *The Aftermath of Partition in South Asia* (London: Routledge).

Tambiah, Stanley J. (1997) *Leveling Crowds – Ethnonationalist Conflicts and Collective Violence in South Asia* (Delhi: Vistaar Publications).

Tamir, Y. (1991) 'The Right of national Self-Determination', *Social Research*, 58.

Tanham, George K. (1992) *Indian Strategic Thought – An Interpretative Essay* (Santa Monica, CA: National Defense Research Institute).

Taylor, Charles (1989) 'Cross-purposes – The Liberal Communitarian Debate' in Nancy L. Rosenblum (ed.), *Liberalism and the Moral Life* (Cambridge, MA: Harvard University Press).

Taylor, Charles (1992) *Multiculturalism and the Politics of Recognition* (Princeton, NJ: Princeton University Press).

Taylor, Charles (1996) 'Why Democracy Needs Patriotism' in Martha C. Nussbaum et al, *For Love of Country – Debating the Limits of Patriotism* (Boston: Beacon Press).

Teloken, Stefan (1998) 'It's Long Way to Harmonization', *Refugees*, 2 (113).

Thakur, Sankarshan and others, (1999) *Guns and Yellow Roses: Essays on the Kargil War* (New Delhi: Viking).

Thornberry, Patrick (1991) *International Law and the Rights of Minorities* (Oxford: Clarendon Press).

Thranhardt, D, ed. (1992) *Europe – A New Immigration Continent* (Lit, Munster).

Tillema, Herbert K. (1991) *International Armed Conflict since 1945 – A Bibliographic Handbook of Wars and Military Interventions* (Boulder, Colorado: Westview Press).

Tilly, Charles (1986) *The Contentious French – Four Centuries of Popular Struggle* (Cambridge, MA: Belknap Press).

Tilly, Charles (1998) 'Where Do Rights Come From?' in Theda Skocpol (ed.), *Democracy, Revolution, and History* (Ithaca: Cornell University Press).

Tilly, Charles (1998) *Durable Inequality* (Berkeley: University of California Press).

Tilly, Charles (1999) 'Power – Top Down and Bottom Up', *Journal of Political Philosophy*, 7.

Tinker, Hugh (1974) *A New System of Slavery – The Export of Indian Labour Overseas, 1830-1920* (Oxford: Oxford University Press).

Tuitt, Patricia (1996) has commented in *False Images – Law's Construction of the Refugee* (London: Pluto Press).

Van Hear, Nicholas (1998) *New Diasporas – The Mass Exodus, Dispersal and Regrouping of Migrant Communities* (London: UCL Press).

Varsori, A., ed. (1995) *Europe 1945-1990s – The End of an Era?* (London: Macmillan).

Vayrymen, Raimo (2000) 'Anti-Globalization Movements at the Crossroads', The Joan B. Kroc Institute for International Peace Studies, University of Notre Dame, Policy Brief (4), November.

Verwoerd, Wilhelm (2000) 'Towards the Recognition of Past Injustices' in Charles Villa-Vicencio and Wilhelm Verwoerd, *Looking Back Reaching Forward – Reflections on the Truth and Reconciliation Commission of South Africa* (London and Cape Town: Zed Press and University of Cape Town Press).

Vijayakumar, Veerabhadran (2000) 'A Critical Analysis of Refugee Protection in South Asia', *Refuge – Canada's Periodical on Refugees*, 19 (2), December.

Villa-Vicencio, Charles and Wilhelm Verwoerd, (2000) *Looking Back Reaching Forward – Reflections on the Truth and Reconciliation Commission of South Africa* (London and Cape Town: Zed Press and University of Cape Town Press).

Virilio, Paul (1997) *Open Sky* (London: Verso).

Walton, Prichard and Robert Mckersie (1965) *A Behavioral Theory of Labor Negotiations – An Analysis of Social Interaction System* (New York: McGraw Hill).

Walzer, Michael (1977) *Just and Unjust Wars – A Moral Argument with Historical Illustrations* (New York: Basic Books).

Walzer, Michael (1983) *Spheres of Justice* (New York: Basic Books).

Warner, Daniel (1992) 'We are all Refugees', *International Journal of Refugee Law*, 4 (3).

Warner, Daniel (1999) 'Searching for Responsibility/Community in International Relations' in David Campbell and Michael J. Shapiro (eds.), *Moral Spaces – Rethinking Ethics and World Politics* (Minneapolis: University of Minnesota Press).

Warner, Daniel (1999) 'The Refugee State and State Protection' in Frances Nicholson and Patrick Twomey (eds.), *Refugee Rights and Realities – Evolving International Concepts and Regimes* (Cambridge: Cambridge University Press).

Weber, Max (1946) 'Politics as a Vocation' in *From Max Weber – Essays in Sociology*, eds, Hans Gerth and C. Wright Mills (New York: Oxford University Press).

Weerakoon, Bradman (1998) 'Government of Sri Lanka (President Premadasa) and the LTTE Peace Negotiations 1989-1990' in Kumar Rupesinghe (ed.), *Negotiating Peace in Sri Lanka – Efforts, Failures and Lessons* (London: International Alert).

Wehr, Paul (1979) *Conflict Regulation* (Boulder, Colorado: Westview Press).

Weiner, Myron, (1995) 'Security, Stability and International Migration' in Sean M. Lynn-Jones and Steven E. Miller (eds.), *Global Dangers – Changing Dimensions of International Security* (Cambridge, MA: The MIT Press).

Weiss, Linda (1998) *The Myth of the Powerless State* (Cambridge: Polity Press).

White, Brian, Richard Little and Michael Smith, eds., (1997) *Issues in World Politics* (London: Macmillan).

Wilkinson, Ray (1998) 'The Deabate over Asylum', *Refugees*, 2 (113).

Wirsing, Robert G. (1994) *India, Pakistan, and the Kashmir Dispute* (New York: St. Martin's Press).

Withol de Venden, Catherine (1998) 'How Can One be Muslim – The French Debate on Allegiance, Intrusion and Transnationalism', *International Review of Sociology*, 8 (2).

Withol de Venden, Catherine (1999) 'Post-Amsterdam Migration Policy and European Citizenship', *European Journal of Migration and Law*, 1.

Wohlforth, W.C. (1994-95) 'Realism and the End of the Cold War', *International Security*, 19 (3).

Wolf, Eric R (1982) *Europe and The People without History* (Berkeley, California: UCLA).

Wolpert, Stanley (1993) *Zulfi Bhutto of Pakistan: His Life and Times* (Delhi: Oxford University Press).

Wood, Ellen Meiksins (1996) 'Capitalism, Merchants and bourgeois Revolution – Reflections on the Brenner Debate and Its Sequel', *International Review of Social History*, Vol. 41, part 2, August.

Wriggins, W. Howard (1960) *Ceylon – Dilemma of a New Nation* (Princeton, N.J.: Princeton University Press).

Yaftachel, Oren (1997) 'Israel: Metropolitan Integration or "Fractured Regions"?' *Cities*, 14 (6).

Yaftachel, Oren (2000) 'Ethnocracy and Its Discontents: Minorities, Protest and the Israeli Polity', *Critical Inquiry* 26 (4), Summer.

Yans-MacLaughlin, Virginia, ed., (1990) *Immigration Reconsidered – History, Sociology and Politics* (New York: Oxford University Press).

Yiftachel, Oren (2000) 'Ethnocracy and Its Discontents: Minorities, Protest and the Israeli Polity', *Critical Inquiry* 26 (4), Summer.

Yiftachel, Oren (2001) '"Rights-Sizing" or "Right-Shaping"? Politics, Ethnicity, and Territory in Plural States' in Brendan O'Leary, Ian S. Lustick, and Thomas Callaghy (eds.), *Right-sizing the State – The Politics of Moving Borders* (Oxford: Oxford University Press).

Yiftachel, Oren and Avinoam Meir, (1998) *Ethnic Frontiers and Peripheries – Landscapes of Development and Inequality in Israel* (Boulder, Colorado: Westview Press).

Yiftachel, Oren and Avinoam Meir, eds. (1998) *Ethnic Frontiers and Peripheries – Landscapes of Development and Inequality in Israel* (Boulder, Colorado: Westview Press).

Yonou, Asoso (1974) *The Rising Nagas* (Delhi: Vikas).

Young, James E. (1990) *Writing and Rewriting the Holocaust – Narrative and the Consequences of Interpretation* (Bloomington: Indiana University Press).

Zartman, I. William (1986) 'Ripening Conflict, Ripe Moment, Formula, and Mediation' in Diane Bendahamane and John McDonald Jr., *Prospective on Negotiation – Four Case Studies and Interpretations* (Washington, D.C.: Foreign Service Institute).

Zartman, I. William, ed. (1995) *Elusive Peace – Negotiating an End to Civil Wars* (Washington D.C.: Brookings Institution).

Zbigniew, Brzezinski (1993) *Out of Control – Global Turmoil on the Eve of Twentieth Century* (New York: Collier Books).

Zelin, Elizabeth and Eric Hershberg, (1996) *Constructing Democracy – Human Rights, Citizenship and Society in Latin America* (Boulder, Colorado: Westview Press).

Index